Economic Strategies and Policy Suggestions of Agricultural Sustainable Food Production

Economic Strategies and Policy Suggestions of Agricultural Sustainable Food Production

Editors

Roberto Henke
Filiberto Altobelli

MDPI

Basel • Beijing • Wuhan • Barcelona • Belgrade • Novi Sad • Cluj • Manchester

Editors
Roberto Henke
Research Centre for Agricultural Policies and Bioeconomy
Rome
Italy

Filiberto Altobelli
Research Centre for Agricultural Policies and Bioeconomy
Rome
Italy

Editorial Office
MDPI
St. Alban-Anlage 66
4052 Basel, Switzerland

This is a reprint of articles from the Special Issue published online in the open access journal *Agriculture* (ISSN 2077-0472) (available at: https://www.mdpi.com/journal/agriculture/special_issues/4RSZQV08K4).

For citation purposes, cite each article independently as indicated on the article page online and as indicated below:

Lastname, A.A.; Lastname, B.B. Article Title. *Journal Name* **Year**, *Volume Number*, Page Range.

ISBN 978-3-7258-0959-2 (Hbk)
ISBN 978-3-7258-0960-8 (PDF)
doi.org/10.3390/books978-3-7258-0960-8

Contents

About the Editors

Roberto Henke

Roberto Henke is Director of Research at CREA-PB (Council for Agricultural Research and Economics–Research Centre for Agricultural Policies and Bioeconomy), Italy. From June 2017 to December 2021, he served as Director at CREA-PB. He has over 25 years of experience in the areas of agricultural economics and policies, rural development. agri-food systems, and agri-food trade, with many scientific publications in international and national journals, and contributions to several monographies and books. He holds a PhD in agricultural economics and policies from the University of Naples Federico II.

Filiberto Altobelli

Filiberto Altobelli, PhD, is a Senior Researcher at CREA-PB (Council for Agricultural Research and Economics–Research Centre for Agricultural Policies and Bioeconomy), Italy. He is involved in research on the sustainability of natural resources in agriculture, cropping systems, irrigation and new technologies to adopt in sustainable field crop management (precision farming and modelling), agricultural meteorology, as well as natural resource policy and economy focused at sustainable agricultural productivity growth. He is involved in national and international projects on these fields. He is a member of the Scientific Board of the Italian Association of Agrometeoroloy (AIAM), the Italian Society of Agronomy, and the Italian Society of Soil Science (SISS). He is also head of secretariat for the International Union of Soil Sciences (IUSS). He has more than 70 publications (scientific articles appearing in exteso in national and international congresses, and peer-reviewed scientific articles).

Preface

The success of a Special Issue is confirmed not only by the number of submissions but also by the variety of relevant issues treated and the geographical coverage of the authors. These elements are at the base of this Reprint, which we are honored to introduce and serve as Guest Editors. The theme of sustainable food production is key in understanding and analyzing the complexity of the agri-food systems all over the world, in developed regions as well as in emerging areas and less developed countries. The concept itself of sustainability assumes different meanings and nuances according to the context, requiring specific actions and policies along its three main dimensions: economic, social, and environmental. In other words, sustainability, like development, is not a static concept but evolves and calls for different measures according to the status and the features of the territories, the population, the levels of welfare, and the natural resources. The agri-food system is always a good indicator of the conditions of sustainability at any territorial level, since it necessarily involves economic, social, and environmental issues, but has also to do with the cultural involvement of people in what they eat and how they produce and make their own food. The contents of this Reprint stem from Europe to Asia, from the North to the South of the world, and it hosts several outstanding and distinguished authors, whom we sincerely thank for their precious contributions. The twelve papers present a variety of approaches with interesting methodologies and different scales of analysis, from micro to macro, with the common thread of keeping together economic strategies, social inclusion, and environmental stewardship in sustainable food production.

Roberto Henke and Filiberto Altobelli
Editors

agriculture

Editorial

Economic Strategies and Policy Suggestions of Agricultural Sustainable Food Production

Filiberto Altobelli and Roberto Henke *

Council for Research in Agriculture and Economics (CREA), Research Centre for Agricultural Policies and Bioeconomy, Via Barberini 36, 00187 Rome, Italy; filiberto.altobelli@crea.gov.it
* Correspondence: roberto.henke@crea.gov.it

Sustainability is increasingly becoming a keyword for viable agriculture and food production. Alongside Agenda 2030, sustainability is acknowledged as a multidimensional issue involving three main spheres of knowledge and action: environmental, economic, and social. Agriculture and food are strongly influenced by climate change, increasing scarcity of natural resources, and changes in land availability and use. At the same time, the agri-food system is at the center of relevant economic interests, both at the global and local level, which rise relevant social conflicts involving local communities, city dwellers, and institutions. This calls for a new approach to be established regarding primary agricultural activities and food production, and a different perspective for studies and projects is also needed.

A sustainable approach to agriculture, in the context of global governance, can lead to an improvement of ecosystems, a reduction of waste of food and natural resources, equitable access to food, and require a new set of policies able to overcome the trade-offs among objectives and searching for win–win solutions.

The articles in this Special Issue contribute to the investigation, discussion (at a scientific level), and dissemination (at an international level) of the possible economic strategies and policies for implementing sustainable agricultural systems and food products, and making rural areas more attractive, thus reducing unbalances concerning urban areas. The result is an Issue rich of interesting innovative approaches and challenging methodologies, including contributions from all over the world, in the right spirit of the free circulation of ideas and research.

The 12 articles in this Special Issue of *Agriculture*, entitled "Economic Strategies and Policy Suggestions of Agricultural Sustainable Food Production", include contributions from a variety of researchers from various countries, following different scientific approaches and methodologies, but all aimed to investigate the complex relationships between the three dimensions of sustainability. Their full list is presented in Table 1.

Firstly, a paper by Italian researchers [1] on consumer evaluations and attitudes towards new genome editing techniques emphasizes the importance of communication and dissemination activities, where clarity and broad appeal are key to assessing knowledge levels and determining how consumers' backgrounds, including social and demographic characteristics, affect their knowledge levels.

Next, a study by Chinese researchers is included, which, through a spatial correlation network structure of and factors influencing technological progress in citrus-producing regions in China, shows that Chinese mandarin and tangerine production is experiencing technological progress, with a gradual slowdown [2]. However, mandarin production technology is advancing faster than tangerine production technology. Overall, network structures are rather dense and complex, with spatial spillovers. Economically developed eastern regions have a higher status and stronger control in spatially correlated networks. Key factors influencing technological progress in citrus production include education, informatization, economic development, innovation, and financial support.

Citation: Altobelli, F.; Henke, R. Economic Strategies and Policy Suggestions of Agricultural Sustainable Food Production. *Agriculture* **2024**, *14*, 504. https://doi.org/10.3390/agriculture14030504

Received: 12 March 2024
Accepted: 18 March 2024
Published: 20 March 2024

A work by Ecuadorian researchers analyzes how family farming contributes to food sovereignty [3], using the community of Guarainag (in the canton of Paute, in the province of Azuay-Ecuador) as a case study. This work responds to the need to explain the elements that influence food sovereignty in the current food crisis in Latin America, specifically in Ecuador, in the search for self-sufficiency in healthy food and local culture.

Another paper by Chinese researchers [4] investigates the reasons for the "Mezzogiorno trap" (inspired by a well-known economic paradox regarding the Italian Southern regions) in the Chinese agricultural economy and presents constructive recommendations based on the research findings. The research process shows that this methodology is better suited to studying regional disparities in specific economic sectors, and the results obtained are more stable and reliable.

As reported by an Italian research group in an analysis of preference weights and priority setting by users of irrigation extension services based on the analytical hierarchy process, "ensuring economic sustainability" is the most important criterion [5]. The contribution of this study is twofold: First, it presents the application of a methodology that involves converting farmers' linguistic judgment into a matching weight. Second, it addresses the decision-making process to improve the use of IAS by evaluating the preferences expressed by stakeholders. Irrigation extension services can play a vital role in helping users adopt new techniques and technologies for more efficient water use and increased production.

The aim of the South African authors in their work was to influence smallholder farmers' perceptions of the adoption of digital technologies in the Eastern Cape Province of South Africa [6]. Their study aimed to identify factors influencing smallholder farmers' perceptions towards the adoption of digital technologies. It used a purposively selected sample of 250 smallholder farmers who were interviewed cross-sectionally in the local municipalities of Port St Johns and Ingquza Hill in South Africa. This study recommends the provision of low-cost digital technologies that promote indigenous knowledge, targeting youth and young farmers with lower educational attainment who live in small households and who are full-time farmers with medium to high incomes and are part of farmer groups/organizations.

Another study on citrus was conducted by Chinese researchers [7] on the spatial-temporal evolution and spatial convergence analysis of total factor productivity of citrus in China. This study shows that from the perspective of time series evolution, the growth rate of total factor productivity of mandarin and tangerine in China slowed down year by year after reaching its maximum value in 2008. Technological progress was the main factor influencing the total factor productivity of citrus fruits. The total factor productivity growth of mandarin was more stable than that of tangerine. Moreover, the pure technical efficiency index and the scale efficiency change index of mandarin and tangerine were not stable.

A study carried out by Italian researchers [8] proposes a business model (BM) as a tool for scaling up irrigation advisory services (IASs) within a business perspective, with the aim of promoting the diffusion of this technology while enhancing the associated environmental and social benefits; BM provides a detailed revenue strategy that guarantees the financial sustainability of IASs. The Business Model Canvas © was adopted for the design and presentation of our BM. In conclusion, an innovative and well-structured BM has the potential to make IASs profitable and capable of ensuring environmental and social sustainability.

Research in Saudi Arabia and Egypt tested an extended model of the Theory of Planned Behavior (TPB) to investigate the determinants of green investment intentions in the Saudi food industry [9]. This study, using PLS-SEM, found that a positive attitude, perceived behavioral control, green investment knowledge, and commitment to green consumption significantly influence potential investors' green investment intentions. Conversely, subjective norms had a negative impact, while religiosity played a moderating role in these relationships. The implications highlight the importance for academics and policymakers in higher education to prioritize instilling positive attitudes, enhancing behavioral control, and providing green investment knowledge to graduates in order to promote environmentally responsible investments.

Table 1. Summary of the twelve articles included in this Special Edition and their contributions to Agricultural Sustainable Food Production.

	Authors	Article	Focus	Outcome
1	Romeo Lironcurti, S.; Demaria, F.; D'Annolfo, R.; Sardone, R.	Consumer Evaluations of and Attitudes towards New Genome Editing Techniques: An Italian Case Study	The research aims are twofold: (a) to assess the level of knowledge and (b) to determine how consumer background, including social and demographic characteristics, affects their level of knowledge.	Emphasize the importance of communication and dissemination activities, in which clarity and a broad appeal are key.
2	Gu, Y.; Qi, C.; He, Y.; Liu, F.; Luo, B.	Spatial Correlation Network Structure of and Factors Influencing Technological Progress in Citrus-Producing Regions in China	The research has a dual focus: firstly, assessing technological progress in key mandarin and tangerine-producing regions in China between 2006 and 2021. Secondly, examining the network structures of spatial correlations in citrus production technology progress, both overall and individually. The study utilizes the quadratic assignment procedure to analyze factors impacting the spatial network.	Chinese mandarin and tangerine production is experiencing technological progress, with a gradual slowdown. Mandarin-production technology is advancing faster than tangerine technology. Overall network structures are denser and more complex, displaying spatial spillover effects. Economically developed eastern regions have a higher status and stronger control in spatial correlation networks. Key factors influencing citrus-production technology progress include education, informatization, economic development, innovation, and financial support.
3	Verdugo, G.; Cuadrado, G.; Castillo, Y.	Family Farming as a Contribution to Food Sovereignty, Case Guarainag Parish	The objective of this research is to analyze how family farming contributes to food sovereignty; the Guarainag parish of the Paute canton in the province of Azuay-Ecuador is taken as a case of study. This work responds to the necessity to explain the elements that impact food sovereignty in the existing food crisis in Latin America and specifically in Ecuador in search of self-sufficiency for healthy food products and people's own local culture.	The research has a correlational and explanatory scope; quantitative methods were used to measure food sovereignty through a binary logit regression model, which provided an answer to the hypothesis of the research, which consisted of testing the influence of family farming on food sovereignty. Furthermore, to collect the information, a survey was applied to 372 small farmers with the support of digital mapping and the Kobol Tulboox software version 1.27.3. The result was a Food Sovereignty Index of 59.79%, which, according to the scale used, places the territory in a high average.
4	Li, X.; Yang, P.; Zou, Y.	An Empirical Investigation of the "Mezzogiorno Trap" in China's Agricultural Economy: Insights from Data Envelopment Analysis (2015–2021)	This paper proposes a methodology for identifying the "Mezzogiorno Trap". By employing this approach and combining panel data on Chinese agriculture from 2015 to 2021, it is discovered that despite the overall development of the Chinese agricultural economy during this period, the "Mezzogiorno Trap" still exists.	The paper analyzes the reasons behind the "Mezzogiorno Trap" in the Chinese agricultural economy and presents constructive recommendations based on the research findings. The research process demonstrates that this methodology is better suited for studying regional disparities in specific economic sectors, and the obtained results are more stable and reliable.

Table 1. *Cont.*

	Authors	Article	Focus	Outcome
5	Donati, I.; Viaggi, D.; Srdjevic, Z.; Srdjevic, B.; Di Fonzo, A.; Del Giudice, T.; Cimino, O.; Martelli, A.; Dalla Marta, A.; Henke, R.; Altobelli, F.	An Analysis of Preference Weights and Setting Priorities by Irrigation Advisory Services Users Based on the Analytic Hierarchy Process	The present study has two objectives. The first is to individuate the priorities of the preferences expressed by the stakeholders. The second objective is to carry out a ranking of the weights of the criteria by case study, ranking the groups and their associated properties among farmers' profiles.	The results show that "assuring economic sustainability" was the most important criterion. The contributions provided by this study are twofold: firstly, it presents an application of a methodology that involves the conversion of a linguistic judgement of farmers in a correspondence weight. Secondly, it tackles decision making regarding improving the use of IASs, evaluating the preferences expressed by the stakeholders. Irrigation advisory services can play a key role in assisting users to adopt new techniques and technologies for more efficient water use and increased production.
6	Bontsa, N.; Mushunje, A.; Ngarava	Factors Influencing the Perceptions of Smallholder Farmers towards Adoption of Digital Technologies in Eastern Cape Province, South Africa	The objective of the study was to determine the factors that influence the perceptions of smallholder farmers towards the adoption of digital technologies. A purposively selected sample of 250 smallholder farmers who were cross-sectionally surveyed from Port St Johns and Ingquza Hill Local Municipalities in South Africa was used in the study.	there are economic, social justice, and traditional perceptions towards digital technologies by smallholder farmers, with socio-economic factors affecting the perceptions. The study recommends providing low-cost digital technologies that promote Indigenous Knowledge, which should target the youth and young farmers with less education in small households who are full-time farmers with moderate-to-high incomes and are part of farmer groups/organisations
7	Gu, Y.; Qi, C.; Liu, F.; Lei, Q.; Ding, Y.	Spatiotemporal Evolution and Spatial Convergence Analysis of Total Factor Productivity of Citrus in China	In this study, the DEA–Malmquist index method was used to measure the total factor productivity of citrus in seven major mandarin-producing provinces and seven major tangerine-producing provinces in China from 2006 to 2020.	The results show that from the perspective of time series evolution, the growth rate of total factor productivity of mandarin and tangerine in China slowed down year by year after reaching the maximum value in 2008. Technological progress was the main factor affecting the total factor productivity of citrus. The total factor productivity growth of tangerine was more stable than that of mandarin, and the pure technical efficiency index and scale efficiency change index of mandarin and tangerine were not stable.
8	Santini, A.; Di Fonzo, A.; Giampietri, E.; Martelli, A.; Cimino, O.; Dalla Marta, A.; Annosi, M.; Blanco-Velázquez, F.; Del Giudice, T.; Altobelli, F.	A Step toward Water Use Sustainability: Implementing a Business Model Canvas for Irrigation Advisory Services [8]	This paper proposes a business model (BM) as a tool for scaling up IASs within a business perspective, with the aim of encouraging the diffusion of this technology while enhancing the associated environmental and social benefits.	BM provides a detailed revenues strategy that guarantees the financial sustainability of IASs. To design and represent our BM, the "Business Model Canvas ©" has been adopted. Concluion swonn an innovative and well-structured BM has the potential to leave the IASs profitable and capable to ensure environmental and social sustainability

Table 1. *Cont.*

	Authors	Article	Focus	Outcome
9	Aliedan, M.; Alyahya, M.; Elshaer, I.; Sobaih, A.	Who Is Going Green? Determinants of Green Investment Intention in the Saudi Food Industry	This research tests an expanded model of the Theory of Planned Behaviour (TPB) to investigate the determinants of green investment intention in the Saudi food industry. A questionnaire survey was electronically directed to 550 fresh agricultural and food sciences graduates in public KSA universities.	This study, utilizing PLS-SEM, revealed that positive attitudes, perceived behavioral control, green investment knowledge, and commitment to green consumption significantly influence potential investors' green investment intentions. Conversely, subjective norms exhibited a negative impact, while religiosity played a moderating role in these relationships. The implications emphasize the importance for scholars and policymakers in higher education to prioritize instilling positive attitudes, enhancing behavioral control, and imparting green investment knowledge to graduates for fostering environmentally conscious investments.
10	Xiuling, D.; Qian, L.; Lipeng, L.; Sarkar, A	The Impact of Technical Training on Farmers Adopting Water-Saving Irrigation Technology: An Empirical Evidence from China	The study takes 707 farmers who grow watermelons and muskmelon in Yuncheng and Xian City of Shanxi and Shaanxi provinces as the research object to analyse the influence of risk aversion and technical training and their interaction terms on farmers' WSIT adoption behaviour. The study uses the Probit and moderating effect models to outline the findings	The empirical analysis reveals the following outcomes: (i) 27.44% of the sample farmers adopt water-saving irrigation technology, indicating that the current adoption rate and the enthusiasm for adoption are relatively low; (ii) risk aversion has a significant negative impact on farmers' adoption of WSIT; (iii) both online and offline technical training have a significant positive impact on farmers' adoption of WSIT; (iv) significant group differences exist in the effects of risk aversion, online technical training, offline technical training and interaction items on farmers' WSIT adoption behaviour.
11	Borsotto, P.; Cagliero, R.; Giarè, F.; Giordani, G.; Iacono, R.; Manetti, I.; Sardone, R.	Measuring Short Food Supply Chain Sustainability: A Selection of Attributes and Indicators through a Qualitative Approach	This paper presents the results of a participatory analysis conducted within the agro BRIDGES H2020 project, with the aim of defining a list of economic, social, and environmental attributes and indicators to assess the sustainability of SFSC, Short food supply chains, and set up a decision-making tool to support producers in self-assessing their sustainability level and choosing the most appropriate business model (BM) from those identified within the project.	Early results highlighted three main issues: indicator calculation feasibility, business model categorization, and the simplicity of the framework for sustainability self-assessment. Some recommendations are made, including the importance of using a participatory process in building an evaluation framework on SFSC sustainability and the necessity of its adaptation to territorial contexts and needs.

Table 1. *Cont.*

	Authors	Article	Focus	Outcome
12	Sudomo, A.; Leksono, B.; Tata, H.; Rahayu, A.; Umroni, A.; Rianawati, H.; Asmaliyah; Krisnawati; Setyayudi, A.; Utomo, M.; Pieter, L.; Wresta, A.; Indrajaya, Y.; Rahman, S.; Baral, H.	Can Agroforestry Contribute to Food and Livelihood Security for Indonesia's Smallholders in the Climate Change Era?	In Indonesia, smallholders have historically practiced agroforestry, which warrants examination in terms of food and livelihood security within sustainable community forest frameworks. Based on a literature review, we analyzed these two forms of security related to smallholder agroforestry practices.	Main findings indicate diverse agroforestry systems, with 88% focusing on non-timber forest products (NTFPs) and 12% on timber. While 42% prioritize direct food supply, 58% emphasize income generation through product sales. However, agroforestry that does not produce food for direct consumption by smallholders generates revenue for purchasing food necessities. Agroforestry supports both food needs (46–61%) and income (51–54%) for smallholders, surpassing traditional agriculture (13%). Semi-commercial agroforestry (57%) is a predominant livelihood prospect. The remaining 27% are purely subsistence, and 15% are purely commercial. However, the commercialization of agroforestry that focuses only on high-value commodities results in a negative impact on biodiversity.The research directly related to food security and ecosystem services quantification remains limited, necessitating further investigation. Policy support and incentives are essential for smallholders practicing complex agroforestry for climate adaptation and mitigation.

Australian and Chinese researchers are evaluating the impact of technical training on farmers' adoption of water-saving irrigation technology [10]. In this work, an in-depth analysis of the impact of risk aversion, technical training, and their interaction on farmers' adoption of WSIT will help the government to promote WSIT to facilitate agricultural resource conservation and sustainable development. Their study suggests that the role of technical training in the diffusion of WSIT should be strengthened and that differentiated technical training for various types of farmers should be implemented to reduce the degree of farmers' risk aversion.

Subsequently, Italian researchers carried out a study of the Short Food Supply Chain (SFSC), which can be understood as a supply chain with a minimum number of intermediaries [11]. Although they have been shown to bring economic, social, and environmental benefits, they still represent a niche phenomenon in the agri-food market.

Finally, "Can agroforestry contribute to food and livelihood security for Indonesia's smallholder farmers in the era of climate change?" is the research question investigated by Indonesian and Chinese researchers [12]. They found that research directly related to food security and the quantification of ecosystem services is still limited and needs further investigation. Hence, policy support and incentives are essential for smallholders practicing complex agroforestry for climate adaptation and mitigation.

Author Contributions: Conceptualization, F.A. and R.H.; writing—original draft preparation, F.A. and R.H.; writing—review and editing, F.A. and R.H.; visualization, F.A. and R.H. All authors have read and agreed to the published version of the manuscript.

Acknowledgments: We would like to sincerely thank all authors who submitted papers to this Special Issue of *Agriculture*, titled "Economic Strategies and Policy Suggestions of Agricultural Sustainable Food Production"; the reviewers of these papers for their constructive comments and thoughtful suggestions; and the editorial staff of *Agriculture*.

Conflicts of Interest: The authors declare no conflicts of interest.

References

1. Borsotto, P.; Cagliero, R.; Giarè, F.; Giordani, G.; Iacono, R.; Manetti, I.; Sardone, R. Measuring Short Food Supply Chain Sustainability: A Selection of Attributes and Indicators through a Qualitative Approach. *Agriculture* **2023**, *13*, 646. [CrossRef]
2. Xiuling, D.; Qian, L.; Lipeng, L.; Sarkar, A. The Impact of Technical Training on Farmers Adopting Water-Saving Irrigation Technology: An Empirical Evidence from China. *Agriculture* **2023**, *13*, 956. [CrossRef]
3. Aliedan, M.; Alyahya, M.; Elshaer, I.; Sobaih, A. Who Is Going Green? Determinants of Green Investment Intention in the Saudi Food Industry. *Agriculture* **2023**, *13*, 1047. [CrossRef]
4. Santini, A.; Di Fonzo, A.; Giampietri, E.; Martelli, A.; Cimino, O.; Dalla Marta, A.; Annosi, M.; Blanco-Velázquez, F.; Del Giudice, T.; Altobelli, F. A Step toward Water Use Sustainability: Implementing a Business Model Canvas for Irrigation Advisory Services. *Agriculture* **2023**, *13*, 1081. [CrossRef]
5. Gu, Y.; Qi, C.; Liu, F.; Lei, Q.; Ding, Y. Spatiotemporal Evolution and Spatial Convergence Analysis of Total Factor Productivity of Citrus in China. *Agriculture* **2023**, *13*, 1258. [CrossRef]
6. Bontsa, N.; Mushunje, A.; Ngarava, S. Factors Influencing the Perceptions of Smallholder Farmers towards Adoption of Digital Technologies in Eastern Cape Province, South Africa. *Agriculture* **2023**, *13*, 1471. [CrossRef]
7. Donati, I.; Viaggi, D.; Srdjevic, Z.; Srdjevic, B.; Di Fonzo, A.; Del Giudice, T.; Cimino, O.; Martelli, A.; Dalla Marta, A.; Henke, R.; et al. An Analysis of Preference Weights and Setting Priorities by Irrigation Advisory Services Users Based on the Analytic Hierarchy Process. *Agriculture* **2023**, *13*, 1545. [CrossRef]
8. Li, X.; Yang, P.; Zou, Y. An Empirical Investigation of the "Mezzogiorno Trap" in China's Agricultural Economy: Insights from Data Envelopment Analysis (2015–2021). *Agriculture* **2023**, *13*, 1806. [CrossRef]
9. Verdugo, G.; Cuadrado, G.; Castillo, Y. Family Farming as a Contribution to Food Sovereignty, Case Guarainag Parish. *Agriculture* **2023**, *13*, 1827. [CrossRef]
10. Sudomo, A.; Leksono, B.; Tata, H.; Rahayu, A.; Umroni, A.; Rianawati, H.; Asmaliyah; Krisnawati; Setyayudi, A.; Utomo, M.; et al. Can Agroforestry Contribute to Food and Livelihood Security for Indonesia's Smallholders in the Climate Change Era? *Agriculture* **2023**, *13*, 1896. [CrossRef]
11. Gu, Y.; Qi, C.; He, Y.; Liu, F.; Luo, B. Spatial Correlation Network Structure of and Factors Influencing Technological Progress in Citrus-Producing Regions in China. *Agriculture* **2023**, *13*, 2118. [CrossRef]
12. Romeo Lironcurti, S.; Demaria, F.; D'Annolfo, R.; Sardone, R. Consumer Evaluations of and Attitudes towards New Genome Editing Techniques: An Italian Case Study. *Agriculture* **2024**, *14*, 51. [CrossRef]

agriculture

MDPI

Article

Consumer Evaluations of and Attitudes towards New Genome Editing Techniques: An Italian Case Study

Simona Romeo Lironcurti *, Federica Demaria, Raffaele D'Annolfo and Roberta Sardone

CREA—Research Centre for Agricultural Policies and Bioeconomy, 00187 Rome, Italy; federica.demaria@crea.gov.it (F.D.); raffaele.dannolfo@crea.gov.it (R.D.); roberta.sardone@crea.gov.it (R.S.)

* Correspondence: simona.romeo@crea.gov.it; Tel.: +39-0647856312

Abstract: The ever-increasing development of assisted evolution technologies (AETs) in agriculture has boosted crop improvement. The commercialization of improved biotech crops can be promoted by modern gene editing instead of conventional genetic modification, which is a cheaper and faster approach that can help address future agriculture challenges, such as food security, environmental sustainability, and climate change. However, the use of these technologies is still sensitive and debated in many countries. Each region promotes a different approach, depending on regulatory policies, and adopting these technologies requires knowledge of consumer views and stakeholder acceptance. For this purpose, we conducted a survey of 564 Italians regarding their knowledge of genetic techniques, informational tools, purchase preferences, environmental sustainability, and food safety issues. The research aims are twofold: (a) to assess the level of knowledge and (b) to determine how consumer background, including social and demographic characteristics, affects their level of knowledge. Our findings emphasize the importance of communication and dissemination activities, in which clarity and a broad appeal are key.

Keywords: assisted evolution technologies (AETs); consumer attitude; environmental risk and food safety; Italian consumer sample

Citation: Romeo Lironcurti, S.; Demaria, F.; D'Annolfo, R.; Sardone, R. Consumer Evaluations of and Attitudes towards New Genome Editing Techniques: An Italian Case Study. *Agriculture* **2024**, *14*, 51. https://doi.org/10.3390/agriculture14010051

Academic Editors: Xiangping Jia and Maurizio Lanfranchi

Received: 13 October 2023
Revised: 28 November 2023
Accepted: 21 December 2023
Published: 27 December 2023

1. Introduction

In recent decades, interest in food security and environmental sustainability has steadily increased as the global population grows. Addressing these issues requires continuous improvement in food production methods, more sustainable systems, and higher product yields.

Advances in science and technology have significantly affected the food industry, improving food production, increasing crop resistance to diseases and drought, and increasing food nutritional content.

Consumers are increasingly interested in learning about their food sources and how different food production systems impact the environment and food safety. In particular, there is growing interest in understanding genetically modified food crops.

In Europe, as in many other places around the world, the related legislation represents the main obstacle to the use of these new technologies. Since 2001, assisted evolution technologies (AETs) have developed rapidly. In response, in 2019, the European Council asked the European Commission to assess the adequacy of existing legislation on the development and potential application of New Genomic Techniques. The resulting Commission study on NGTs (2021) found that the existing legislation was not fit for purpose and was ineffective in terms of risk assessment. The study argued that risk assessment should have requirements adapted to the characteristics and risk profile of a plant and stated that such assessment should be deemed unnecessary for plants produced through conventional plant breeding or classical mutagenesis. The study concluded that there are strong indications that the current EU GMO legislation is not fit to regulate NGT plants obtained by targeted mutagenesis or cisgenesis, nor products (including food and feed) derived from them, and consequently, legislation should be adapted on the basis of scientific and technical progress [1–3].

This raises the problem of different applications of national laws between countries. Indeed, the various national legal frameworks show divergences in biotechnology regulation, which contributes to limiting biotechnological development [4]. Several GM food issues have also become controversial at different levels, from food security to environmental issues, including risks and benefits.

Assisted evolution technologies, including NBTs, are an alternative to traditional genetic selection tools and represent a potential response to the important challenge of sustainably increasing food production. Differently from GMOs, NBTs also allow the introduction of new characteristics without involving genes from other species, keeping the genetic heritage unaltered. Consequently, the resulting plant product is free from foreign genes and would not be distinguishable from the product generated by conventional breeding techniques [5].

Myths and facts guide consumer choices. One popular myth is that GM foods are not natural food products and, thus, are unsafe for consumption, as most of the relevant literature highlights. The fact is that GM foods are safe, and almost everything we eat is genetically modified; just think of "pachino" tomatoes, apples, or corn. The literature has also shown that consumers can have both positive and negative perceptions of GM products that are offset by personal values and ethics [6]. Scholars also stress that the perception and acceptance of GM foods could be influenced by the values held by specific groups, including overall concerns about global food and food security, climate change, and ethical beliefs. Moreover, empirical evidence suggests that more education in terms of quality information would allow consumers to make purchasing decisions that accurately reflect their beliefs [7].

The level of consumer acceptance of GM products is also guided by trust in institutions, scientific knowledge, and pricing. This means that the government's decision to ban or approve GM crop cultivation, as well as the communication strategies used, can influence consumer attitudes and choices [8,9]. Literature findings reveal that perceived benefits and risks play a significant role in determining consumer behavior toward GM food.

Finally, gender differences and income levels also seem to play a role in consumer perceptions. Bellows et al. [10] found that women were more strongly inclined to purchase non-GM foods, and income level impacts attitudes toward GMOs.

In light of the above, the present study aims to understand Italian consumers' feelings towards and perceptions of AET. By using a descriptive analysis, we study the correlation between the consumer's level of knowledge of AET and sample demographic characteristics. The correlation between consumer backgrounds and knowledge levels has been confirmed here. We used an original Italian sample administered via the Appinio agency platform. The relevance of this research is linked to the reluctance of Italian consumers to accept genetically modified foods. The results of this study reveal that consumer opinions are heavily linked to the information tools used, where information accessed via social media, press, and television is of poor quality, and scientific communications are aimed at a very small audience.

2. Literature Review

The development of new genomic techniques has been an incentive to launch a revision of the legal framework governing the use of agricultural biotechnologies. The use of advanced breeding techniques will likely be required by the Green Deal, which seeks to transform the EU into a modern, resource-efficient, and competitive economy.

These techniques could provide the key to addressing modern agricultural challenges (climate change, food supply, environmental sustainability). However, understanding consumer opinions is essential since they are the end users of products. Consumer knowledge of genetic techniques is low [11]. McGarry et al. [12] compared consumer knowledge levels in the United States, Japan, and Italy, finding that US consumers were more likely to be familiar with GMOs, with a percentage of 40.9% compared to 33.3% in Japan and only 28% in Italy.

Lusk et al. [13] find heterogeneity in terms of knowledge and that some consumers recognize there are differences between GMOs and genome editing (GE). Consumers pay attention to risk analyses in terms of health and the environment rather than the process used to create new products. Consumer knowledge affects their attitudes toward GM food and other consumer goods. Knowledge of GM technology is also linked to consumers' perceptions of the benefits and risks of genetically modified food and is considered a significant factor in correcting distorted perceptions and intentions regarding GMOs [14].

Another branch of the literature points out that one of the main obstacles to consumer acceptance is attributable to the type of information tools used [7,15,16].

The concerns related to food safety remain, although there is some consensus on the contribution of biotechnology to sustainable food systems, particularly in creating opportunities for environmental protection and generating good performance in the agricultural and food sectors [17–21].

Some scholars agree that consumers perceive transgenic products with the same properties as conventional ones and, consequently, the same probability of being harmful to humans. These consumers also believe GM foods positively address global food insecurity [17,19,22]. Conversely, many consumers still believe that genetically modified plants are the result of an artificial crossing of naturally incompatible genes; hence, a source of high risk for human health. These consumers believe GM foods alter the characteristics of native foods, causing harmful health effects. Many uncertainties remain, and consumers show strong resistance to food safety dimensions, feeding the scientific debate on the use of biotechnology in the agri-food field [8,23–27].

De Marchi et al. [28] investigate the motivations for consumer acceptance of cisgenic products. By comparing four information treatments related to basic information, naturalness, health, and environment, the authors show that information on health-related benefits and, particularly, environmental benefits helps generate a positive communication landscape around cisgenic food. The results suggest the need to develop food policies and new communication strategies aimed at increasing consumer acceptance of edited food.

The introduction of GM food should be accompanied by adequate policies to guarantee consumer safety; this would allow a decrease in consumer-perceived risk concretely relating to health [29,30].

Controversies linked to sociocultural factors, lack of public education, deficits in science communication, and ethical issues remain.

It is necessary to develop scientific evidence-based decision-making models and assess the risks and benefits of agricultural biotechnology with the highest scientific rigor, considering agricultural biotechnology could reasonably be considered practicable when certain conditions are met: consumers have a good knowledge of the products; products are accepted globally and by the communities in which they are developed; there is potential to directly benefit farmers; and there is responsible design to limit and minimize risks [31].

Nonetheless, disagreement regarding GM foods persists on several levels, including whether it is safe, whether it should be labeled and if so, how, and whether agricultural biotechnology is needed to address world hunger now or in the future. The empirical evidence has demonstrated that innovations based on these new research techniques provide substantial economic benefits and environmental improvements in critical ecosystems with important overall contributions to sustainable development, specifically in terms of higher yields, lower pesticide use, increased farm incomes, reduced pollution, and increased resilience to weather [32].

Many consumers have no deep knowledge of GM foods and could overestimate their objective knowledge. Moreover, McGarry et al. [12] show that European consumers have a strong preference for organic foods, which could affect their attitudes toward GM products. The US and EU have two contrasting approaches to GMO legislation in terms of approval. The US advocates for GMOs through approval and production, but in the EU, the precautionary principle toward GMO use is being applied. These legal norms are closely related to various social values, directly or indirectly [33]. The European Commission is taking action to rebuild the current EU GMO legislation in order to exclude NGTs from the current detection, labeling, and approval requirements with the aim of informing the public about the benefits of NGTs and breaking the resistance to GM crops.

Scholars have validated that social trust can play a prominent role, in particular, trust in regulatory agencies and procedures in use [31].

Increased knowledge through communication and education might affect public perceptions about agricultural biotechnology, in particular, building trusting relationships between experts, policymakers, regulators, and the public.

3. Materials and Methods

3.1. Survey Description

The purpose of this document is to examine, via a questionnaire, consumer's feelings about genetic modification techniques. In particular, we explore the relationship between level of objective and subjective knowledge and levels of income and education. The dataset used is original and derived from a sample analysis involving 564 men and women residing in Italy who answered 15 questions in Italian (see Appendix B). The survey was conducted by the Appinio research agency (Appinio DEU, Hamburg, Germany) using a CAWI methodology (computer-assisted web interviewing), which is a survey and data collection method that relies on web-based technology to administer questionnaires and collect responses from participants. Data was automatically collected on Appinio's own survey platform. Subjects came from two main sources: from Appinio's own panel, i.e., people who registered to the Appinio App and expressed their consent and availability to complete surveys, or from partner panels, which work as an extension of the Appinio panel.

In both cases, upon registration, panelists were invited to provide their basic socio-demographic information, mainly gender, age, location, educational level, employment status, civil status, and household income. Thanks to this information, Appinio can direct the survey invitations exclusively to respondents that match specific survey requirements (e.g., if the target group consists of people with a university degree, only people who previously declared having a degree would receive the notification). This methodology allows the Appinio data to be of high accuracy and quality—meeting exactly the quotas needed for each study.

After delivery of surveys, APPINIO collects the answers given by the end users and provides them in anonymous, aggregated form, together with anonymous demographic and statistical data on the client in the form of an evaluation for market research purposes.

Panelists provide us with this data voluntarily and to a self-determined extent. The data provided are aggregated and made available anonymously in a statistical evaluation that does not allow any conclusions to be drawn about individual users.

The information on the specific use of data that Appinio collects and provides is available in Appendix C.

The sample collected and used in this study is representative and stratified by sex and age. A high standard of responses was granted from Appinio quality controls, with a low margin of error and a high confidence level (e = 4.38% | 95% C.I.).

3.2. Survey Structure

The questionnaire was divided into four sections, as described in Figure 1. The first stage of the questionnaire includes a self-assessment of knowledge with questions related to subjective and objective knowledge; it explores what consumers think they know about genetically modified techniques (GMT) and what they really know. The next questions relate to the tools used to acquire information on GMTs. The third section comprises topics related to the environment, food safety, and willingness to purchase GM products.

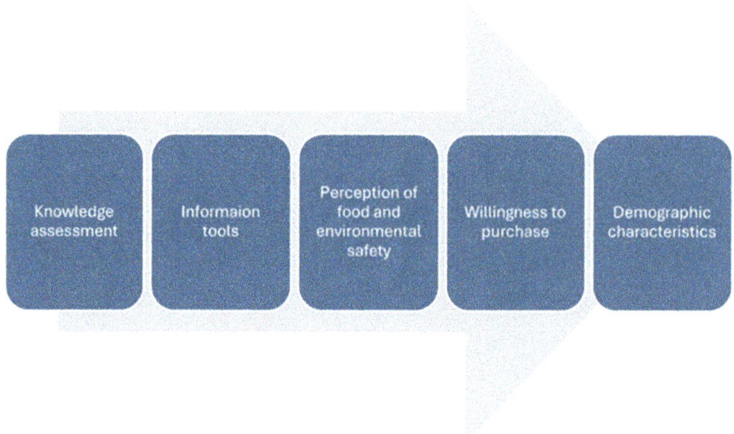

Figure 1. Survey structure.

Some of the responses were based on a 5-point Linkert scale; others were based on single or multiple-choice. Some of these responses were recorded based on their distribution.

In the first part of the survey, regarding knowledge, respondents answered four questions on GM food; three of these were related to subjective facts, while one was related to objective facts, with only one correct answer option (see Appendix B, questions F1 to F4 in the survey). Participants were asked what genetic improvement methods they know with multiple response options, mainly distinguishing between techniques that are used regardless of genetic modification and new techniques, like GMOs or NBTs (see Appendix B, question F1 in the survey). Moreover, the knowledge of the differences between NBTs and GMOs and the purpose of genetic modification has been investigated (see Appendix B, questions F2 and F3 in the survey). The last question of the first section regards the "objective" or real knowledge and investigates the consumer's knowledge of the most cultivated agricultural species, such as GMO seeds (see Appendix B, question F4 in the survey). A score ranging from 1 to 5 indicates the level of objective knowledge on the topic. Respondents who chose soy and cotton have a high level of objective knowledge. Based on the answer to question 1 (see Appendix B, question F1 in the survey), we generated a new variable to capture the subjective knowledge, which is the sum of the methods known by respondents. This variable captures the level of presumed knowledge. We also investigate what respondents think about the scope of genetic modification by evaluating both the single and multiple-choice responses (see Appendix B, question F3 in the survey).

Consumer knowledge is key in the consumer's perception and propensity to purchase GM products. Nevertheless, the literature highlights the existence of a gap between objective and subjective knowledge. The first one refers to what consumers really know about GM techniques, while the latter concerns what consumers think they know. Fernbach et al. [34] pinpoint that consumers who are opposed to GM products have a high subjective knowledge but low real knowledge. Therefore, it is important to understand this gap to address the propensity of the public to negatively regard NBTs and GMOs, which are often considered to be the same.

Section two includes a question related to the use of information tools (see Appendix B, question F5 in the survey). The variable related to information tools is scored from 1 to 5, and it is also coded as the sum of the total number of information tools used to acquire materials on GM food crops and products.

Section three includes questions related to the perception of food and environmental safety (see Appendix B, questions from F6 to F8 in the survey). Participants were asked to choose from the following options: (A) I believe GM foods are unsafe (to eat or for the environment), (B) I believe they are a little safe, (C) I don't know, (D) I believe they are fairly safe, (E) I believe they are absolutely safe. Moreover, participants were asked whether they believe GM foods or crops could contribute to environmental sustainability. These variables are both re-coded between 1 and 3, where 1 corresponds to "No, they don't"; 2, "I don't know"; and 3, "Yes, they do".

Finally, the survey explores consumers' attitudes toward purchasing GM products (see Appendix B, questions from F9 to F10 in the survey). Specifically, it investigates the willingness to purchase GM foods. Also, this score, initially ranging between 1 and 5, has been re-coded in a new variable with 1 for "I don't buy", 2 for "I don't know", and 3 for "I do buy". Additionally, we investigate the willingness to buy depending on the product's origin.

Table 1 shows the demographic characteristics (The survey, reported in Appendix B due to space constraints, includes only questions related to income, educational degree, and geographical area. Gender and age were used for the sample stratification.) of the sample regarding age, gender, income, education level, and region. The respondents, men and women between 18 and 65 years old, were prescreened by gender, education, household, region, and income to ensure the representativeness of the statistical population. The distribution of the sample by geographical area is illustrated in Figure 2, with a greater component in northeastern Italy, equal to 14.5% of the women and 14.7% of the men.

Table 1. Demographic characteristics (*N* = 564).

Gender	*N*	**(%)**
Male	275	48.7
Female	288	51.1
No response	1	0.2
Total	***564***	***100***
Income		
0–15,000 Euros	160	28.3
15,000–30,000 Euros	249	44.3
30,000–50,000 Euros	99	17.6
>50,000 Euros	55	9.8
Total	***564***	***100***
University Degree		
Doctorate Degree	27	4.8
Master	98	17.4
First Degree	77	13.7
High school diploma	304	53.9
Middle school diploma	55	9.8
No qualification	3	0.4
Total	***564***	***100***
Number of family members		
1	85	15.1
2	158	28.0
3	175	31.0
4	115	20.4
5	27	4.8
+6	3	0.5
No response	1	0.2
Total	***564***	***100***

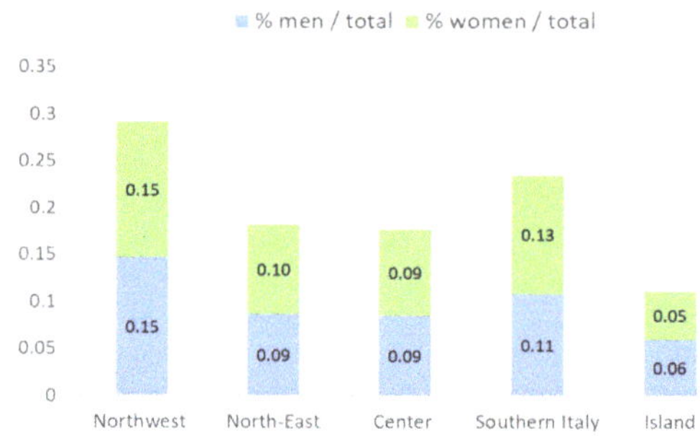

Figure 2. Sample distribution by geographical area.

As shown in Table 1, respondents comprise 275 men (48.8%) and 288 (51.1%) women. More than half of the sample reported having a high school diploma, 13.7% a bachelor's degree, and 17.8% a master's degree. Forty-four percent of respondents earn between 15,000 and 30,000 euros per year (249 participants out of 564, or 44.3%), while the remainder are divided between those who earn more than 30,000 euros and those who earn less than 15,000 euros.

4. Results

4.1. A Preliminary Analysis of the Survey Results

Among all techniques, most people (68%) stated that the most improved methods they know are the GM methods, followed by in vitro culture techniques (47%). Specifically, the vitro culture technique is more known by women than men (W: 51% vs. M: 42%). Only 7.1% are informed about new breeding techniques; we can say that Italian consumers have heard little about NBTs, and the percentage of respondents is really informative in that sense. The responses to this question highlight the knowledge gap in the Italian population. This is strictly related to the other two questions on the differences between GMOs and NBTs and on the purpose of genetic modification. Concerning the differences between GMOs and NBTs, 42% stated they know that there are differences between all the new breeding techniques, but only 15% really know what these differences are. In contrast, 2.8% claimed to know that differences between these two kinds of techniques exist (Figure 3a). Regarding the scope of genetic improvement, resistance to pests, diseases, and herbicides was recognized as the main purpose of genetic modification (48%), especially by respondents belonging to the age group 55–65. Adaptation to climate change was the second most selected option (Figure 3b).

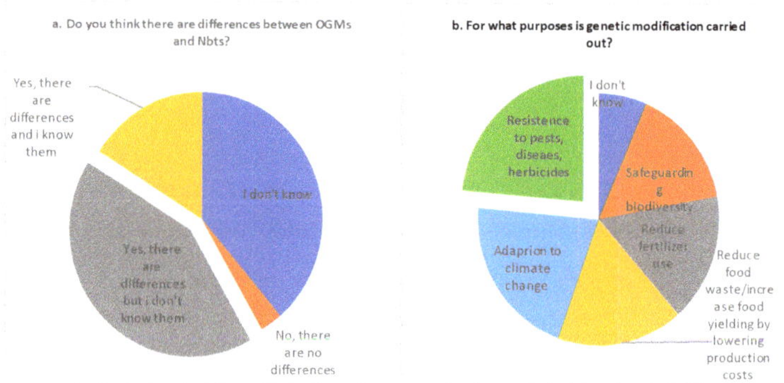

Figure 3. (**a**,**b**) Knowledge level of genetic improvement.

4.1.1. Consumers' Subjective and Real Knowledge

As the literature points out, consumers' knowledge affects their attitudes toward GM food. Consumers' knowledge of GM techniques is also related to their perceptions of the aims, risks, and benefits of GMOs. However, the level of knowledge is strictly related to educational level. Below, we analyze the subjective and real knowledge of the level of income and education.

Concerning the real knowledge, a question related to the most commonly cultivated GM seed was posed. Thirty-five percent of respondents thought this was maize, 21% wheat, while only 19% of respondents correctly selected soya, and 5% selected cotton. These results emphasize a distortion in consumers' real knowledge. This suggests that the level of consumer knowledge is relatively low. Because of this, confidence in GM products is undermined, and consumers may delay or avoid making decisions because they feel anxiety or uncertainty about their purchase consequences [14].

The second section of the survey deepens the relationship between genetic editing and environmental and food sustainability. Participants answered questions on two issues: whether they feel the GM products are safe to eat and whether they consider AET safe for the environment (Figure 4). Regarding GM food safety, 34% of respondents responded in the affirmative. Among them, men were more likely to be confident with them (average—M: 3.3% vs. W: 3%), and most of these respondents hold a doctoral degree.

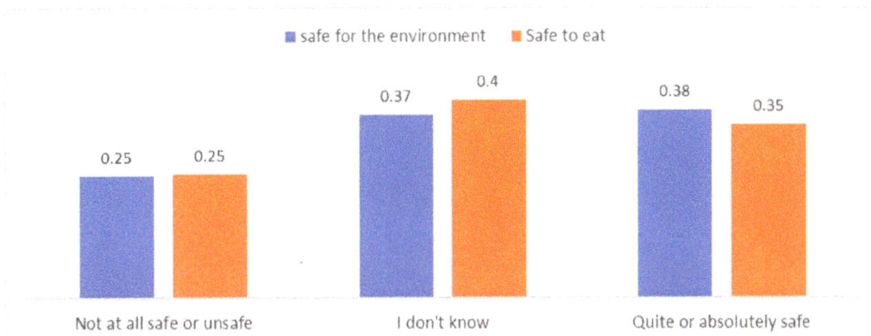

Figure 4. Food and environmental safety.

A degree of skepticism among Italian people towards genetically modified organisms has been confirmed [12]. However, when they are interviewed about the possible purposes/benefits of AET for the environment, their opinions improve. Thirty-eight percent of the sample believe that AETs are fairly or very safe for the environment, while more than half of the sample (55%) felt that AETs could contribute to environmental sustainability (Figure 4). Younger generations are more likely to identify the preservation of biodiversity as a goal. Twenty-seven percent were undecided on the subject. Men recognized the contribution of AET to sustainability more than women (M: 61% vs. W: 49%). Figure 4 shows that people have similar feelings in terms of food and environmental issues.

Scholars suggest people generally support cisgenic application to reduce pesticide residues but maintain more negative perspectives on GM foods [35].

4.1.2. Consumer Information Tools

Notably, when asked about the tools used to obtain information on GM products, 35.8% of respondents indicated social media, 33.7% scientific publications, 46.6% press and television, and 16.1% friends and acquaintances, while 12.3% do not use any kind of information tool and 1.4% selected "other". This result highlights how our sample relies on Press/Television and social media (with a percentage of 84%). This finding reflects the fact that confidence in scientific publications plays a lesser role in comparison to the other tools used by our sample. Unsurprisingly, this outcome is confirmed by other empirical work, such as [36], which suggests that the Press and Television can often create confusion that affects levels of the acceptance of GMOs. On the other hand, social media plays an intermediate role, more popular among young people [19,20].

4.2. Consumer Behavior

In the survey, we included a question related to product origin. Four in ten Italians would buy a product from GM crops. Consumers are more confident in products coming from Western Europe (62%) and North America (50%) than those from African (24%) and Asian (26%) countries (Figure 5).

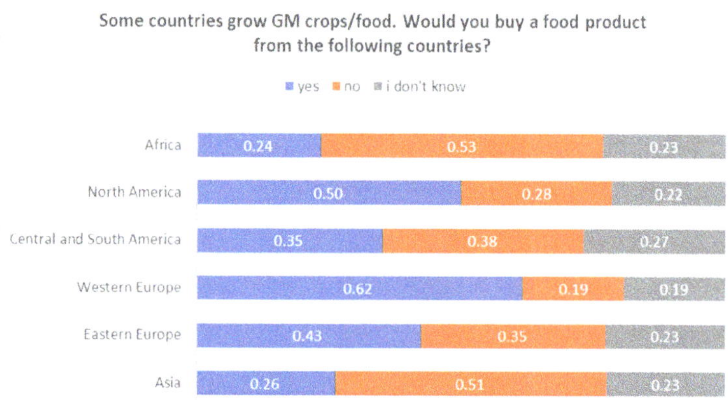

Figure 5. Geographical purchasing preferences.

4.3. Descriptive Analysis

Two statistical techniques were selected to analyze qualitative and quantitative variables extracted from the survey: (i) contingency tables have been used to explore the relationships between categorical variables (e.g., real knowledge, gender); (ii) a correlation matrix has been adopted to examine correlations between quantitative variables collected in the questionnaire (e.g., food security, environmental safety). The data were analyzed using R software (version 2023.06.1+524), and visualizations (e.g., tables, bar charts) have also been provided accordingly [37].

4.3.1. Consumer Analysis of Real and Subjective Knowledge

Figure 6 illustrates the conditional distribution of real knowledge for each subjective knowledge category. It sheds light on how people's self-assessments align with their understanding of the most grown GM crops. It is evident that a significant percentage of respondents (39.6%) confidently identify corn as the most cultivated GM crop, followed by wheat (27.5%), soy (20.9%), and cotton (8.8%) (see Table A1—Appendix A). Although many respondents consider themselves knowledgeable about GM crops and believe that corn is the most cultivated, in fact, the correct response is cotton and soybean.

Figure 7 offers insights into the conditional distribution of subjective knowledge across different levels of education, revealing how individuals' self-assessed knowledge varies based on their level of education. Among individuals with a high school diploma, which is the most prominent group, representing 54% of the overall sample, a significant proportion (52.7%) acknowledged a lack of knowledge by responding with "I don't know". In contrast, 7.3% confidently asserted "No" to having knowledge about GMOs. Approximately 31% expressed uncertainty with "Yes, I don't know", while 9.1% selected "Yes, I know". This pattern indicates that a notable portion of those with a high school diploma tend to admit uncertainty in their knowledge of GMOs.

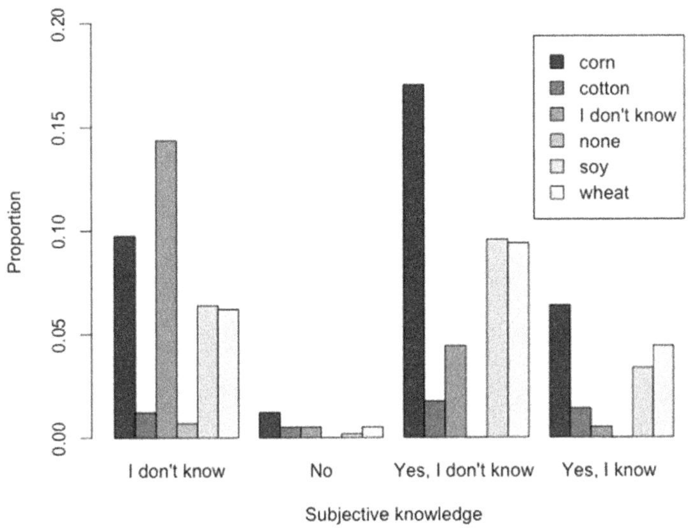

Figure 6. Conditional distribution of real knowledge by subjective knowledge category (proportion).

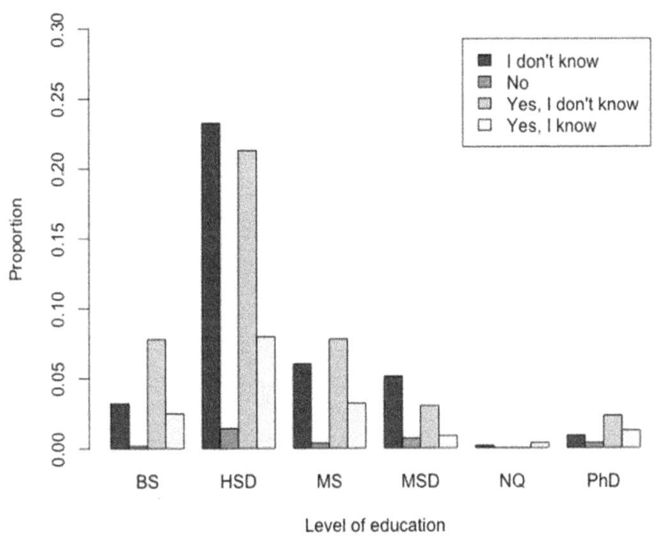

Figure 7. Conditional distribution of subjective knowledge by level of education (proportion). Note: bachelor's degree (BS), high school diploma (HSD), master's degree (MS), middle school diploma (MSD), no qualification (NQ), and doctoral degree (PhD).

Figure 8 examines the distribution of real knowledge across different educational backgrounds. As we can see from Figure 8, the following percentages indicate the respondents who have identified corn as the most cultivated crop: 39% (bachelor's degree), 30.6% (high school diploma), 44.9% (master's degree), 32.7% (middle school diploma), and 66.7% (no qualification). Individuals with doctoral degrees exhibited a lower percentage, with 25.9%, when identifying corn as the most cultivated crop. Conversely, they reported soy with 33.3% as the most cultivated GM crop, indicating a relatively greater real knowledge compared to other educational categories (see Table A2—Appendix A).

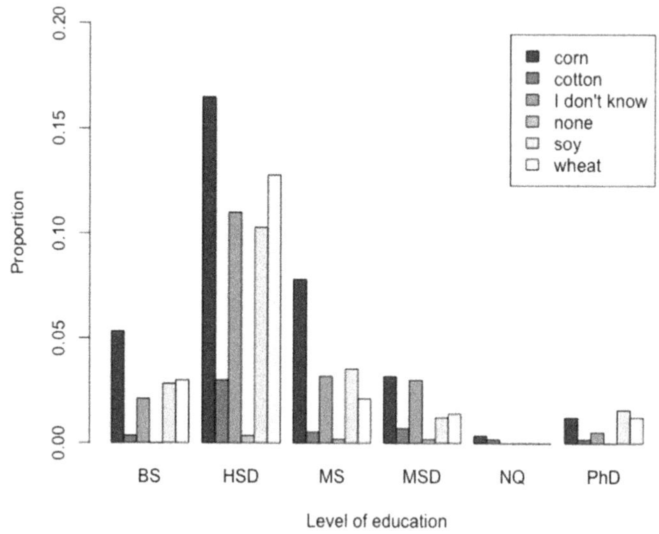

Figure 8. Conditional distribution of real knowledge by level of education (proportion). Note: bachelor's degree (BS), high school diploma (HSD), master's degree (MS), middle school diploma (MSD), no qualification (NQ), and doctoral degree (PhD).

Figure 9 shows the distribution of real knowledge across individuals based on gender. Male respondents displayed a higher response rate regarding corn at 38.9%, while female respondents had a lower percentage at 30.1%. Conversely, women exhibited a higher response rate regarding soy, with 21.1%, compared to men, who had a slightly lower percentage at 17.8%. Additionally, women tended to express more uncertainty, with 21.8% responding "I don't know", while men had a slightly lower percentage at 17.8% (see Table A3—Appendix A).

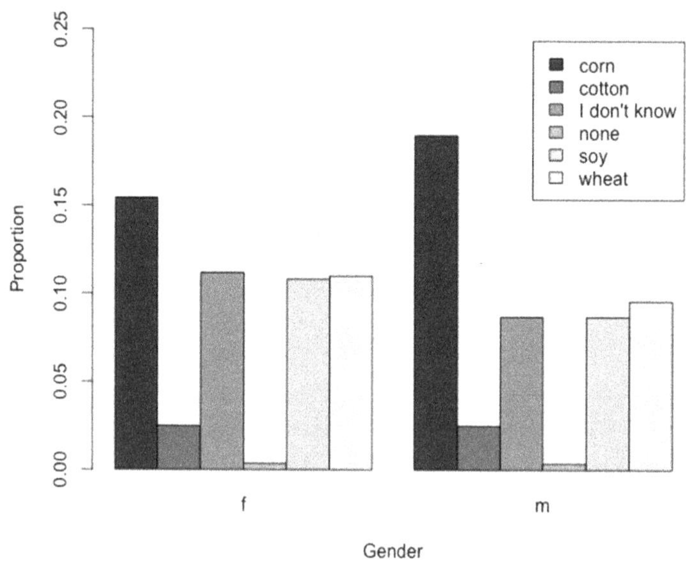

Figure 9. Conditional distribution of real knowledge by gender (proportion).

4.3.2. Consumers' Subjective Knowledge and Environmental Sustainability According to Income Levels

Figure 10 displays the conditional distribution of subjective knowledge among individuals with varying income levels. People with incomes between EUR 15,000–30,000 and between EUR 30,000–50,000 exhibited similar knowledge patterns. However, 50% of those with incomes below EUR 15,000 expressed uncertainty, responding with "I don't know", indicating a remarkable level of doubt regarding their knowledge of GMOs. Conversely, individuals with incomes exceeding EUR 50,000 displayed the highest percentage of having no knowledge of this topic, with 11% (see Table A4—Appendix A).

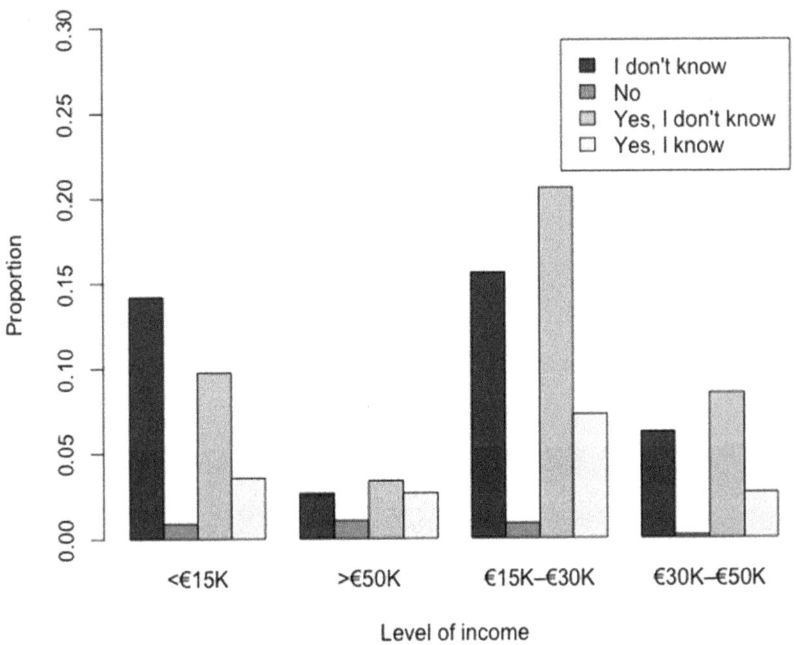

Figure 10. Conditional distribution of subjective knowledge by level of income (proportion).

Figure 11 illustrates variations in perceptions of the environmental sustainability of GMOs across diverse income levels. Individuals with incomes ranging from EUR 30,000 to 50,000 exhibited significant support for environmental sustainability, with 62.6% responding affirmatively with a "Yes". Furthermore, among those with incomes exceeding EUR 50,000, a substantial majority of 60% distinctly embraced a positive perspective on GMO's environmental sustainability. Within the income bracket of EUR 15,000 to 30,000, a significant 55.6% of individuals endorsed environmental sustainability, highlighting substantial support for GMO-related environmental sustainability in this income category. Lastly, individuals with incomes below EUR 15,000 also demonstrate substantial support, with approximately 46.9% expressing a favorable view of environmental sustainability (see Table A5—Appendix A).

4.3.3. Consumers' Real Knowledge According to Geographical Area

Figure 12 sheds light on how residents in the central, northeastern, northwestern, and southern areas understand the main GM crop cultivations. Individuals in all surveyed regions provided a high response rate related to corn, with percentages ranging from 30.1% to 40%. When looking at soy, the northeastern region stood out with 28.2%. In contrast, the central and northwestern regions exhibited rates of 20% and 23%, respectively, while the southern region presented a notably lower response rate on soy with 11.7%. Across all geographical regions, responses regarding cotton were relatively modest, ranging from 3% to 5.8% (see Table A6—Appendix A).

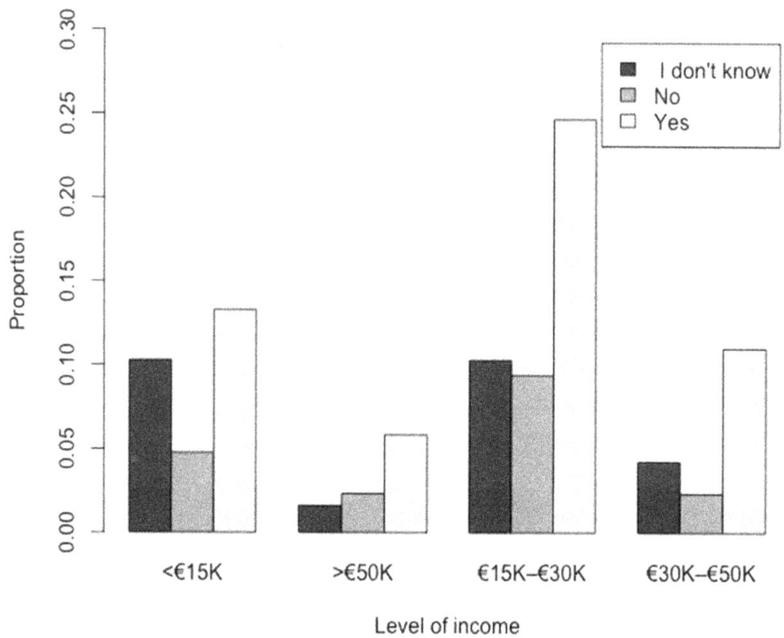

Figure 11. Conditional distribution of environmental sustainability by level of income (proportion).

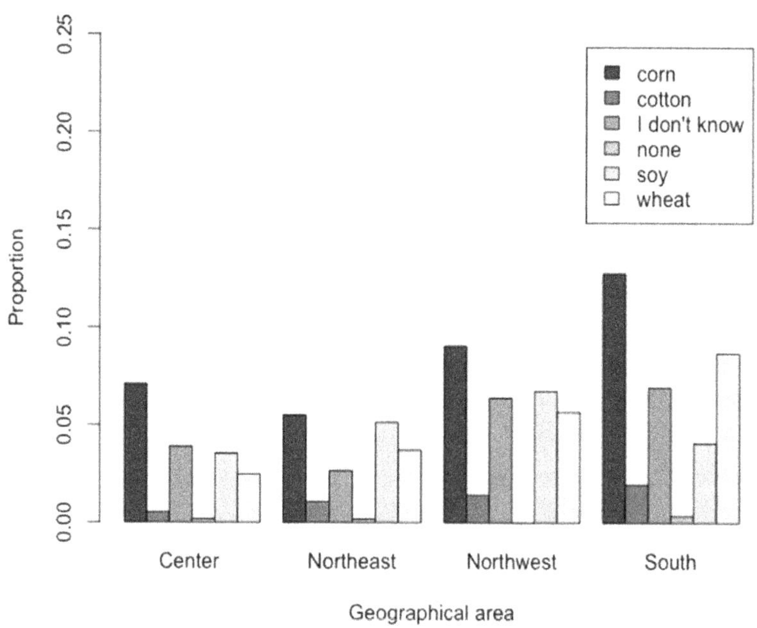

Figure 12. Conditional distribution of real knowledge distribution by geographical area (proportion).

4.3.4. Correlation Analysis: Food Security, Environmental Safety, and Consumer Purchase Propensity for GMOs

Figure 13 displays the level of correlation based on the Pearson correlation coefficient across key responses provided on the Likert scale: food security, environmental safety, and consumer purchase propensity for GMOs. The correlation analysis indicates a strong relationship between the analyzed variables. Firstly, a significant positive correlation of 0.8 was observed between food security and environmental safety responses. This suggests that as levels of food security perception increase, the perception of GMO environmental safety increases as well. Secondly, there was a positive correlation of 0.7 between food security and consumer purchase propensity. It indicates higher levels of food security responses associated with a greater purchasing intention. Thirdly, a robust positive correlation of 0.8 exists between environmental safety and consumers' attitudes toward purchasing GM products. This correlation implies that individuals who consider GM crops or foods environmentally safe may also be more inclined to purchase GM products.

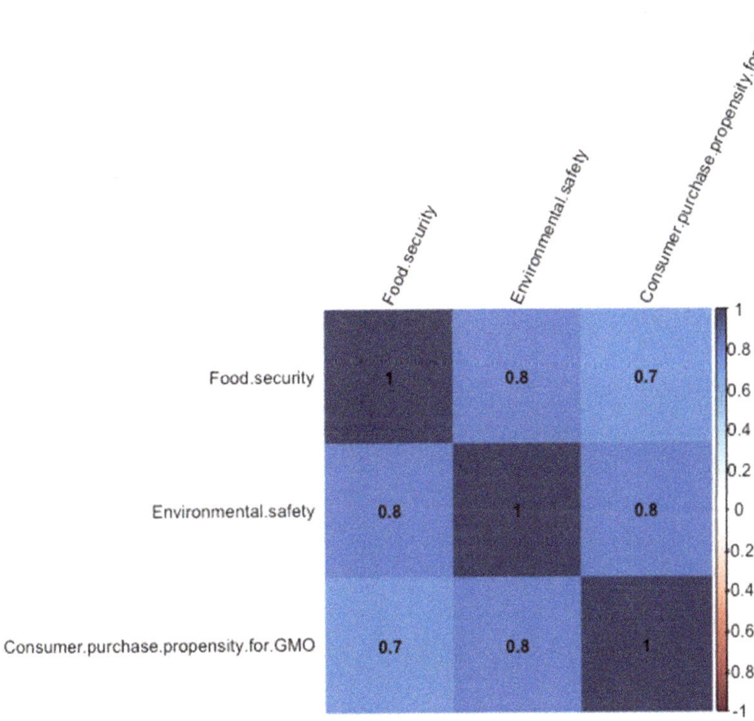

Figure 13. Heatmap: food security, environmental safety, and consumer purchase propensity for GMOs. Note: The color intensity represents the value of the data point, with lighter colors indicating lower values and darker colors indicating higher values.

5. Discussion and Conclusions

This study offers fresh insights into the knowledge of AET, the information tools employed, and the behavioral intentions of a sample of Italian consumers.

The analysis reveals that only a low percentage of consumers know NBTs; conversely, most respondents know in vitro culture techniques and GMOs. Only 16% of the sample stated that there are differences between GMO products and products obtained through the use of new breeding techniques, and they know what these are.

Additionally, consumers feel more confident if products originated in the most developed countries (Western Europe and the US). These results align with the research conducted by Hwang and Nam [14] in South Korea, which focuses on the influence of consumer knowledge on perceptions and purchase intentions toward GM food. The authors show that higher levels of education, income, and food involvement affect knowledge level by producing an overestimated effect.

The education level guides the subjective and real knowledge. Consumers with a high school diploma recognize a lack of precise knowledge of these techniques. Surprisingly, what is generally expected does not match the level of education. These results could be explained by the different information tools used to improve consumers' knowledge. Indeed, individuals often prefer non-scientific information tools, such as the Internet and media, which may contain not only expert sources but also consumers' feelings and unsupported claims. Consequently, consumers most exposed to negative feelings or information are more likely to overestimate their knowledge level. This has also been found by other researchers in different countries [38,39].

The literature has highlighted that a better understanding of GM foods is associated with a positive consumer attitude and purchase intention. However, this connection could be beneficial if the information is tailored and based on scientific approaches. Furthermore, subjective knowledge is linked to an income level where the low- and high-income categories seem to guide consumer knowledge. The role of the scientific community is a key point: researchers, together with other relevant stakeholders (e.g., government agencies, regulatory authorities, and biotechnology companies), should cooperate in developing communication strategies and dissemination activities to inform consumers about GM foods and crops and how the new genetic techniques differ from the traditional ones. The right communication could play a role in informing people and changing attitudes for a future consensus and acceptance of GM food products.

This study shows that both real and subjective consumer knowledge is relatively low, and there is a need to reduce the potential and imbalance gap between these two knowledge levels. Of course, this result could be reached through consumer education as part of educational curricula at school in the case of young people and with the right transparent and understandable communication for the other consumer category.

We are aware of the limitations of the current study, which is mainly based on descriptive analysis. This is a preliminary work that suggests the need for further investigation using econometric analysis. Our future research will entail: (a) enlarging the sample size and including other countries (EU and non-EU); (b) customizing the survey by adding questions on who should legislate AET, labeling, risk perception, and the willingness to pay for GM food; and (c) reinforcing the scope of some questions already present in the survey. Finally, we go further to deeply understand and test the real knowledge level and the intention to purchase.

Author Contributions: Conceptualization: S.R.L., F.D. and R.S.; methodology: S.R.L., F.D. and R.D.; validation: R.S., F.D. and R.D.; resources: S.R.L. and F.D.; data curation: S.R.L. and F.D.; writing—original draft preparation: S.R.L., F.D., R.D. and R.S.; funding acquisition: S.R.L. All authors have read and agreed to the published version of the manuscript.

Funding: This research received no external funding.

Institutional Review Board Statement: Not applicable.

Data Availability Statement: The data supporting the findings of this study are available from the corresponding author on request.

Acknowledgments: The authors would like to express their sincere gratitude to the two anonymous reviewers for their valuable comments and suggestions.

Conflicts of Interest: The authors declare no conflict of interest.

Appendix A

Table A1. Conditional distribution of real knowledge for each subjective knowledge category (percentage).

		Subjective Knowledge				
		I Don't Know	No	Yes, I Don't Know	Yes, I Know	Total
Real knowledge	corn	25.2	41.2	40.3	39.6	34.4
	cotton	3.2	17.7	4.2	8.8	5
	I don't know	37.2	17.7	10.5	3.3	19.9
	none	1.8	0	0	0	0.7
	soy	16.5	5.9	22.7	20.9	19.5
	wheat	16.1	17.7	22.7	27.5	20.6
	Total	100	100	100	100	100

Note: Pearson chi-squared test (8) = 28.5365 Pr = 0.000. We used the χ^2 test of association to verify whether or not subjective and real knowledge are independent or associated. The p-value suggests the variables are not independent; therefore, a strong relationship between these two categorical variables (subjective knowledge and real knowledge) exists.

Table A2. Conditional distribution of real knowledge for each level of education (percentage).

		Level of Education						
		Bachelor's Degree	High School Diploma	Master's Degree	Middle School Diploma	No Qualification	Doctoral Degree	Total
Real knowledge	corn	39	30.6	44.9	32.7	66.7	25.9	34.4
	cotton	2.6	5.6	3.1	7.3	33.3	3.7	5
	I don't know	15.6	20.4	18.4	30.9	0	11.1	19.9
	none	0	0.7	1	1.8	0	0	0.7
	soy	20.8	19.1	20.4	12.7	0	33.3	19.5
	wheat	22.1	23.7	12.2	14.5	0	25.9	20.6
	Total	100	100	100	100	100	100	100

Note: Pearson chi-squared test (20) = 31.9910 Pr = 0.043. The χ^2 test of association suggests that the two categorical variables (level of education and real knowledge) are not independent.

Table A3. Conditional distribution of real knowledge by gender (percentage).

		Gender		
		Female	Male	Total
Real knowledge	corn	30.1	38.9	34.4
	cotton	4.8	5.1	5
	I don't know	21.8	17.8	19.9
	none	0.7	0.7	0.7
	soy	21.1	17.8	19.5
	wheat	21.5	19.6	20.6
	Total	100	100	100

Note: Pearson chi-squared test (4) = 5.2681 Pr = 0.261. The χ^2 test value shows that real knowledge and gender are independent.

Table A4. Conditional distribution of subjective knowledge for each level of income (percentage).

		Level of Income (EUR)				
		<15K	>50K	15K–30K	30K–50K	Total
Subjective knowledge	I don't know	50	27.3	35.2	35.4	38.7
	No	3.1	10.9	2	1	3
	Yes, I don't know	34.7	34.6	46.4	48.5	42.2
	Yes, I know	12.5	27.3	16.4	15.2	16.1
	Total	100	100	100	100	100

Note: Pearson chi-squared test (9) = 32.4408 Pr = 0.000. The χ^2 value claims a strong association between income level and subjective knowledge.

Table A5. Conditional distribution of environmental sustainability for each level of income (percentage).

		Level of Income (EUR)				
		<15K	>50K	15K–30K	30K–50K	Total
Environmental sustainability	I don't know	36.3	16.4	23.2	24.2	26.4
	No	16.9	23.6	21.2	13.1	18.8
	Yes	46.9	60	55.6	62.6	54.8
	Total	100	100	100	100	100

Note: Pearson chi-squared test (6) = 15.8169 Pr = 0.015. Even in the case of environmental sustainability and income level, a relationship is confirmed by the χ^2 test.

Table A6. Conditional distribution of real knowledge by geographical area (percentage).

		Geographical Area				
		Center	Northeast	Northwest	South	Total
Real knowledge	corn	40	30.1	30.9	36.7	34.4
	cotton	3	5.8	4.9	5.6	5
	I don't know	22	14.6	21.8	19.9	19.9
	none	1	1	0	1	0.7
	soy	20	28.2	23	11.7	19.5
	wheat	14	20.4	19.4	25	20.6
	Total	100	100	100	100	100

Note: Pearson chi-squared test (12) = 20.3211 Pr = 0.061. The χ^2 test reveals once again an association between real knowledge and geographical areas.

Appendix B

Hamburg, 05.01.2023

Knowledge Genetic Improvement

Question No.	Survey	Question Type
F1	What genetic improvement methods do you know? Please select all the methods you have heard about. A: Crossing and selection B: Induced genetic mutation C: Assisted selection with molecular markers D: In vitro culture techniques E: New breeding techniques (Nbts) F: GMO G: None of these (not randomized)	Multiple Choice (Answers randomized)

Question No.	Survey	Question Type
F2	Do you think there are differences between GMO products and products obtained through the use of New breeding techniques (Nbts)? A: Yes, there are differences and I know them B: Yes, there are differences but I don't know them C: No, there are no differences Q: I don't know	Single Choice
F3	For what purposes is genetic modification carried out? Select, from the following options, what you think are the purposes of genetic modification. A: Resistance to pests, diseases, herbicides B: Adaptation to climate change C: Reduce food waste/increase food yield by lowering production costs Q: Reduce fertilizer use E: Safeguarding biodiversity F: I don't know (not randomized)	Multiple Choice (Answers randomized)
F4	Based on your knowledge, which of these agricultural species is the most cultivated as GMO seed? A: Soyabean B: Corn C: Wheat D: Cotton E: I don't know (not randomized) F: None of the above (not randomized)	Single Choice (Answers randomized)
F5	What information tools do you use to obtain information on GM products? A: Friends and acquaintances B: Press and television C: Social Media D: Scientific and information publications E: None of these (not randomized) Other tools: (Freetext)	Multiple Choice (Answers randomized)
F6	Do you think GM food products are safe to eat? A: Not at all safe B: Not very safe C: I don't know Q: Quite safe E: Absolutely safe	Likert
F7	In your opinion, could AET contribute to environmental sustainability? A: Yes B: No C: I don't know	Single Choice
F8	How environmentally safe do you think AETs are? A: Not at all safe B: Unsafe C: I don't know Q: Quite safe E: Absolutely safe	Likert
F9	How likely would you be to buy a product from GM crops? A: No chance B: Low probability C: I don't know Q: Some probability E: High probability	Likert
F10	Some countries grow GM crops/food. Would you buy a food product from the following countries? Answer: A: yes B: No C: I don't know Items: A: Africa B: North America C: Central and South America D: Western Europe E: Eastern Europe F: Asia	Matrix (Items randomized)

Question No.	Survey	Question Type
F11	What region do you live in? A: Valle d'Aosta B: Piedmont C: Liguria D: Lombardy E: Trentino Alto-Adige F: Veneto G: Friuli-Venezia Giulia H: Emilia Romagna I: Tuscany J: Umbria K: Marche L: Abruzzo M: Lazio N: Molise O: Campania P: Basilicata Q: Puglia R: Calabria S: Sardinia T: Sicily	Single Choice
F12	Please indicate the highest degree you have earned (degree you have actually completed—not the degree you are currently studying for). A: No qualification (primary school) B: Secondary school C: High school diploma D: Bachelor's degree E: Master's degree F: Doctorate	Single Choice
F13	What is your gross annual income? A: Less than 15,000 euros B: Between 15,000 and 30,000 euros C: Between 30,000 and 50,000 euros Q: More than 50,000 euros	Single Choice

Appendix C

Appendix C.1. Data Protection at Appinio

Data protection and the associated protection of privacy are of central importance to Appinio as a market research company. We are committed to our clients and panelists to ensure at all times that all data is protected in accordance with the legal requirements of the GDPR.

For Appinio, data protection and transparency about how data are handled are principles we follow throughout our product development, also known as Privacy by Design. As a market research company, we have a special position and responsibility with respect to data privacy, as we are always in the middle between clients and survey participants.

Below, we list resources that are available to you as an Appinio customer to inform you about our privacy practices. At the same time, this information will help you to also act in a privacy-compliant manner and to ensure that your own privacy, as well as that of the survey participants, is protected at all times.

Appendix C.2. Data Minimisation and Data Retention

Part of our data protection concept is data economy. According to our data retention policy, we only store the data we really need and only for as long as necessary. Data that users delete on Appinio will be removed within a maximum of 30 days. The only exceptions to this are very specific data, such as invoices and log files, which we need to keep in order to comply with our own legal obligations. For more information, please see our privacy policy.

To be able to guarantee these standards, we train our employees in the responsible handling of data in accordance with legal requirements and our internal guidelines. In addition, we have appointed a Data Protection Officer to oversee DSGVO-compliant data use at Appinio.

Our goal is not only to comply with legal requirements and our internal policies. We want to ensure that when you use Appinio, whether through our platform, in interactions with our employees, or as an Appinio panelist, you always feel good about your privacy as well. Therefore, we are committed to protecting your data and privacy.

Appendix C.3. Data Security

Encryption: All sensitive data (this includes all customer data, panelist data, and survey/response data) are secured via Sha265 encryption. All servers are also SSL secured, which means all data transfer is encrypted.

Server location: All servers are located in Germany (Frankfurt). Appinio uses the infrastructure of AWS (Amazon Web Services) to guarantee the highest security standards and, at the same time, the highest accessibility of the systems. All survey data are stored here and made available exclusively to you, our customer, via our platform.

Appendix C.4. Data Usage and Documents

The following is information on the specific use of data that Appinio collects and provides. These data processing operations are also listed in our privacy policy and terms and conditions. All documents can be accessed in their original form at:

Customer T&Cs (for use of the survey platform) https://research.appinio.com/#/en/tos (accessed on 4 July 2023) Customer Privacy Policy (for website, platform, webapp) https://www.appinio.com/en/privacy (accessed on 4 July 2023).

App Panelist T&Cs (accessible via app/play store and Appinio app) https://link.appinio.com/#/en/tos (accessed on 4 July 2023).

App Panelist Privacy Policy (accessible via app/play store and Appinio app) https://link.appinio.com/#/en/privacy (accessed on 4 July 2023).

The app privacy policy lists what happens to our panelists' data. The Website, Platform, and Web App Privacy Policy lists how we handle data from our customers and web app participants.

This information is intended to help understand and be transparent about the exact data processing procedures. Our privacy statements contain all the data processing provisions our customers need to be DSGVO compliant. In addition, our technical and organizational measures explain what security measures are in place internally to ensure data protection.

After delivery of surveys, APPINIO collects the answers given by the end users and provides them in anonymous, aggregated form, together with anonymous demographic and statistical data, to the client in the form of an evaluation for market research purposes.

Panelists provide us with this data voluntarily and to a self-determined extent. The data provided are aggregated and made available anonymously in a statistical evaluation that does not allow any conclusions to be drawn about individual users.

Appinio GmbH • Handelsregister Amtsgericht Hamburg HRB 133 503 • VAT-No/USt.-ID-Nr: DE 29 67 45 664

Bankkonto Kto: 100 223 5222 BLZ: 200 505 50 • Geschäftsführer: Jonathan Kurfess, Kai Granaß, Max Honig, Dr. Tizian Bonus SEPA Bankkonto IBAN: DE24200505501002235222 • BIC: HASPDEHHXXX

References

1. European Commission. *Explanatory Memorandum to COM(2023)411—Plants Obtained by Certain New Genomic Techniques and Their Food and Feed*; European Commission: Brussels, Belgium, 2023.
2. *EFSA Annual Activity Report 2022*; European Food Safety Authority: Parma, Italy, 2022. [CrossRef]
3. *EFSA Annual Activity Report 2021*; European Food Safety Authority: Parma, Italy, 2021. [CrossRef]
4. Turnbull, C.; Lillemo, M.; Hvoslef-Eide, T.A. Global regulation of genetically modified crops amid the gene edited crop boom–a review. *Front. Plant Sci.* **2021**, *12*, 630396. [CrossRef] [PubMed]
5. EPSO. Opinion on the SAM Explanatory Note on New Techniques in Agricultural Biotechnology. 2015. Available online: https://epsoweb.org/ (accessed on 18 September 2023).

6. Akbari, M.; Fozouni Ardekani, Z.; Pino, G.; Valizadeh, N.; Karbasioun, M.; Padash, H. Consumer Attitude towards Genetically Modified Foods in Iran: Application of Three-Dimensional Model of Corporate Social Responsibility. *Foods* **2023**, *12*, 1553. [CrossRef] [PubMed]
7. Wunderlich, S.; Smoller, M. Consumer awareness and knowledge about food sources and possible environmental impact. *Int. J. Environ. Impacts* **2019**, *2*, 85–96. [CrossRef]
8. Sendhil, R.; Nyikab, J.; Yadavc, S.; Mackolild, J.; Ge, R.P.; Workief, E.; Ragupathyg, R.; Ramasundaram, P. Genetically modified foods: Bibliometric analysis on consumer perception and preference. *GM Crop. Food* **2022**, *13*, 65–85. [CrossRef]
9. Marangon, F.; Troiano, S.; Carzedda, M.; Nassivera, F. Consumers' acceptance of genome edited food and the role of information. *Ital. Rev. Agric. Econ.* **2021**, *76*, 5–21.
10. Bellows, A.C.; Alcaraz, G.; Hallman, W.K. Gender and food, a study of attitudes in the USA towards organic, local, U.S. grown, and GM-free foods. *Appetite* **2010**, *55*, 540–550. [CrossRef]
11. McFadden, B.R.; Lusk, J.L. What consumers don't know about genetically modified food, and how that affects beliefs. *FASEB J.* **2016**, *30*, 3091–3096. [CrossRef] [PubMed]
12. McGarry Wolf, M.; Bertolini, P.; Shikama, I.; Bergerd, A. A Comparison of Attitudes Toward Food and Biotechnology in the U.S., Japan, and Italy. *J. Food Distrib. Res.* **2012**, *43*, 103–112. [CrossRef]
13. Lusk, J.L.; McFadden, B.R.; Wilson, N. Do Consumers Care How a Genetically Food Was Created or Who Created It? *Food Policy* **2018**, *78*, 81–90. [CrossRef]
14. Hwang, H.; Nam, S.-J. The influence of consumers' knowledge on their responses to genetically modified foods. *GM Crop. Food* **2021**, *12*, 146–157. [CrossRef]
15. Ishii, T.; Araki, M. Consumer acceptance of food crops developed by genome editing. *Plant Cell Rep.* **2016**, *35*, 1507–1518. [CrossRef] [PubMed]
16. Lucht, J.M. Public acceptance of Plant Biotechnology and GM crops. *Viruses* **2015**, *7*, 4254–4281. [CrossRef] [PubMed]
17. Noleppa, S.; Cartsburg, M. *The Socio-Economic and Environmental Values of Plant Breeding in the EU and for Selected EU Member States*; Hffa Research Paper; HFFA Research GmbH: Berlin, Germany, 2021.
18. Ferrari, L. Farmers' attitude toward CRISPR/Cas9: The case of blast resistant rice. *Agribusiness* **2022**, *38*, 175–194. [CrossRef]
19. Ferrari, L.; Baum, C.M.; Banterle, A.; De Steur, H. Attitude and labelling preferences towards gene-edited food: A consumer study amongst millennials and Generation Z. *Br. Food J.* **2021**, *123*, 1268–1286. [CrossRef]
20. Demaria, F.; Zezza, A. Scientific information and cognitive bias in the case of New Breeding Techniques: Exploring Millennials behaviour in Italy. *Ital. Rev. Agric. Econ.* **2022**, *77*, 41–60. [CrossRef]
21. Lassoued, L.; Phillips, P.W.B.; Smyth, S.J.; Hesseln, H. Estimating the cost of regulating genome edited crops: Expert judgment and overconfidence. *GM Crop. Food* **2019**, *10*, 44–62. [CrossRef]
22. Meldolesi, A. Can CRISPR Modify Attitudes on Food Crops? Nature Italy. 2021. Available online: https://www.nature.com/articles/d43978-021-00070-1 (accessed on 18 September 2023).
23. Ortega, D.L.; Lusk, J.L.; Lin, W.; Caputo, V. Predicting responsiveness to information: Consumer acceptance of biotechnology in animal products. *Eur. Rev. Agric. Econ.* **2020**, *47*, 1644–1667. [CrossRef]
24. Bain, C.; Lindberg, S.; Selfa, T. Emerging sociotechnical imaginaries for gene edited crops for foods in the United States: Implications for governance. *Agric. Hum. Values* **2020**, *37*, 265–279. [CrossRef]
25. Sheldon, I.M. Regulation of biotechnology: Will we ever 'freely' trade GMOs? *Eur. Rev. Agric. Econ.* **2002**, *29*, 155–176. [CrossRef]
26. Qaim, M. Role of New Plant Breeding Technologies for Food Security and Sustainable Agricultural Development. *Appl. Econ. Perspect. Policy* **2020**, *42*, 129–150. [CrossRef]
27. Bunge, J.; Marcus, A.D. Is this tomato engineered? Inside the coming battle over gene-edited food. *Wall Str. J.* **2018**. Available online: https://www.wsj.com/articles/is-this-tomato-engineered-inside-the-coming-battle-over-gene-edited-food-1523814992 (accessed on 18 September 2023).
28. De Marchi, E.; Cavaliere, A.; Banterle, A. Consumer Choice Behavior for Cisgenic Food: Exploring the Role of Time Preferences. *Appl. Econ. Perspect. Policy* **2021**, *43*, 866–891. [CrossRef]
29. Martinez-Poveda, A.; Molla-Bauza, M.B.; del Campo Gomis, F.J.; Martinez, L.M. Consumer-perceived risk model for the introduction of genetically modified food in Spain. *Food Policy* **2009**, *34*, 519–528. [CrossRef]
30. Bawa, A.S.; Anilakumar, K.R. Genetically modified foods: Safety, risks and public concerns—A review. *J. Food Sci. Technol.* **2012**, *50*, 1035–1046. [CrossRef] [PubMed]
31. Harfouche, A.L.; Petousi, V.; Meilan, R.; Sweet, J.; Twardowski, T.; Altman, A. Promoting Ethically Responsible Use of Agricultural Biotechnology. *Trends Plant Sci.* **2021**, *26*, 546–559. [CrossRef] [PubMed]
32. Smith, V.; Wesseler, J.H.; Zilberman, D. New Plant Breeding Technologies: An Assessment of the Political Economy of the Regulatory Environment and Implications for Sustainability. *Sustainability* **2021**, *13*, 3687. [CrossRef]
33. Woźniak-Gientka, E.; Agata, T.; Milica, P.; Anna, B.; Dennis, E.; Nick, V.; Godelieve, G.; Selim, C.; Naghmeh, A.; Tomasz, T. Public perception of plant gene technologies worldwide in the light of food security, GM Crop. *Food* **2022**, *13*, 218–241.
34. Fernbach, P.M.; Light, N.; Scott, S.E.; Inbar, Y.; Rozin, P. Extreme opponents of genetically modified foods know the least but think they know the most. *Nat. Hum. Behav.* **2019**, *3*, 251–256. [CrossRef]
35. McFadden, B.R.; Smyth, S.J. Perceptions of Genetically Engineered Technology in Developed Areas. *Trends Biotechnol.* **2019**, *37*, 447–451. [CrossRef]

36. Araki, M.; Ishii, T. Towards social acceptance of plant breeding by genome editing. *Trends Plant Sci.* **2015**, *20*, 145–149. [CrossRef]
37. R Core Team. *R: A Language and Environment for Statistical Computing*; R Foundation for Statistical Computing: Vienna, Austria, 2023; Available online: https://www.R-project.org/ (accessed on 19 September 2023).
38. Kim, B.R. Consumer Attitude of Risk and Benefits toward Genetically Modified (GM) Foods in South Korea: Implications for Food Policy. *Inz. Ekon. Eng. Econ.* **2012**, *23*, 189–199. [CrossRef]
39. Rzymski, P.; Królczyk, A. Attitudes toward genetically modified organisms in Poland: To GMO or not to GMO? *Food Sec.* **2016**, *8*, 689–697. [CrossRef]

agriculture

Article

Spatial Correlation Network Structure of and Factors Influencing Technological Progress in Citrus-Producing Regions in China

Yumeng Gu [1,2,3], Chunjie Qi [1,2,*], Yu He [1,2,3], Fuxing Liu [1,3] and Beige Luo [4]

1 College of Economics and Management, Huazhong Agriculture University, Wuhan 430070, China; guyumeng@webmail.hzau.edu.cn (Y.G.); he785676854@webmail.hzau.edu.cn (Y.H.); liufuxing@webmail.hzau.edu.cn (F.L.)
2 Institute of Horticultural Economics, Huazhong Agriculture University, Wuhan 430070, China
3 Hubei Rural Development Research Center, Wuhan 430070, China
4 Yiling District Administration for Rural Revitalization, Yichang 443100, China; luobeigenice@163.com
* Correspondence: qichunjie@mail.hzau.edu.cn; Tel.: +86-158-7178-2598

Abstract: In this study, the transcendental logarithmic cost function model was used to measure the rate of technological progress in seven major mandarin-producing regions and seven major tangerine-producing regions in China from 2006 to 2021. The modified gravity model was used to establish spatial correlation networks. The social network analysis method was used to analyze the characteristics of the overall network structure and the individual network structure of the spatial correlation networks of citrus-production technology progress, and the quadratic assignment procedure was used to analyze the factors influencing the spatial network. The results show the production of Chinese mandarins and tangerines is in the stage of technological progress in general, but the rate of progress is slowing down gradually, and the rate of mandarin-production technology progress is higher than that of tangerine-production technology progress. In terms of the overall network structure characteristics, the spatial networks of technological progress related to Chinese mandarin and tangerine production are becoming increasingly dense and complex, with obvious spatial spillover effects, but the network structure is relatively loose, and the polarization of the tangerine network is more serious. In terms of individual network structure characteristics, the relatively economically developed eastern regions have a higher status in terms of the spatial correlation network and a stronger role in controlling and dominating the resource elements needed for citrus-production technology progress. Education, informatization, economic development, innovation support, and financial support are important factors influencing the formation of the spatial association network of citrus-production technology progress in China.

Keywords: citrus; technological progress; spatial correlation network structure; transcendental logarithmic cost function; social network analysis

Citation: Gu, Y.; Qi, C.; He, Y.; Liu, F.; Luo, B. Spatial Correlation Network Structure of and Factors Influencing Technological Progress in Citrus-Producing Regions in China. *Agriculture* **2023**, *13*, 2118. https://doi.org/10.3390/agriculture13112118

Academic Editors: Roberto Henke and Filiberto Altobelli

Received: 27 September 2023
Revised: 3 November 2023
Accepted: 7 November 2023
Published: 8 November 2023

1. Introduction

Citrus is one of the most important cash crops in the world and the largest category of fruits in the world [1]; it is the largest category in China in terms of planted area and production [2]. China's citrus industry ranks first in the world, and production accounts for about one-third of the world's production [3]. According to the China Rural Statistical Yearbook, in 2021, China's citrus planting area was 2.922 million ha and the production was 55.956 million tons, accounting for 22.82% of China's fruit planting area and 25.81% of the production. China's citrus industry has been developing rapidly, especially in the past 45 years, since the reform and opening up. In China, citrus varieties have been enriched, the spatial layout of citrus production has been optimized, citrus quality has been improved, farmers' enthusiasm for planting is high [4], and the promotion of the healthy development of the citrus industry has become one of the most important methods for boosting industrial prosperity and realizing the revitalization of the countryside [5]. According to the UN Comtrade Database, China's citrus export was 917,700 tons in 2021, accounting for only 5.96% of the world's total citrus exports. So, some scholars say that though China is the world's major citrus producer, it is not a powerhouse of citrus production and trade [6,7]. Compared with developed countries, China's citrus-production efficiency is low [8], and citrus production per unit area is lower than the world average [2]. According to FAO data, China's citrus production per unit area in 2021 was 15.37 t/ha, which is much lower than Indonesia's production per unit area, which is the highest in the world, with a production rate of 38.53 t/ha. China's citrus industry urgently needs to accelerate the innovation-driven transformation of the development mode from "extensive" to "intensive" [1] to improve citrus production per unit area, and the improvement of production per unit area is driven by technological progress [9]. Under the role of factor flow and market and government support mechanisms, technological progress among major citrus-producing provinces does not exist independently but shows a certain spatial correlation [10]. At the same time, in plant taxonomy, mandarins and tangerines belong to the same family and the same genus but are different species of woody plants. Mandarins and tangerines are often collectively referred to as "citrus." There are differences in the mandarin and tangerine planting areas in China, and mandarins and tangerines differ in terms of scientific and technological strength [11]. So, the rates of technological progress [12] and the characteristics of the spatial network structure are also different. Therefore, what is the level of citrus-production technology progress in China? What kind of changing trends exist in mandarin- and tangerine-production technology progress? What are the differences in the technological advances related to mandarins and tangerines? Are they spatially correlated? What are the characteristics of the spatial association network structure? What are the factors affecting the structural formation of citrus spatial association networks? Answering these questions is of great practical significance for optimizing the allocation of resource factors, promoting the technological progress in mandarin and tangerine production, improving production efficiency, and promoting the high-quality development of the citrus industry.

2. Literature Review

Technological progress is the use of a certain amount of input to produce more output, or, conversely, the use of less input to produce a certain amount of output [13]. Theoretical research on technological progress began in the early 19th century. In 1957, Solow created an economic growth accounting model to clarify the contribution of technological progress to economic growth [14]. Scholars at home and abroad began research on economic development and technological progress. Arrow put forward the concept of "learning by doing" and believed that the skills of workers would be continuously improved in production, which led to technological progress, and tried to endogenize technological progress for the first time [15]. Based on the neoclassical investment theory, through the selection of the transcendental logarithmic production function, Christensen et al. concluded that technological progress is the main reason for productivity change [16].

With the continuous progress in agricultural technology, productivity has greatly improved, and a large number of scholars have emerged in the field of agricultural technological progress research. The methods of measuring agricultural technological progress are mainly divided into two categories: the parametric method and the non-parametric method. Tan believes that the overall technological progress in agriculture can be divided into spontaneous technological progress and induced technological progress, and many scholars have followed suit to conduct separate research on spontaneous technological progress and induced technological progress [17]. Mao et al. used data envelopment analysis (DEA) to analyze the total factor productivity of Chinese agriculture in the period 1984–1993 and found it to be the main reason for the change in productivity [18]. Da Silva et al. measured the technological progress in Brazilian agriculture in 1976–2016 and analyzed the efficiency of factor input use in different periods [19]. Tan et al. investigated the relationship between agricultural technological progress, agricultural insurance, and factor input use and concluded that both agricultural technology progress and agricultural insurance have a positive impact on farmers' income [20]. Chen et al. measured different types of environmentally friendly technological progress in Chinese agriculture from 2000 to 2010 and analyzed the spatial spillover effect [21].

In the study of citrus-production technology progress, He et al. measured citrus technical efficiency and technological progress index in 20 cities in Sichuan, China, from 2009 to 2020 and concluded that it was on the low side, which led to low productivity [22]. Gu et al. measured and decomposed the total factor productivity of citrus in China from 2006 to 2020 using the DEA-Malmquist index method and concluded that technological progress is the main factor affecting the total factor productivity of citrus [2]. Xiang et al. analyzed the technical efficiency of citrus cultivation, the time series development law, and the influencing factors from 2007 to 2015 by using the beyond logarithmic production function and concluded that the overall average technical efficiency of tangerine production is higher than that of citrus, and there are regional differences in the technical efficiency of citrus production and cultivation [23].

As spatial analysis methods have improved, many scholars have used social network analysis to study the spatial correlation network structure and the factors influencing it, i.e., agricultural total factor productivity [24], agricultural green total factor productivity [25], agro-ecological efficiency [26], and green science and technology innovation efficiency [27]. In the study of the spatial correlation network structure of technological progress, Wang et al. concluded that there are obvious spatial correlation and spillover effects in the development of agricultural science and technology innovation in China and presented the shape of the spatial correlation network structure [28]. He et al. concluded that agricultural location centrality and intermediary centrality have a significant positive moderating effect on technological progress [29].

After combing through the literature, we found that there are abundant research results on agricultural technology progress and spatial correlation network structure in academic circles, which are of great reference value for this study, but there are still the following deficiencies: first, the existing studies focus on agricultural technological progress and there is a lack of studies on what kind of change characteristics citrus-production technology progress presents. Second, existing studies have failed to reveal the structural characteristics of the spatial correlation between citrus-production technology progress and the factors affecting the formation of its spatial correlation network structure.

In view of this and taking into account the differences in mandarin and tangerine cultivation and related technological progress, in this paper, based on the panel data of factor prices, inputs, and outputs of seven main mandarin-producing areas and seven main tangerine-producing areas from 2006 to 2021, we adopted the beyond logarithmic cost function to measure the citrus-production technology progress in China's main citrus-producing areas and construct a modified gravity model to determine the spatial correlation of citrus-production technology progress. Following this, the overall network structure and individual network characteristics of citrus-production technology progress were systematically analyzed using social network analysis (SNA), and the factors driving the spatial correlation of citrus-production technology progress in China were identified using quadratic assignment procedure (QAP) regression analysis with a view to providing certain reference and informative value for further exploration of the potential of citrus production in China and promoting the coordinated sustainable development of China's citrus industry.

3. Materials and Methods

3.1. Materials

To measure technological progress in citrus production, one output variable and five input variables were selected for this paper, taking into account the availability of data [11], as shown in Table 1. The output variable was the output of the main products per ha of citrus. The selected input variables were: labor price, expressed as labor cost per workday; land price, expressed as the land input cost per ha (including the rent of the transfer land and the folding rent of the self-camp); fertilizer price, obtained by dividing the fertilizer cost per ha by the amount of fertilizer per ha (including nitrogen fertilizer, phosphorus fertilizer, potash fertilizer, and compound fertilizer); and pesticide prices, obtained by dividing the cost of pesticides per ha by the amount of pesticides per mu. Other direct or indirect costs, such as the cost of farm fertilizer, drainage and irrigation, fuel and power, and marketing costs per acre, were used to measure other material and service price inputs, and since data on the volume of other material and service inputs were not available, the index of agricultural production materials was used in place of the other material and service prices.

Table 1. Descriptive statistics of input and output variables.

	Variable	Observed Value	Unit	Minimum Value	Maximum Value	Average Value	Standard Deviation
Output	Output of main products	224	kg	498.56	4633.70	1834.93	690.56
Input	Labor prices	224	CNY/workday	17.46	85.13	43.12	15.38
	Land prices	224	CNY/ha	387.50	6642.20	1965.78	69.18
	Fertilizer prices	224	CNY/kg	3.04	43.21	5.66	3.06
	Pesticide prices	224	CNY/kg	16.37	1441.29	273.51	238.37
	Other prices	224	—	100.00	184.97	147.07	22.48

In order to further explore the factors influencing the spatial correlation network structure of citrus-production technology progress in China, considering that the formation of this network is related to the mechanisms of factor flow, market drive, and government support and with reference to previous studies [24,25], this paper selected seven factors closely related to technological progress as research variables, namely industrial structure (str), level of informatization (inf), level of education (edu), level of economic development (eco), strength of innovation support (inn), strength of financial support (fin), and rate of agricultural disaster (dis), as shown in Table 2. To construct a matrix of regional differences in industrial structure, the proportion of the value added of the primary industry in the regional GDP as a proxy variable for industrial structure was used; to construct a matrix of regional differences in the level of informatization, the number of regional internet broadband access ports to characterize was used; to construct a matrix of regional differences in the level of education, the level of education of the rural workforce in each region was used; to construct a matrix of regional differences in economic development levels, the GDP per capita was used; to construct a matrix of regional differences in innovation support, the proportion of the total fiscal expenditures each region spends on science and technology as a proxy for innovation support was used; to construct a matrix of regional differences in fiscal support, the proportion of expenditure on agriculture, forestry, and water affairs in the total fiscal expenditure in each region as a proxy variable for financial support was used; and to construct a matrix of regional differences in the rate of agricultural disasters, the proportion of the agricultural disaster area in the affected area in each region was used.

Considering the differences in technological progress in mandarin and tangerine production and the availability of data, we divided citrus into two categories (mandarin and tangerine) in the measurement of technological progress. We selected the panel data of the seven major mandarin-producing areas, i.e., Guangdong, Guangxi, Jiangxi, Hubei, Hunan, Fujian, and Chongqing, and the seven major tangerine-producing areas, i.e., Guangdong, Jiangxi, Zhejiang, Hubei, Hunan, Fujian, and Chongqing, from 2006 to 2021. In 2021, the citrus production of these areas accounted for 84.02% of China's citrus production, making the data highly representative. The data were obtained from the database of the National Bureau of Statistics, the China Population and Employment Statistical Yearbook, and the National Compilation of Cost and Benefit Information of Agricultural Products. Some of the missing data were obtained by consulting the statistical yearbooks of the region for the relevant years and using the moving average method. In order to ensure data consistency and comparability, each value variable was deflated using the corresponding price index, with 2006 as the base period.

Table 2. Influencing factors and variables of spatial correlation network of citrus-production technology progress.

Influencing Factors	Variable Code	Calculation Methods and Explanations	Data Sources
Industrial structure	str	Value added of primary sector/GDP	Database of the National Bureau of Statistics
Informatization level	inf	Number of internet broadband access ports	Database of the National Bureau of Statistics
Education level	edu	Educational level of the rural labor force	China Population and Employment Statistical Yearbook
Economic development level	eco	GDP per capita	Database of the National Bureau of Statistics
Innovation support	inn	Science and technology expenditure/ Total fiscal expenditure	Database of the National Bureau of Statistics
Financial support	fin	Expenditure on agriculture, forestry, and water affairs/Total fiscal expenditure	Database of the National Bureau of Statistics
Agricultural disaster rate	dis	Area damaged/Area affected	Database of the National Bureau of Statistics

3.2. *Methods*

3.2.1. Transcendental Logarithmic Cost Function

In 1973, the transcendental logarithmic cost function was first proposed by L. Christensen et al. [30] and compared with the commonly used DEA method, the Malmquist index method, and the traditional C-D function. Its functional form is in line with the setting of economic theory, and it allows for the use of dummy and proxy variables according to theoretical and real needs so as to enable accurate economic explanations [31]. So, in this paper, we selected the beyond logarithmic cost function model to measure the technological progress in Chinese citrus production. In general, for the second-order Taylor expansion of the transcendental logarithmic cost function, we do not need to set up a special functional form or define the substitution relationship between elements. However, the method also has two drawbacks: first, the time-varying nature of the estimated coefficients cannot be observed, and second, the correlation between some variables cannot be well verified [9]. Therefore, in this paper, the third-order Taylor expansion of the transcendental logarithmic cost function was chosen in the following form:

$$\begin{aligned}\ln C_t = {} & \alpha_0 + \sum_{m=1}^{k} \alpha_m \ln P_{mt} + \alpha_y \ln Y_t + \alpha_T T + \tfrac{1}{2} \sum_{m=1}^{k} \sum_{n=1}^{k} \alpha_{mn} \ln P_{mt} \ln P_{nt} \\ & + \sum_{m=1}^{k} \alpha_{my} \ln P_{mt} \ln Y_{mt} + \sum_{m=1}^{k} \alpha_{mT} T \ln P_{mt} + \tfrac{1}{2}\alpha_{yy} (\ln Y_t)^2 + \alpha_{yT} T \ln Y_t \\ & + \tfrac{1}{2}\alpha_{yy} T (\ln Y)^2 + \tfrac{1}{2}\alpha_{TT} T^2 + \tfrac{1}{2} \sum_{m=1}^{k} \sum_{n=1}^{k} \alpha_{mnT} \ln P_{mt} \ln P_{nt} + \sum_{m=1}^{k} \alpha_{myT} T \ln Y_t \ln P_{mt}\end{aligned} \tag{1}$$

In Equation (1), C_t denotes the total cost of production at time t; P_{mt} denotes the price of factor m at time t; m = 1, 2,..., k denotes the number of factors; P_{nt} denotes the price of a factor of production except factor m at time t; n = 1, 2,..., $k - 1$ denotes the number of factors; Y_t denotes the level of output at time t; and T denotes the time trend. Because there are too many independent variables in Equation (1), if direct estimation is carried out, the problem of covariance will arise. Therefore, instead of the cost function being estimated directly, the Shephard lemma is usually used to bias the prices of the factors of production to construct the cost share equation in the following form:

$$S_{mt} = \alpha_m + \sum_{m=1}^{k} \alpha_{mn} \ln P_{nt} + \sum_{n=1}^{k} \alpha_{mnT} T \ln P_{nt} + \alpha_{my} \ln Y_t + \alpha_{nyT} T \ln Y + \alpha_{mT} T \tag{2}$$

The transcendental logarithmic cost function is quadratically differentiable with respect to the logarithm of input prices, and the Hessian matrix of this function is symmetric. So, the restrictions are as follows:

$$\alpha_{mn} = \alpha_{nm}, \alpha_{mnT} = \alpha_{nmT}, \forall m \neq n \tag{3}$$

$$\sum_{m=1}^{k} \alpha_m = 1, \sum_{m=1}^{k} \alpha_{mn} = \sum_{m=1}^{k} \alpha_{my} = \sum_{m=1}^{k} \alpha_{mT} = 0, \sum_{m=1}^{k} \alpha_{mnT} = \sum_{n=1}^{k} \alpha_{mnT} = \sum_{m=1}^{k} \alpha_{myT} = 0 \tag{4}$$

For a known output, progress in production technology is accompanied by a reduction in factor inputs at constant factor prices or a reduction in factor prices at the same factor inputs, all of which lead to a continuous decline in production costs as technology progresses. In this way, the total rate of technological progress (TP) can be obtained by using the total cost of production taken logarithmically and derived for time. If $TP > 0$, then there is technological progress, and if $TP < 0$, then there is technological regression. The formula is as follows:

$$\begin{aligned} TP &= -\partial \ln C/\partial T \\ &= -\alpha T - \sum_{m=1}^{k} \alpha_m T \ln P_{mn} - \alpha y T \ln Yt - \tfrac{1}{2}\alpha_{yyT}(\ln Y)^2 - \alpha_{TT} T \\ &\quad -\tfrac{1}{2}\sum_{m=1}^{k}\sum_{n=1}^{k} \alpha_{mnT} \ln P_{mt} \ln P_{nt} - \sum_{m=1}^{k} \alpha_{myT} \ln Y_t \ln P_{mt} \end{aligned} \tag{5}$$

3.2.2. Modified Gravity Model

The analysis of spatial correlation networks first needs to establish a spatial correlation network matrix, and existing research has mainly adopted two methods to construct a spatial correlation network matrix, namely the VAR model and the gravitational model. Because the VAR model cannot portray the trend of change in the spatial correlation network and is too sensitive to the selection of a lagging order, it will reduce the accuracy of the network structure characterization to a certain extent [32], while the gravitational model is constructed based on the principle of distance decay and the law of gravity, which can combine technological progress and geographical distance to better characterize spatial correlation [27]. Therefore, in this paper, referring to the study of Wang et al. [33], we introduced a modified gravitational model to measure the gravitational strength of the spatial correlation of citrus-production technology progress in the major citrus-producing areas in China and constructed a spatial correlation matrix with the following formula:

$$S = K\frac{TP_i \cdot TP_j}{\omega_{ij}^2/(g_i - g_j)^2}, K = \frac{TP_i}{TP_i + TP_j} \tag{6}$$

In Equation (6), S is the strength of the correlation of citrus-production technology progress between the main citrus-producing areas i and j; TP_i and TP_j denote the rate of citrus-production technology progress in provinces i and j, respectively; K denotes the contribution of main production area i to S; ω_{ij} is the spherical distance between the two main production areas; and g_i and g_j are the values of the real per capita GDP of the two main production areas, respectively.

By modifying the gravity model to measure the correlation strength of mandarin- and tangerine-production technology progress in each citrus-producing region, a 7 × 7 correlation strength matrix was constructed, and the average value of each row of the matrix was taken as the threshold. If the correlation strength was greater than the threshold, it was recorded as 1, indicating that there is a spatial spillover of citrus-production technology progress from the main production region in that row to the main production region in that column. If the correlation strength was less than the threshold, it was recorded as 0, indicating that there is no spatial spillover of citrus-production technology progress from the main production region in that row to the main production region in that column, and a final 7 × 7 oriented spatial correlation 0–1 type asymmetric matrix was formed.

3.2.3. Social Network Analysis

Social network analysis (SNA) is an important research paradigm that uses graph theoretic tools and algebraic modeling techniques to explore the relationship between members in a network structure and has been widely used in many fields, such as sociology, economics, and management [34,35]. The social network analysis method usually involves research on the overall network association structure and the individual network structure characteristics.

(1) Overall network correlation structure analysis: In this paper, we used four indicators, namely network density, network correlation degree, network level, and network efficiency, as follows:

Network density is measured according to the ratio of the number of real connections in the network to the theoretical maximum number of connections that can be carried in the network, which reflects the closeness of the spatial association network of citrus-production technology progress: the greater the density of the network, the closer the connection between mandarin- and tangerine-production technology progress in each main production area, and the greater the impact of the network on technological progress in citrus production in each main citrus-producing area. The value of network density is between 0 and 1. The specific measurement is shown in Equation (7), where D_n is the network density, n is the number of connections that actually exist, and N is the number of network nodes. So, the maximum number of carrying connections in the directed network graph is $N(N-1)$.

$$D_n = n/[N \times (N-1)] \tag{7}$$

The network correlation degree is an indicator of the robustness and vulnerability of a network structure. If there is a correlation between any two citrus-producing regions in the correlation network, this network structure is robust. When many lines are connected to only one or two citrus-producing regions, then the dependence of the citrus-production technology progress association network on that region is high, and once that region is excluded, the network may collapse, and its network relevance is low. The value of network relevance is between 0 and 1. Measured as shown in Equation (8), R is the network relevance and Z is the number of unreachable nodes in the network.

$$R = 1 - Z[N(N-1)/2] \tag{8}$$

The network level reflects the hierarchical structure of each citrus-producing region in the association network, and there is a degree of two-by-two asymmetric arrival between each production region in the network. The value of network level is between 0 and 1. Measured as shown in Equation (9), H is the network level degree, V is the number of pairs of symmetric reachable points in the network, and max(V) is the number of pairs of symmetric reachable points that can be carried by the theoretically existing network.

$$H = 1 - V/\max(V) \tag{9}$$

Network efficiency reflects the degree of existence of redundant correlations between main production areas in the spatial correlation network of citrus-production technology progress. A lower network efficiency indicates that citrus-production technology progress has more spatial spillover channels and that the correlations between citrus-production technology progress in the main production areas are closer and the spatial correlation network is more stable. The value of network efficiency is between 0 and 1. Measured as shown in Equation (10), E is the network efficiency, M is the number of redundant links in the network, and max(M) is the maximum number of redundant links that can be carried by the theoretically existing network.

$$E = 1 - M/\max(M) \quad (10)$$

(2) Individual network structure characterization: This paper adopted three indicators, namely point degree centrality, proximity centrality, and intermediary centrality, to conduct centrality analysis and reveal the role of each citrus-producing region in the network, as follows:

Point degree centrality reflects the degree to which a certain main production area is in the center of the association network structure. The larger the point degree centrality is, the more the main production area is connected with other main production areas in the association network, and the more prominent the center position of the main production area is in the association network. Measured as shown in Equation (11), C is the point degree centrality, q is the number of regions directly associated with a certain main production area in the association network, and Q is the maximum number of directly associated regions that the main production area can carry.

$$C = q/(Q - 1) \quad (11)$$

Proximity centrality reflects the ability of citrus-producing regions to be free from the control of other production regions in the correlation network, which is the sum of the shortcut distances between a citrus-producing region and other production regions in the network. The larger the proximity centrality is, the more direct the spatial correlations between the production region and other production regions are, and the easier it is for the region to play the role of a "central actor" in the correlation network. Measured as shown in Equation (12), d_{ij} is the shortcut distance between the main production area i and main production area j.

$$C_{APi}^{-1} = \sum_{j=1}^{n} d_{ij} \quad (12)$$

Intermediary centrality reflects the "bridge" and "intermediary" role of each citrus-producing region in the correlation network. The larger the degree of intermediary centrality, the greater the bridge and intermediary role of the citrus-producing region in the correlation network. Assuming that the number of shortcuts between the main citrus-producing areas a and b in the association network is L_{ab}, the number of shortcuts of the third main production area i between the main production areas a and b is $L_{ab}(i)$, and the probability that the main production area i exists between the main production areas a and b is $P_{ab}(i) = (L_{ab}(i))/L_{ab}$. The intermediary centrality measure is given in Equation (13).

$$C_{abi} = \sum_{a}^{N}\sum_{b}^{N} P_{ab}(i), (a \neq b \neq i) \quad (13)$$

(3) Quadratic assignment procedure (QAP) model: The QAP model is a non-parametric method to explore the relationship between matrices by comparing different matrix data with permutation [34], which usually includes two stages: QAP correlation analysis and QAP regression analysis. This method does not need to assume that the explanatory variables are independent of each other, which can effectively solve the endogeneity problem of relational data, and the regression results are more stable [36]. QAP correlation analysis compares the correlation between two matrices by looking at the matrices as long vectors containing $n(n - 1)$ numbers and then similarly comparing the correlations between the two variables and calculating the correlation coefficients of the two vectors [37,38]. QAP regression analysis is the study of regression relationships between multiple matrices and one matrix by performing a regular regression analysis on the long vector elements corresponding to the independent and dependent variable matrices

and then performing a regression on the rows and columns of the dependent variable. The variables are replaced, the regression is repeated, all coefficient values are saved, and the value of R^2 is determined [37]. The QAP model is constructed as follows:

$$TPM = f(str, \inf, edu, eco, inn, fin, dis) \quad (14)$$

In Equation (14), TPM is the spatial correlation matrix of citrus-production technology progress, and the independent variables are the regional difference matrix of industrial structure (*str*), the regional difference matrix of informatization level (*inf*), the regional difference matrix of education level (*edu*), the regional difference matrix of economic development level (*eco*), the regional difference matrix of innovation support (*inn*), the regional difference matrix of financial support (*fin*), and the regional difference matrix of agricultural disaster rate (*dis*).

4. Characterization of Changes in Citrus Production Technology Progress

Using stata16 (StataCorp LLC, College Station, TX, USA), based on citrus input and output variables, the technological progress in mandarin and tangerine production in China from 2006 to 2021 was obtained through the transcendental logarithmic cost function, as shown in Figure 1.

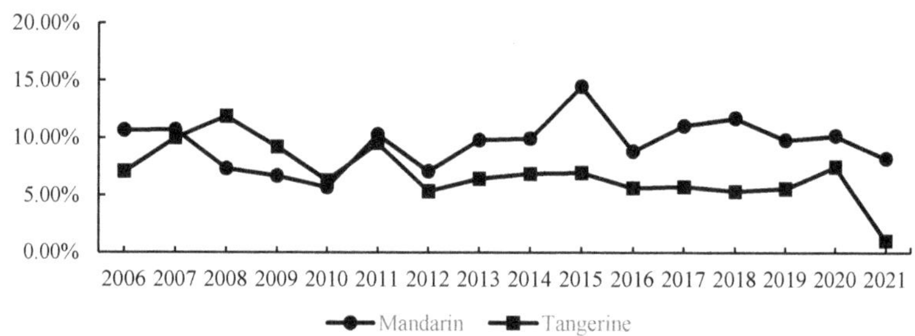

Figure 1. Changes in TP in China's mandarin and tangerine production from 2006 to 2021.

As can be seen from Figure 1, the rate of technological progress in mandarin and tangerine production in China was positive from 2006 to 2021, indicating that the production of mandarins and tangerines in China, in general, has been in the stage of technological progress. In terms of the overall level of technological progress, the rate of technological progress in mandarin production has been higher than that in tangerine production since 2011, and the gap was the largest in 2015, with the rate of technological progress in mandarin production being 7.5% higher than that in tangerine production. From the trend of change, the technological progress rate of mandarin and tangerine production fluctuated within the range of 6–12% from 2006 to 2011. The technological progress rate of mandarin production remained stable from 2006 to 2011 and fluctuated from 2011 to 2021, while the technological progress rate of tangerine production showed a fluctuating and decreasing trend and decreased to a minimum of 1.05% in 2021, which may be because tangerine-production technology was internalized and the new technological breakthroughs were difficult [2].

The rate of technological progress in each main mandarin- and tangerine-producing area is shown in Table 3. Regarding the technological progress in mandarin production, the average technological progress rate of Guangdong was the highest, being 11.79% higher than that of Hunan, which was the lowest. The technological progress rate of Fujian tangerine production had the largest increase, which was 9.92% in 2021 compared to 2006. The rate of technological progress in the five main mandarin-producing regions of Guangxi, Chongqing, Hubei, Jiangxi, and Hunan showed a fluctuating downward trend, with the rate of technological progress slowing down. Regarding the technological progress in tangerine production, the average technological progress rate of Hunan was the highest,

11.56% higher than that of Hubei, which was the lowest. Only the technological progress rate of Chongqing showed a rising trend, and the technological progress in the other six main tangerine-producing areas showed different degrees of slowdown, among which, the technological progress rate of Hunan decreased the most, by 21.29%. In summary, it can be seen that, first, the mandarin- and tangerine-production technology progress in both Guangdong and Fujian is at a medium-high level because Fujian and Guangdong have sufficient precipitation, heat, and light and, at the same time, the government has enough financial resources to carry out technological research, development, and diffusion [39]. Second, there are some regional differences in the technological progress in the main production areas of mandarins and tangerines, and the mandarin- and tangerine-production technology progress levels vary in the same area, which may be related to the levels of scientific research in the main production areas of mandarins and tangerines [11]. Third, except for a few years when some of the main production areas may have had negative values for the technological progress rate due to climate and pests and diseases, the rate of citrus-production technology progress in most of the years is still positive, which indicates that all the main production areas are in the stage of technological progress.

Table 3. TP for the major mandarin- and tangerine-producing areas in China from 2006 to 2021.

Classification	Areas	2006	2009	2012	2015	2018	2021	Average Value
Mandarin	Guangdong	12.57%	14.76%	12.77%	16.75%	15.26%	21.96%	15.99%
	Fujian	3.62%	6.85%	10.58%	20.52%	17.06%	13.54%	12.66%
	Guangxi	10.80%	8.62%	7.88%	15.51%	11.84%	8.45%	10.69%
	Chongqing	12.59%	8.50%	7.36%	10.10%	10.57%	6.76%	9.55%
	Hubei	18.72%	8.97%	6.64%	4.87%	6.48%	8.98%	8.15%
	Jiangxi	17.08%	3.97%	2.17%	2.85%	3.09%	3.39%	6.46%
	Hunan	10.33%	3.57%	1.64%	6.26%	3.03%	3.82%	4.20%
Tangerine	Hunan	29.61%	20.98%	16.79%	6.78%	14.18%	8.32%	15.22%
	Zhejiang	13.35%	14.60%	4.84%	8.62%	4.31%	6.51%	9.90%
	Guangdong	10.42%	8.06%	4.45%	7.28%	11.21%	6.57%	9.05%
	Jiangxi	14.77%	9.09%	5.14%	12.94%	6.99%	5.56%	8.59%
	Fujian	6.59%	5.13%	3.16%	8.09%	5.52%	0.61%	6.18%
	Chongqing	3.34%	1.68%	0.97%	3.79%	6.80%	7.42%	4.25%
	Hubei	6.61%	6.72%	5.38%	4.00%	−0.28%	0.49%	3.66%

Note: Due to space constraints, results for other years are not reported and are available from the authors upon request.

5. Characterization of the Spatial Correlation Network Structure

In order to visualize the structural shape and evolution of the spatial correlation network of citrus-production technology progress in China, this paper maps the structure of the spatial correlation networks of mandarins and tangerines in 2006 and 2021 by using Ucinet6.212 (University of California, Irvine, CA, USA) and ArcGIS10.2.2 (Environmental Systems Research Institute, Redlands, CA, USA) (Figures 2 and 3).

As can be seen from Figures 2 and 3, the citrus-production spatial correlation network of technological progress in China presents the structural characteristics of multi-relationships and multi-directions, and the network correlations are increasingly strengthened. From the structure of the spatial correlation network of mandarin-producing areas, it can be seen that in 2006, Guangdong and Jiangxi had the highest number of correlations and were at the core of the spatial correlation network, forming a dual-core structure, while other mandarin-producing areas also had more correlations and were at the sub-core position. In 2021, the closeness of the technological progress correlation among the main citrus-producing areas was enhanced and the position of Hubei and Fujian in the spatial correlation network was significantly improved, with Guangdong, Hubei, and Fujian occupying the core positions. At the same time, the association links of other main production areas also increased significantly.

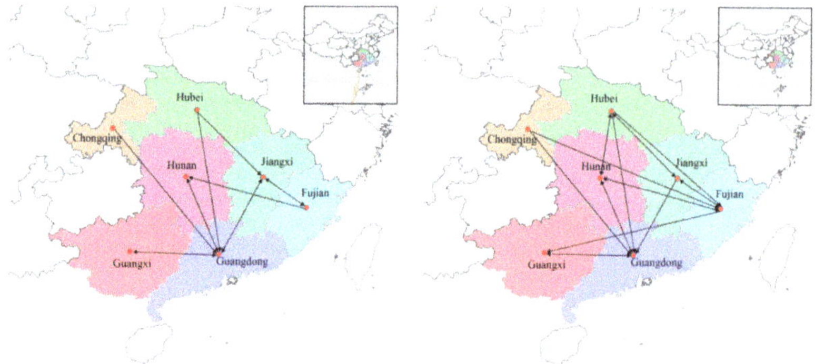

(**a**) 2006 (**b**) 2021

Figure 2. Spatial correlation network of the TP in China's mandarin production.

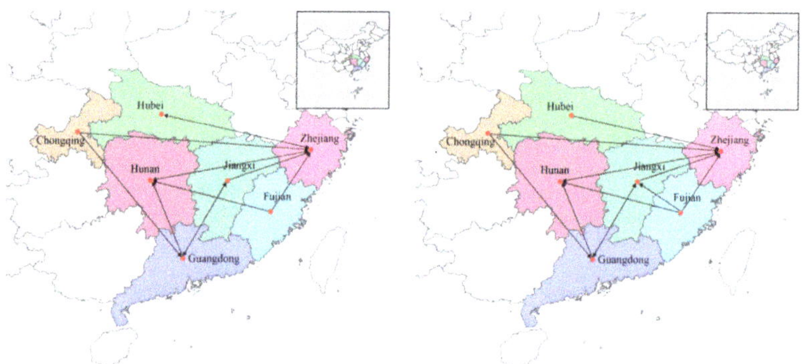

(**a**) 2006 (**b**) 2021

Figure 3. Spatial correlation network of the TP in China's tangerine production.

From the structure of the tangerine spatial correlation network, it can be seen that in 2006, Zhejiang, Hunan, and Guangdong had the highest number of correlations and were at the core of the spatial correlation network, forming a triple-core structure, and the other tangerine-producing regions were in the sub-core positions. In 2021, the positions of Jiangxi and Fujian in the spatial correlation network gradually improved, because of which the spatial correlation network of the tangerine-production technology progress presents a multi-core structure characteristic. The main reason is that with the implementation of the strategy of agricultural power and regional coordinated development, under the dual roles of market mechanism and government macro-control, the mobility of inter-regional citrus-production factors has been enhanced, mutual exchanges and cooperation among main citrus-producing areas have been strengthened, the frequency of interaction has increased, the spatial interaction of citrus-production technology progress has been strengthened, and the stability of the spatial network has improved.

5.1. Characteristics of the Overall Network Structure

In order to grasp the overall structure of the spatial correlation network of China's mandarin- and tangerine-production technology progress in more depth, Ucinet6.212 (University of California, Irvine, CA, USA) was used to examine and analyze the overall network structure in four aspects, namely network density, network correlation degree, network level, and network efficiency, and the results are shown in Figures 4 and 5.

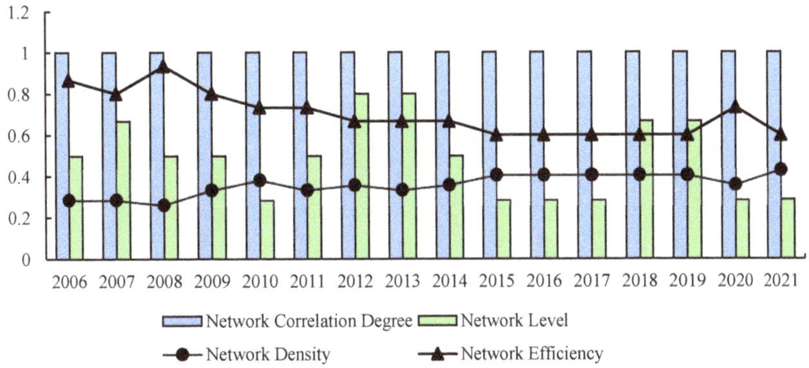

Figure 4. Overall network structure characteristics of the TP in China's mandarin production from 2006 to 2021.

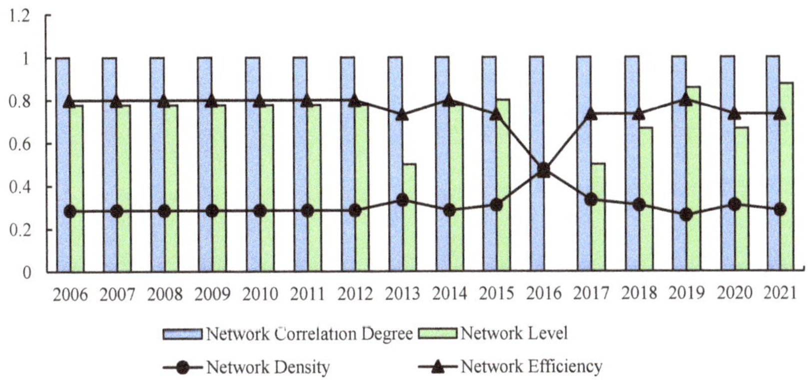

Figure 5. Overall network structure characteristics of the TP in China's tangerine production from 2006 to 2021.

First, with regard to network density, the spatial network density of mandarin-production technology progress showed a fluctuating upward trend, and the spatial network density of citrus-production technology progress reached the maximum value of 0.4286 in 2021. The spatial association network density of tangerine-production technology progress reached the maximum value of 0.4762 in 2016 and then gradually declined. However, the overall mandarin and tangerine network densities were not large, indicating that at present, the degree of closeness of the spatial correlation relationship of citrus-production technology progress in China is not high, the network structure is relatively loose, and the spatial cooperation and interaction of technological progress need to be strengthened. Second, regarding the network correlation degree, the network correlation degree of both mandarins and tangerines was 1, which indicates that the network correlation structure of technological progress related to mandarin and tangerine production in China has good connectivity and robustness, all the main production areas of mandarins and tangerines are in the spatial association network of citrus-production technology progress, there is no isolated main production area detached from the network, and the spatial spillover effect of the network is obvious. Third, regarding the network level, except for 2016, when the network level of tangerines was 0, the spatial network level of technological progress related to both mandarin and tangerine production in other years was not 0. Therefore, the spatial association network of technological progress related to both mandarin and tangerine production needs to be further optimized. In contrast, the spatial network rank of technological progress related to mandarin production was lower than that related to tangerine production, indicating that the gradient characteristics of the spatial network of technological progress related to tangerine production is stronger than that related to mandarin production, and the spatial network of technological progress related to tangerine production is more polarized, with more main tangerine-producing areas at the edge of the network. Fourth, regarding network efficiency, the spatial network efficiency of technological progress in relation to both mandarin and tangerine production showed a decreasing trend, indicating that there is an increase in the number of connecting lines in the technological progress correlation network, an increase in the stability of the network, and the existence of multiple superimposed spillover channels.

Overall, through the overall network structure characterization, there are significant spatial correlation and spillover paths in the spatial association network of technological progress related to Chinese citrus production, the phenomenon of synergistic development is obvious, and a more stable spatial association network of technological progress has been formed; however, the network structure is relatively loose and there are strong gradient characteristics. Thus, improving the tightness of the network and decreasing the degree of the network level are the key points for promoting technological progress in citrus production in China.

5.2. Characteristics of the Individual Network Structure

In order to examine the position and role of each main citrus-producing area in the spatial correlation network of technological progress in a more detailed way and to grasp the characteristics of its individual network structure, this paper measured the centrality of the main mandarin- and tangerine-producing areas, and the results are shown in Table 4.

Table 4. Network centrality of China's mandarin- and tangerine-production technology progress in 2021.

Classification	Areas	Point Degree Centrality			Proximity Centrality	Intermediary Centrality
		Degree of Point-Out	Degree of Point-Entry	Degree of Centrality		
Mandarin	Guangdong	2	5	83.333	85.714	25.556
	Guangxi	1	2	33.333	60	1.333
	Jiangxi	3	2	50	66.667	1.333
	Hubei	4	2	66.667	75	3.556
	Hunan	3	3	50	66.667	1.333
	Fujian	3	4	83.333	85.714	25.556
	Chongqing	2	0	33.333	60	1.333
Tangerine	Guangdong	2	2	50	60	8.889
	Jiangxi	2	2	50	66.667	7.778
	Zhejiang	1	5	83.333	85.714	48.889
	Hubei	1	0	16.667	50	0
	Hunan	1	3	50	66.667	7.778
	Fujian	3	0	50	66.667	2.222
	Chongqing	2	0	33.333	60	4.444

In terms of the point degree centrality, in the spatial correlation network of technological progress in Chinese-mandarin-producing areas, the point degree centrality rankings in decreasing order were Guangdong, Fujian, Hubei, Jiangxi, Hunan, Guangxi, and Chongqing, among which Guangdong and Fujian ranked first, which indicates that these two main production areas are in the "core position" of the spatial correlation network of technological progress in mandarin-producing areas and play an important role in the network. The rankings of the point-out degree in decreasing order were Hubei, Jiangxi, Hunan, Fujian, Guangdong, Chongqing, and Guangxi, indicating that the technological progress related to mandarin production in Hubei has a greater influence on other main mandarin-producing regions. The rankings of the point-entry degree in decreasing order were Guangdong, Fujian, Hunan, Guangxi, Jiangxi, Hubei, and Fujian, indicating that the technological progress in Guangdong mandarin production is more influenced by other main mandarin-producing areas, among which the point-entry degree of Chongqing was 0, which indicates that Chongqing belongs to the technological spillover main production areas in the spatial correlation network of technological progress in China's mandarin production and is influenced to a limited extent in the overall network. In terms of the spatial correlation network of the technological progress in the major Chinese mandarin-producing areas, the rankings for the point degree of centrality in decreasing order were Zhejiang, Guangdong, Jiangxi, Hunan, Fujian, Chongqing, and Hubei, of which Zhejiang ranked first, indicating that it has the highest correlation relationship with other major mandarin-producing areas and is in the "core position" of the spatial correlation network of technological progress in major mandarin-producing areas. The rankings of the point-out degree in decreasing order were Fujian, Guangdong, Jiangxi, Chongqing, Zhejiang, Hubei, and Hunan, indicating that the technological progress in Fujian tangerine production has a greater influence on other tangerine-producing regions. The rankings of the point-entry degree in decreasing order were Zhejiang, Hunan, Guangdong, Jiangxi, Hubei, Fujian, and Chongqing, indicating that the technological progress in Zhejiang tangerine production is greatly influenced by other tangerine-producing regions, and the point-entry degree of Hubei, Fujian, and Chongqing was 0, which indicates that these three tangerine-producing regions belong to the technological spillover type of the spatial correlation network of the technological progress in tangerine production, and the degree of their influence in the overall network is limited.

In terms of proximity centrality, the average value of proximity centrality in the spatial correlation network of the technological advancement of major mandarin-producing regions in China was 71.39, and there were two major mandarin-producing regions for which the values of proximity centrality were higher than this average value, namely Guangdong and Fujian, which indicates that Guangdong and Fujian can more quickly generate intrinsic connections with other major production regions in the spatial correlation network of technological advancement of major mandarin-producing regions. In other words, Guangdong and Fujian play the role of central actors in the network. In the spatial correlation network of technological progress in Chinese-tangerine-producing regions, the average value of the proximity centrality of tangerine-producing regions was 65.10, and there were four tangerine-producing regions that had higher proximity centrality than this average value, namely Zhejiang, Jiangxi, Hunan, and Fujian, indicating that these four regions are able to connect with other tangerine-producing regions more quickly in the spatial correlation network of technological progress and they play the role of a central actor.

In terms of intermediary centrality, the average value of intermediary centrality in the spatial correlation network of technological advancement of major mandarin-producing regions in China was 8.57, and the major mandarin-producing regions with intermediary centrality values higher than this average value were Guangdong and Fujian, which indicates that Guangdong and Fujian have a stronger ability to control the technological exchange among other major mandarin-producing regions in the spatial correlation network of technological advancement, are at the core of the network, and play the role of intermediary and bridge. In the spatial correlation network of technological progress in major tangerine-producing areas in China, the average value of the intermediary centrality of major tangerine-producing areas was 11.43, and the major tangerine-producing area with a higher average intermediary centrality value was Zhejiang, which indicates that Zhejiang has a stronger ability to control technological exchanges among other major tangerine-producing areas in the spatial correlation network of technological progress in these areas, is at the core of the network, and plays the role of intermediary and bridge. The degree of intermediary centrality of Guangdong, Fujian and Zhejiang was much higher than that of other mandarin-producing regions because these three production regions belong to the eastern developed region, which has a strong driving ability and plays a controlling and dominating role over other production regions.

Taken together, the analysis results of point centrality, proximity centrality, and mediation centrality are similar. The spatial network structure of China's citrus-production technology progress shows an obvious Matthew effect, with the relatively economically developed eastern provinces of Guangdong, Fujian, and Zhejiang having a higher status in the spatial network and a stronger dominant role in controlling the resource elements needed for progress in citrus-production technology. However, Chongqing, in the western part of the country, is in a passive position in citrus-production technology exchanges and cooperation.

6. Analysis of Factors Influencing the Spatial Correlation Network

6.1. QAP Correlation Analysis

Clarifying the factors influencing the spatial correlation network of citrus-production technology progress in China and its functioning mechanism is an important foundation for optimizing and regulating the structure of the correlation network of regional citrus-production technology progress. Therefore, the spatial correlation matrices of mandarin and tangerine production were combined, and correlation analysis was carried out using the quadratic assignment procedure (QAP) model based on the influencing factors selected above. As shown in Table 3, differences in education levels and economic development levels passed the 1% significant level test, differences in innovation support passed the 5% significant level test, and differences in informatization levels and financial support passed the 10% significant level test, indicating that these five factors significantly affect the formation of the spatial correlation network structure of citrus-production technology progress in China. Among them, the correlation coefficient of differences in education levels was negative, indicating that similar education levels are an important factor in generating spatial association and spatial spillover of citrus-production technology progress. The correlation coefficients of four variables, namely differences in informatization levels, differences in economic development levels, differences in innovation support, and differences in financial support, were positive, indicating that regional differences in these four variables are conducive to the formation of the spatial correlation network of citrus-production technology progress. The correlation coefficients of differences in industrial structure and differences in agricultural disaster rates were positive, but their significance levels were higher than 10%, indicating that their effects on the spatial correlation of citrus-production technology progress in China are not significant.

6.2. QAP Regression Analysis

In order to avoid multicollinearity between independent variables causing bias in the regression results, this paper set the number of random permutations to times to conduct QAP regression analysis on the model of factors influencing the spatial correlation of the technological progress in citrus production in China.

As can be seen from Table 5, $Adj\ R^2 = 0.983$, indicating that seven factors (i.e., industrial structure differences, informatization level differences, education level differences, economic development level differences, innovation support differences, financial support differences, and agricultural disaster rate differences) can explain approximately 98.3% of the spatial correlation effect of citrus-production technology progress in China. Among them, the regression coefficient of the differences in education levels was significantly negative at the 1% level, indicating that the differences in rural education levels significantly hinder the formation of the spatial correlation of citrus technological advancement in China. This is mainly because similar rural education levels mean that growers in these main production areas have similar abilities to learn the technology, which can help in the mutual exchange of citrus-production technology between the main production areas and thus promote the formation of the spatial correlation of technological advancement. The differences in informatization levels, economic development levels, innovation support, and financial support were significantly positive, at 10%, 1%, 5%, and 10%, respectively, indicating that the spatial correlation of technological advancement is more likely to occur among citrus-producing regions with higher differences in informatization levels, economic development levels, innovation support, and financial support. The higher the level of informatization and economic development, coupled with an increase in government support for innovation and finance, the greater the potential and opportunity for technological innovation in the main citrus-producing areas and the stronger the attraction of talent and capital to the main production areas that are lagging behind in development, making it easier for the resource elements needed for technological progress to flow across regions between the main production areas, thus facilitating the formation of spatial correlation relationships. The regression coefficients of the differences in industrial structure and the differences in the rates of agricultural disasters were positive but not significant ($p > 0.1$), indicating that the regional differences in industrial structure cannot significantly affect the spatial correlation of citrus-production technology progress, the impact of meteorological disasters on the technological progress in citrus production is limited, and Huanglong disease is usually the main reason affecting the production of citrus [40].

Table 5. Factors driving the spatial correlation network of citrus-production technology progress in China.

Influencing Factors	QAP Correlation Analysis		QAP Regression Analysis	
	Correlation Coefficient	*p*-Value of Significance	Coefficient of Regression	*p*-Value of Significance
Industrial structure	0.239	0.184	0.365	0.179
Informatization level	0.467 *	0.085	0.614 *	0.078
Educational level	−0.460 ***	0.002	−0.877 ***	0.004
Economic development	0.932 ***	0.000	2.012 ***	0.001
Innovation support	0.612 **	0.019	1.137 **	0.024
Financial support	0.386 *	0.082	0.865 *	0.081
Agricultural disaster rate	0.127	0.333	0.321	0.323

Note: $R^2 = 0.989$; $Adj\ R^2 = 0.983$; "*", "**", and "***" indicate statistical significance at the 10%, 5%, and 1% levels, respectively.

7. Conclusions and Policy Implications

7.1. Conclusions

Based on the cost–benefit data of China's main mandarin- and tangerine-producing areas from 2006 to 2021, this paper measured the rate of technological progress in China's mandarin and tangerine production using the transcendental logarithmic cost function model and analyzed its changing characteristics. The paper constructed a modified gravity model to determine the spatial correlation relationship of technological progress in the main areas producing mandarins and tangerines and established spatial correlation networks, based on which, it explored the characteristics of the overall network structure and the individual network structure by applying the social network analysis method. The paper further investigated the factors influencing the spatial correlation networks using the QAP regression analysis method and arrived at the following conclusions:

First, regarding the changing characteristics of the technological progress in Chinese citrus production, in general, the production of Chinese mandarins and tangerines is in the stage of technological progress, and the rate of technological progress related to mandarins is higher and more stable than that related to tangerines. In terms of spatial distribution, the average rate of technological progress in mandarin production is the highest in Guangdong, and that in tangerine production is the highest in Hunan. There are obvious inter-regional and intra-regional differences in technological progress in mandarin and tangerine production. In terms of time series development, except for Fujian and Guangdong, the technological progress rates of other major mandarin-producing regions show a fluctuating downward trend and a slowdown. In addition, except for Chongqing, the technological progress rates of other tangerine-producing areas show a slowdown.

Second, in terms of the overall network structure, China's citrus-production technology progress spatial network is becoming denser and more complex, and the stability of the spatial network is constantly improving, presenting a multi-core structure in space. The spatial network density of the technological progress in mandarin and tangerine production is not high, and the network structure is relatively loose. The network correlation degree is 1, there is no isolated main production area outside the network, and the network spatial spillover effect is obvious. Differences in the characteristics of mandarin and tangerine network levels are more obvious, with the mandarin-production technology progress space-network level being overall lower than that of tangerines, and the tangerine-production technology progress space-correlation-network polarization phenomenon is more serious. The spatial network efficiency of technological progress in both mandarin production and tangerine production shows a decreasing trend, the network stability is increasing, and there are multiple superimposed spillover channels.

Third, in terms of individual network characteristics, the point degree centrality of Guangdong and Fujian, the two main production areas, is the highest among the main mandarin-producing areas; that of Zhejiang is the highest among the main tangerine-producing areas; and they are in the core positions in the spatial correlation networks of technological progress in the main mandarin- and tangerine-producing areas, respectively. At the same time, the proximity centrality of Guangdong and Fujian is higher than the national average of mandarin-producing regions, and the proximity centrality of Zhejiang, Jiangxi, Hunan, and Fujian is higher than the national average of tangerine-producing regions. These main regions play the role of central actors in the spatial association network of technological progress. The intermediary centrality of Guangdong and Fujian is also higher than the national average for mandarin-producing regions, and the intermediary centrality of Zhejiang is higher than the national average for tangerine-producing regions, and they play the roles of intermediary and bridge in the spatial correlation network of technological progress. The spatial network of the technological progress in citrus production in China shows a significant Matthew effect, with the relatively economically developed eastern provinces of Guangdong, Fujian, and Zhejiang having a higher status in the spatial network, while Chongqing, in the western part of the country, is in a passive position in citrus-production-related technological exchanges and cooperation.

Fourth, the QAP analysis results show that the differences in education levels have a significant negative impact on the structure of the spatial correlation network of citrus-production technology progress in China. Differences in the levels of information technology, economic development, innovation support, and financial support have a significant positive effect on the spatial network structure of citrus-production technology progress in China. Differences in the industrial structure and agricultural disaster rates do not have a significant effect on the spatial network structure of citrus-production technology progress in China.

7.2. Policy Implications

First, a modern citrus industry science and technology innovation system should be built, increasing citrus-specific scientific research input, innovating and developing citrus-production technologies, and promoting new technologies through demonstration projects to provide a new driving force for China's citrus-production technology progress.

Second, there is a need to grasp the structure of the overall linkage network, implement the strategy of coordinated regional development of citrus science and technology innovation, and promote the construction of spatial spillover channels of citrus technological advancement. For the marginal production regions in the spatial linkage network of citrus technological advancement, the role of the government is necessary in seeking technical assistance, obtaining a guaranteed supply of citrus-production technology in terms of citrus breeding and planting, and promoting the balanced development of China's citrus-production technology in the region.

Third, with regard to the positioning of the network roles of the main production regions, differentiated policy regulation should be implemented to accurately identify the central actors in the spatially linked network and the main production regions that act as intermediaries and bridges in the spatial transmission paths, so as to provide an accurate idea of their vital driving roles.

Finally, there is a need to speed up the development of rural education in the production regions that are trailing behind and reduce the disparity in the quality of rural education between the production regions. To increase the likelihood of forming correlations with other production regions, each production region needs to pay more attention to the effects of information technology, economic development, support for innovation, and financial support.

This study used the cost–benefit panel data of seven main mandarin-producing regions and seven main tangerine-producing regions from 2006 to 2021. Although the data of each production factor of each main production region were collected and processed, the data of other production regions could not be obtained. Thus, to some extent, this study could not accurately reflect the actual level of technological progress in citrus production in China. In the future, with better data information, it is hoped that the data related to the production of all mandarin- and tangerine-producing areas will be collected, with a sufficiently large sample size to make the research conclusions more accurate and representative. Meanwhile, in the future, research can also be conducted on how to optimize the factor allocation of land, labor, and capital to promote citrus-production technology progress; how to introduce technology-embedded governance to promote citrus-production technology progress; and how to implement institutional-embedded governance to innovate institutional mechanisms.

This article belongs to the same series of research as the author's previous article published in this Special Issue, entitled "Spatiotemporal Evolution and Spatial Convergence Analysis of Total Factor Productivity of Citrus in China". The previous article's research concluded that technological progress is the main factor affecting the total factor productivity of citrus. This article further measured the technological progress in citrus production in China by using the transcendental logarithmic cost function model and analyzed the spatial correlation network structure and its influencing factors, which expands on the previous article.

Author Contributions: Conceptualization, Y.G.; methodology, Y.G. and B.L.; software, Y.H.; validation, F.L.; formal analysis, Y.H.; resources, Y.G.; data curation, Y.H. and F.L.; writing—original draft preparation, Y.G.; writing—review and editing, Y.G. and F.L.; visualization, Y.G. and B.L.; supervision, C.Q.; project administration, C.Q.; funding acquisition, C.Q. All authors have read and agreed to the published version of the manuscript.

Funding: This research was funded by the China Agriculture Research System-National Citrus Industry Technology System (CARS-26-06BY), the National Social Science Fund of China (23BJY155), and the Major Consulting Project of Jiangxi Research Institute for the Development Strategy of China Engineering Science and Technology (2022-DFZD-37).

Institutional Review Board Statement: Not applicable.

Data Availability Statement: The data presented in this study are available from the first author upon request.

Conflicts of Interest: The authors declare no conflict of interest.

References

1. Qi, C.J.; Gu, Y.M.; Zeng, Y. Progress of citrus industry economy in China. *J. Huazhong Agric. Univ.* **2020**, *31*, 1–4. (In Chinese)
2. Gu, Y.M.; Qi, C.J.; Liu, F.X.; Lei, Q.Y.; Ding, Y.C. Spatiotemporal Evolution and Spatial Convergence Analysis of Total Factor Productivity of Citrus in China. *Agriculture* **2023**, *13*, 1258. [CrossRef]
3. Deng, X.X. Review and Prospect of 60 Years of Citrus Breeding in China. *J. Hortic.* **2022**, *49*, 2063–2074. (In Chinese)
4. Guo, W.W.; Ye, J.L.; Deng, X.X. Fruit scientific research in New China in the past 70 years: Citrus. *J. Fruit Sci.* **2019**, *36*, 1264–1272. (In Chinese)

5. Huang, W.H.; Qi, C.J. Study on the evolution of spatial layout and driving factors of citrus production in China. *J. Huazhong Agric. Univ. (Soc. Sci. Ed.)* **2022**, *4*, 90–103. (In Chinese)
6. Zhang, Y.; Zhao, Y.; Qi, C. Analysis of world citrus trade pattern. *World Agric.* **2007**, *12*, 26–29. (In Chinese)
7. He, J.; Qi, C. Comparison and reference of Chinese and foreign citrus industry development mode. *Economy* **2010**, *2*, 110–113. (In Chinese)
8. Wu, Y. Assessment of technical efficiency and analysis of influencing factors in Chinese citrus plantation. *China Agric. Resour. Zoning* **2018**, *39*, 94–102. (In Chinese)
9. Chen, S.; Hu, H.; Fu, S. Impacts of factor price changes on agricultural technology progress and factor substitution—Taking corn production as an example. *J. Hunan Agric. Univ. (Soc. Sci. Ed.)* **2018**, *19*, 24–31. (In Chinese)
10. Zhou, H.Y. Research on the Spatial Correlation of Agricultural Technology Progress and Its Influencing Factors in Xinjiang. Master's Thesis, Shihezi University, Shihezi, China, 2022.
11. Fang, G.Z.; Qi, C.J.; Lei, Q.Y. Calculation of total factor productivity of citrus in China and the difference of region—Based on the DEA-Malmquist index method. *Chin. J. Agric. Resour. Reg. Plan.* **2019**, *40*, 29–34. (In Chinese)
12. Li, D.H.; Guo, J.Y.; Zhu, S.B. Total factor productivity, technological progress and efficiency changes in China's citrus industry. *J. Jiangxi Agric. Univ.* **2010**, *9*, 43–47. (In Chinese)
13. Xu, J.C. Research on Technological Progress and Its Causal Factors of Wheat Production in China. Master's Thesis, Huazhong Agricultural University, Wuhan, China, 2020.
14. Solow, R.M. Technical change and the aggregate production function. *Rev. Econ. Stat.* **1957**, *39*, 312–320. [CrossRef]
15. Arrow, K.J. The economic implications of learning by doing. *Rev. Econ. Stud.* **1962**, *29*, 155–173. [CrossRef]
16. Christensen, L.R.; Jorgenson, D.W.; Lau, L.J. Transcendental logarithmic utility functions. *Am. Econ. Rev.* **1975**, *65*, 367–383.
17. Tan, C. *Introduction to Development Economics*; Wuhan University Press: Wuhan, China, 2009. (In Chinese)
18. Mao, W.; Koo, W.W. Productivity growth, technological progress, and efficiency change in Chinese agriculture after rural economic reforms: A DEA approach. *China Econ. Rev.* **1997**, *8*, 157–174. [CrossRef]
19. Da Silva, G.; Gomes, E.G.; de Andrade Alves, E.R.; Garcia Gasques, J. Technological progress in the Brazilian agriculture. *Socio-Econ. Plan. Sci.* **2020**, *72*, 100879.
20. Tan, C.; Tao, J.; Yi, L.; He, J.; Huang, Q. Dynamic Relationship between Agricultural Technology Progress, Agricultural Insurance and Farmers' Income. *Agriculture* **2022**, *12*, 1331. [CrossRef]
21. Chen, G.; Deng, Y.; Sarkar, A.; Wang, Z. An Integrated Assessment of Different Types of Environment-Friendly Technological Progress and Their Spatial Spillover Effects in the Chinese Agriculture Sector. *Agriculture* **2022**, *12*, 1043. [CrossRef]
22. He, Y.; Chen, W. Evaluation of Sustainable Development Policy of Sichuan Citrus Industry in China Based on DEA-Malmquist Index and DID Model. *Sustainability* **2023**, *15*, 4260. [CrossRef]
23. Xiang, Y.; Qi, C.J.; Wang, W.X. Research on factor substitution relationship and growth path of citrus production—An empirical analysis based on panel data of main production areas. *J. China Agric. Univ.* **2017**, *22*, 200–209. (In Chinese)
24. Zhang, F.; Wu, L.; Wang, F.L. Research on the spatial correlation network structure and driving factors of agricultural total factor productivity in China. *Res. Agric. Mod.* **2020**, *41*, 587–598. (In Chinese)
25. Tan, R.H.; Liu, H.M. Evolution of spatial correlation network characteristics and influencing factors of green total factor productivity in Chinese agriculture. *Chin. J. Ecol. Agric.* **2022**, *30*, 2011–2026. (In Chinese)
26. Qu, H.; Yin, Y.; Li, J.; Xing, W.; Wang, W.; Zhou, C.; Hang, Y. Spatio-temporal Evolution of the Agricultural Eco-efficiency Network and Its Multidimensional Proximity Analysis in China. *Chin. Geogr. Sci.* **2022**, *32*, 724–744. [CrossRef]
27. Sun, Z.R.; Fan, J.; Sun, Y. Characteristics of spatial correlation network structure and influencing factors of green technology innovation efficiency in China. *Econ. Geogr.* **2022**, *42*, 33–43. (In Chinese)
28. Wang, F.; Wu, L.; Zhang, F. Network structure and influencing factors of agricultural science and technology innovation spatial correlation network—A study based on data from 30 provinces in China. *Symmetry* **2020**, *12*, 1773. [CrossRef]
29. He, H.; Ding, R. Spatiotemporal heterogeneity effect of technological progress and agricultural centrality on agricultural carbon emissions in China. *Front. Environ. Sci.* **2023**, *10*, 1078357. [CrossRef]
30. Christensen, L.R.; Jorgenson, D.W.; Lau, L.J. Transcendental logarithmic production frontiers. *Rev. Econ. Stat.* **1973**, *55*, 28–45. [CrossRef]
31. Jiang, Q.J.; Wu, M.M. Analysis of the contribution rate of technological progress to China's fishery economic growth—A study based on the beyond logarithmic production function method. *China Fish. Econ.* **2023**, *41*, 1–11. (In Chinese)
32. Wang, X.H.; Yang, Y.Q.; Luo, X.Y.; Wen, T. Spatial correlation network and its mechanism for high-quality economic development in China. *J. Geogr.* **2022**, *77*, 1920–1936. (In Chinese)
33. Wang, M.H.; Su, M.M. Network structure and influencing factors of livelihood capital of interprovincial residents in China. *Econ. Geogr.* **2022**, *42*, 36–44+94. (In Chinese)
34. Shao, H.Q.; Wang, Z.F. Network structure of spatial correlation of habitat environment and its driving factors in city clusters in the middle reaches of the Yangtze River. *Yangtze River Basin Resour. Environ.* **2022**, *31*, 983–994. (In Chinese)
35. Liu, H.J.; Liu, C.M.; Sun, Y.N. Research on the structural characteristics of spatial correlation network of energy consumption and its effects in China. *China Ind. Econ.* **2015**, *5*, 83–95. (In Chinese)
36. Doreian, P.; Conti, N. Social context, spatial structure and social network structure. *Soc. Netw.* **2012**, *34*, 32–46. [CrossRef]
37. Liu, J. *Lectures on Whole Network Approach: A Practical Guide to UCINET*; Truth & Wisdom Press: Beijing, China, 2009.

38. Krackardt, D. QAP partialling as a test of spuriousness. *Soc. Netw.* **1987**, *9*, 171–186. [CrossRef]
39. Jiang, S.Y.; Liu, Y.; Wang, M.M. Research on the dynamic evolution and regional differences of total factor productivity of cereals in China. *Agric. Technol. Econ.* **2016**, *6*, 13–20. (In Chinese)
40. Cheng, C.Z.; Zeng, J.W.; Zhong, Y. Research progress on citrus yellow dragon disease. *J. Hortic.* **2013**, *40*, 1656–1668. (In Chinese)

agriculture

MDPI

Article

Family Farming as a Contribution to Food Sovereignty, Case Guarainag Parish

Graciela Verdugo, Gina Cuadrado * and Yonimiler Castillo

Faculty of Postgraduate, School of Economics, Universidad Católica de Cuenca, Cuenca 010107, Ecuador; gracielav1965@hotmail.com (G.V.); ycastilloo@ucacue.edu.ec (Y.C.)
* Correspondence: gcuadrado@ucacue.edu.ec

Abstract: The objective of this research is to analyze how family farming contributes to food sovereignty; the Guarainag parish of the Paute canton in the province of Azuay-Ecuador is taken as a case of study. This work responds to the necessity to explain the elements that impact food sovereignty in the existing food crisis in Latin America and specifically in Ecuador in search of self-sufficiency for healthy food products and people's own local culture. For this purpose, a Food Sovereignty Index was constructed through ten quantitative and qualitative indicators. The research has a correlational and explanatory scope; quantitative methods were used to measure food sovereignty through a binary logit regression model, which provided an answer to the hypothesis of the research, which consisted of testing the influence of family farming on food sovereignty. Furthermore, to collect the information, a survey was applied to 372 small farmers with the support of digital mapping and the Kobol Tulboox software version 1.27.3. The result was a Food Sovereignty Index of 59.79%, which, according to the scale used, places the territory in a high average. In addition, the hypothesis was verified, concluding that there is a direct relationship among the following elements of family farming such as number of household members, family labor, group of products, type of animals, tillage technology, natural fertilizer, and altitudinal levels with food sovereignty. For future research, it is recommended that the variable of climate change has to be incorporated in order to observe its impact on food sovereignty.

Keywords: family farming; food sovereignty; agricultural; diversification; family labor force

Citation: Verdugo, G.; Cuadrado, G.; Castillo, Y. Family Farming as a Contribution to Food Sovereignty, Case Guarainag Parish. *Agriculture* **2023**, *13*, 1827. https://doi.org/10.3390/agriculture13091827

Academic Editor: Sanzidur Rahman

Received: 20 July 2023
Revised: 18 August 2023
Accepted: 22 August 2023
Published: 18 September 2023

1. Introduction

The food crisis requires that we reformulate the current capitalist agrifood model and build a new relationship that respects the cultural system and natural ecosystems; in this context, family farming represents a new way of looking at economic activity. Hence, the food crisis is an opportunity because it has allowed a debate on the importance of food sovereignty, as a right of the population to obtain their own food system, which allows the consumption of healthy, nutritious food produced under technologies compatible with the environment [1]. Food sovereignty is also focused on defending the rights of peasant communities harmed by the policies that promote economic globalization [2].

The pandemic revealed the significance of this kind of agriculture since the health crisis could not have been managed if the production of food from peasant production had been interrupted. The close relationship between family farming and food sovereignty is also evident.

Agriculture productive units (UPAs) dedicated to family farming produce around 80% of food worldwide, on an average area of 70% of agricultural land globally [3], while in Ecuador, family farming provides 60% of the food demanded by the population [4]. In the same way, in the Guarainag parish, the productive units dedicated to family farming occupy 2.53% (42.74 hectares) of the total agriculture area, the same ones that are located on different bioclimatic floors and adjacent to populated centers.

The concept of family farming in Latin America originated at the end of the 19th century and the beginning of the 20th century, from the publications made by the historians Chayanov and Ayala who call it the "family economic unit" [5], which means: "a farm of sufficient size to provide the livelihood of a family and that in its operation does not require salaried labor, but that could be attended with the labor force of the family itself" [6].

However, it must be emphasized that, from the year 2000, after the end of the dictatorships, the term family farming began to be used with greater emphasis in the region due to the resistance of this segment to economic shocks [7]. The use of this term was also influenced by the declaration of the United Nations General Assembly (UNGA), in which 2014 was nominated as the year of Family Farming.

Likewise, food sovereignty is analyzed as a concept that emerged from the year 1996, promoted by the social organization called Vía Campesina, in the different sceneries of the fight against hunger [1]. It emerges as an opposition to the policies that promote capitalist agricultural activity [8] and as a response to food sovereignty proposed by the Food and Agriculture Organization of the Nations—FAO [9]. It also constitutes a tool that aims to change food systems from the bottom up [10]. It makes it possible to carry out studies on its characterization and contribution to the economy of the region; that is, family farming became the mobilizing axis of a series of events, academic studies, and public policies [11].

In 2017, the Food and Agriculture Organization of the Nations-FAO and the Latin American and Caribbean Parliament (Parlatino) published the Parlatino Family Farming Model Law, as a basis for the formulation of laws and policies in the region, in which the concept of family farming was established as a productive unit and a way of life in which men and women from the same family nucleus work and their production is used for self-consumption, exchange, and commercialization.

Family farming systems in general, and home gardens in particular, are important settings for the implementation of agroecological practices [12]; its administration is in charge of the family and the family resides in it or in a nearby place [11], a way of life of community work which differentiates it from other types of capitalist economy [13].

The Inter American Institute for Cooperation in Latin America (IICA) states that family farming is a social category made up of individuals who live in rural areas, work the land for productive and reproductive purposes, and in which family work predominates, due to which it has been positioned in different countries as a relevant social subject [14].

In addition, family farming is not only a productive activity, but also a way of life of the peasant population [11,15]; it is characterized by the high number of people who work in it [16]. Both men and women from the same family unit [17,18] mainly use family labor and occasionally hired labor [19]. Furthermore, they are owners of their means of production, use low-tech tools and equipment, are friendly to the environment (animal traction, picks, shovels), and they are located on different altitudinal floors where agricultural activities are carried out which allows them to have a high diversification of their production [20].

This is related to the approach of Fabron and Castro [18] and Acebedo [19], who state that family farming is the basis of food sovereignty. It provides healthy and nutritious food with strong identity roots, which gives it elements of multidimensionality. This criterion also coincides with both Byaruhanga and Isgren and [21] Wald and Hill [22], who consider that the multiscale perspective plays an important role in the analysis of food systems and their implementation.

Food sovereignty is focused on the rights of both the State and the people in order to democratically define what to sow in their agricultural production units without external imposition [22]. Food sovereignty is also established as a right that people have to a diet with nutritious, accessible products produced under an agroecological approach that represents their culture, and also to freely choose what they want to produce for their food [23]. Food sovereignty is a political proposal promoted by social organizations at the international level and subsequently assumed by the different States within their regulatory frameworks [9].

From the literature review, it has not been possible to find a study that shows the analysis of the role of family farming in family sovereignty, which allows us to formulate the following research question: How does family farming contribute to food sovereignty in Guarainag Parish? The research hypothesis shows that there is a positive influence of family farming on food sovereignty, according to which was proposed the objective of analyzing how family farming contributes to food sovereignty. The Guarainag parish of the Paute canton in the province of Azuay-Ecuador is taken as a case.

Ecuador is a pioneer country in terms of food sovereignty, incorporating it into the Ecuadorian Constitution of 2008 [24]. In Article 281, it states that: "Food sovereignty constitutes a strategic goal and an obligation of the State to guarantee that individuals, communities, people, and nationalities achieve self-sufficiency in healthy and culturally appropriate food on a permanent basis" [25]. In the case of Ecuador, the Heifer International Foundation in 2018 characterizes family farming as a productive system linked to mainly family labor and it is the family nucleus that decides what to plant and how it organizes work, and manages the transmission of knowledge [26].

In Guarainag, despite the problems caused by migration and the effects of changing land use, family agricultural production is maintained, characterized by its diversification, ancestral techniques in crops, employment of family labor, and the concentration of UPAs around their homes, thereby contributing to food sovereignty through access to health products, for which qualitative indicators and quantitative analysis were used through an inferential statistic.

2. Materials and Methods

The research was carried out in the Guarainag parish, belonging to the province of Azuay of the Paute canton, which borders the cantons: Azogues (Cañar province), Sevilla de Oro, Guachapala, and the Tomebamba parish (Azuay province). The parish is made up of 8 communities: Las Juntas, Llamacón, Selel, Ucumarina, Bella Unión, Rambran, Coyal, and Guarainag Center, as shown in Figure 1.

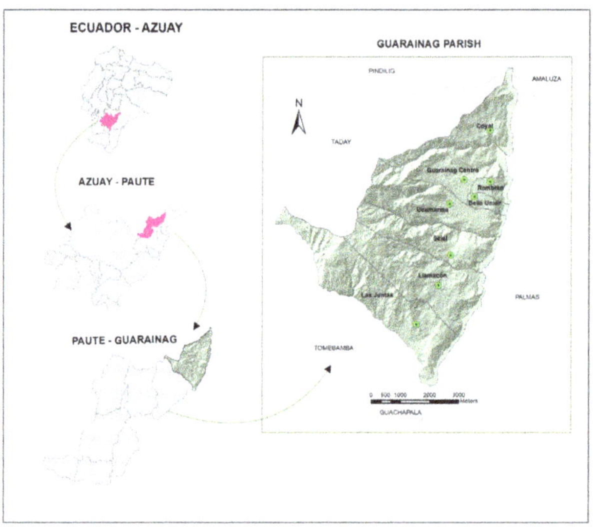

Figure 1. Location of the research area.

The reason why this parish was chosen to carry out the study is that its agricultural productive units present characteristics of family farming based mainly on the use of family labor, product diversification, and ancestral productive practices, among others which allowed us to determine to what extent this family farming contributes to food sovereignty. The study is based on the application of the inductive and deductive methods to understand the theoretical generalities regarding the proposed study, and how family farming contributes to food sovereignty based on estimates of the logistic regression econometric model. The development of the research was carried out in 4 stages according to Figure 2.

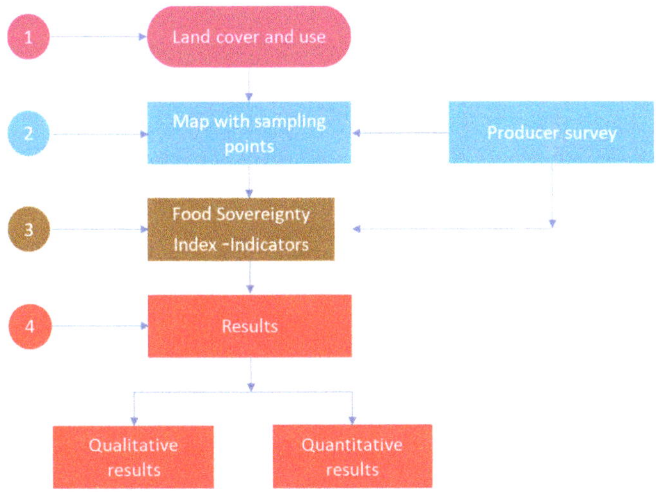

Figure 2. Research Stages.

(1) Compilation of cartographic material at a scale of 1:5000 of the coverage and use of the land of the Guarainag parish.

(2) Cartographic identification of agricultural productive units (UPAs) through a map which provided the number of UPAs dedicated to family farming, and which were validated by the directors of the Parish Government. The lack of knowledge of the exact number of these units was a limitation for the development of this research since there is little updated official information available due to the fact that the last agricultural census in Ecuador was carried out in the year 2000. This problem was solved with cartography updated to 2020, where a total of 387 UPAs dedicated to family farming were identified.

Based on the total number of UPAs, the sample was established with a total of 372 surveys to be collected randomly. The formula selected for the sample responds to the characteristics of the study where the population universe is known. The formula used is described in Equation (1).

The formula for the sample:

$$n = NPQ/(N-1)\left(\frac{E}{K}\right)^2 + PQ \tag{1}$$

Fountain: [27]

Where:

n = sample size

N= population size

P = probability of success

Q = probability of failure

E= error 1%

K = 1.96

$$n = \frac{387(0.5*0.5)}{(387-1)\left(\frac{0.01}{1.96}\right)^2 + (0.5*0.5)} = 372$$

With the number of samples determined, a map was made with the sampling points where the survey was applied. The location of the UPAs to be surveyed was carried out by means of a simple random sampling where each family agricultural productive unit had the same possibility of being included [28].

To carry out the sampling, each of the geo-referenced plots was numbered and the statistical probability calculation function in Excel software version 16.0.4266.1003 was used to select the plots to be surveyed. Figure 3 shows the UPAs where the surveyed households are located in relation to the total.

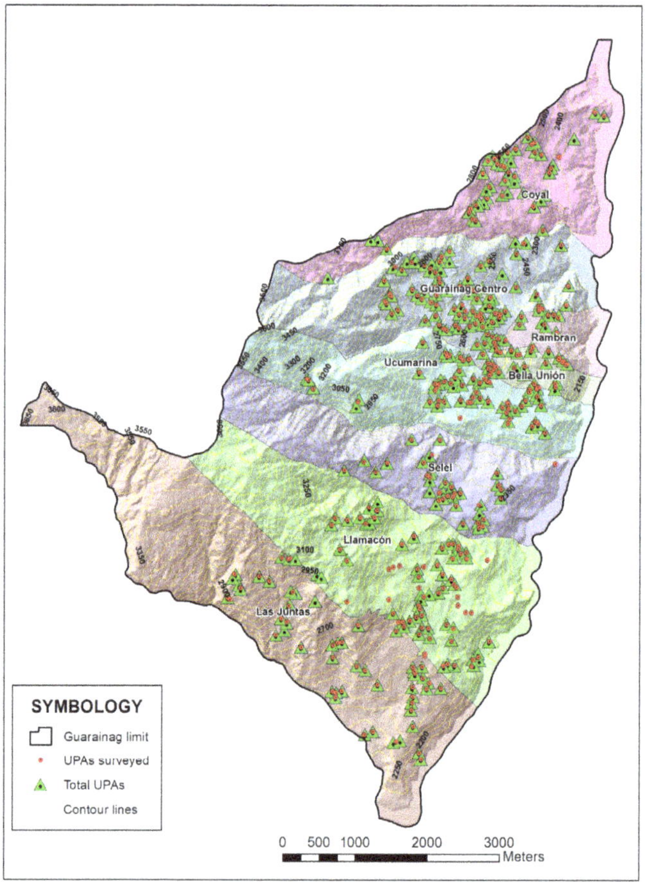

Figure 3. Location of the UPAs surveyed.

(3) A Food Sovereignty Index (IDS) was constructed as a criterion for measuring the food sovereignty variable. The following process was followed to construct the IDS: establish the factors of each variable, assign indicators for each factor, and assign its scale or measurement value [29–32]. The summary of this process is observed in Table 1, where the indicators are expressed as a percentage and the IDS is a simple average of food consumption (products, animals).

Table 1 shows that IDS has been formed based on 10 indicators, which were chosen from Johanes M. Waldmueller y Laura Rodríguez [24], Salgado et al. (2020) [33]. The IDS is the result of the simple average of the 10 indicators that compose it. It was decided to give the same weight to each variable due to the insufficiency of empirical and theoretical evidence as reference for making decisions regarding the weights of these indicators.

Table 1. Indicators of food sovereignty.

Indicators (%)
Average of vegetables you consume from your crops
Average of legumes consumed from their crops
Average of cereals consumed from their crops
Average of tubers consumed from their crops
Average of fruits consumed from their crops
Average of dairy consumed from its production
Average of chickens that you consume from your property
Average of guinea pigs that you eat from your property
Average of cattle consumed from the farm
Average of pigs consumed from your farm
Food Sovereignty Index

This may be a limitation of the present research; however, in the authors' opinion, it constitutes a first approximation that has a practical application in the measurement of food sovereignty in rural parishes. The analysis of the weights of each of the indicators on food sovereignty will be considered for future research.

The Food Sovereignty Index–IDS was established with a stratification; the Likert scale was used in 5 ranges: *Low,* those with <20% IDS; Middle-Low, among the range of >20%–<40%; Half, in the range >40%–<60%, High average, in the range >60%–<80%, and High, between the ranges >80%–<100%. This classification can be seen in Table 2.

Table 2. Ranks of the Food Sovereignty Index.

Ids Ranges	**Rating of the Food Sovereignty Index**
IDS < 20%	Low
20% < IDS < 40%	Middle-low
40% < IDS < 60%	Half
60% < IDS < 80%,	High average
80% < IDS < 100%	High

Table 3 shows the 3 axes and the 12 indicators that were used to measure the family farming variable. The choice of these was based on: Victor Hugo Verdezoto y Jorge Enrique Viera [27], Johanes M. Waldmueller y Laura Rodriguez [24].

Table 3. Indicators of family farming.

Axes	**Indicators**	**Unit**
Labor	Family labor	%
	Hired labor	%
Production units	Products group	Number
	Animals group	Number
	Property tenure	% own % rented
Technology	Ancestral tillage technology	%
	Plowing technology	%
	Associated products	%
	Agricultural calendar	%
	Rotation practices	%
	Access to irrigation	%
	Natural fertilizer	%

(4) The determination of the results where the quantitative indicators that allowed to determinate the association and correlation between variables were analyzed. For the

treatment of the data and the elaboration of the econometric model, the Stata statistical software version 14.0 was used, where 11 predictors were entered for the Logit Model, which validates the correlation between the dependent and independent variables. Likewise, the qualitative indicators that describe and characterize the studied variables were also analyzed [27]. The Logit model was constructed with the IDS as the dependent variable and the 11 predictor variables are those corresponding to the family farming indicators found in Table 4. The specification of the model can be seen in Equation (2).

Specification of the logit model:

$$\Pr(Y = 1|X_1, X_2 \ldots X_{11}) = \frac{1}{1 + \left(\frac{1}{e^{(\beta_0 + \beta_1 X1 + \beta_2 X2 + \cdots + \beta_{12} X11)}}\right)} \quad (2)$$

When the IDS is greater than 60%, it takes the value of 1 and is classified as having food sovereignty, and if the IDS is less than 60%, it does not have food sovereignty, qualifying it as 0. According to the above, 65.68 would be the percentage that had food sovereignty and 34.32% that did not. Hence, the description and categorization of the predictor variables can be seen in Table 4.

Table 4. Variables predictors.

Axes.	Variable	Categories
Independent variables	Number of household members managed by the Productive Unit	Membership
	Labor used UPA	1. Labor is Family 0. Labor is hired
	Group of products that it harvests Type of animals you raise	Number of product groups Animal type number
	Technologies used in tillage	1. Plow management 0. Tractor handling
	Use of the associated sowing technique	1. If they plant associates 0. Do not sow associates
	Use of rotation practices	1. If you do rotation practices 0. Does not carry out rotation practices
	Use of natural fertilizer	1. If you use natural fertilizer 0. Does not use natural fertilizer
	Tenure of the land	1. Own 0. Leased, rented, at first
	Acquisition of seeds	1. Own harvest 0. Another way of acquiring
	Altitudinal floors	1. Altitude of 2140–2499 2. Altitude of 2500–2799 3. Altitude of 2800–2999 4. Altitude of 3000–4000
Dependent variable	Food Sovereignty Index	0 < 60% without food sovereignty 1 $\geq$ 60% with food sovereignty

The bivariate analysis method allows us to analyze the correlation between variables through the Chi-Square test (chi2), the same one that contrasts from the observed frequencies if the differences between the two groups are attributable to chance [28].

3. Results

According to the established methodology, the results are presented below in two sections: qualitative and quantitative indicators.

3.1. Qualitative Indicators

As mentioned above, the family farming variable is measured through the 12 qualitative indicators in Table 3, as well as the food sovereignty variable, measured with the 10 indicators in Table 1. Table 5 shows them for each of the variables, axes, and indicators.

Table 5. Variables of family farming and food sovereignty results.

Variables	Axes	Indicators	Unit	Value
Family farming	Labor	Family labor	%	77.75
		Hired labor	%	22.25
	Production units	Products group	Number	6
		Animals group	Number	4
		Property tenure	% own	95.71
			% rented	4.29
	Technology	Ancestral tillage technology	%	86.06
		Plowing technology	%	13.94
		Associated products	%	93.03
		Agricultural calendar	%	91.42
		Rotation practices	%	97.32
		Access to irrigation	%	6.70
		Natural fertilizer	%	91.69
Food sovereignty		Average of vegetables you consume from your crops	%	58.17
		Average of legumes consumed from their crops	%	38.57
		Average of cereals consumed from their crops	%	82.88
		Average of tubers consumed from their crops	%	15.28
		Average of fruits consumed from their crops	%	23.82
		Average of dairy consumed from its production	%	91.58
		Average of chickens that you consume from your property	%	90.89
		Average of guinea pigs that you eat from your property	%	95.06
		Average of cattle consumed from the farm	%	86.05
		Average of pigs consumed from your farm	%	87.34

Based on the variables identified in Table 1, it can be observed that the labor used in the productive units is closely linked to family labor with 77.75%, and hired labor with 22.25%; the latter is mainly due to the aging of the population which is linked to agricultural production. The surveys conducted showed that family farming in the case of Guarainag has a diversity of products, which are grouped into six categories: vegetables, legumes, cereals, fruits, dairy products, and tubers, and in terms of livestock, they raise small animals (guinea pigs, pigs, poultry) and cattle. This characteristic coincides with what was stated by [6].

Likewise, the characteristics of the Agricultural Productive Units—UPAs dedicated to family farming, their concentration, area, diversity of products, animal husbandry, bioclimatic floors, irrigation systems, and property ownership were analyzed. Regarding the concentration of family UPAs, these are adjacent to the populated centers of the parish communities, as we can see in Figure 4.

The average area of the agricultural area dedicated to family farming is 2893 square meters, where there is a diversity of products grouped into six species: vegetables, legumes, cereals, fruits, dairy products, and tubers. Regarding the raising of animals, there are four types: chickens, guinea pigs, pigs, and cattle. Livestock and crop production are located in four bioclimatic zones. Likewise, 86.33% of the family farms dedicated to family agriculture are located in two bioclimatic floors, ranging from 2140 to 2499 and 2500 to 2799 m above sea level, as shown in Figure 5.

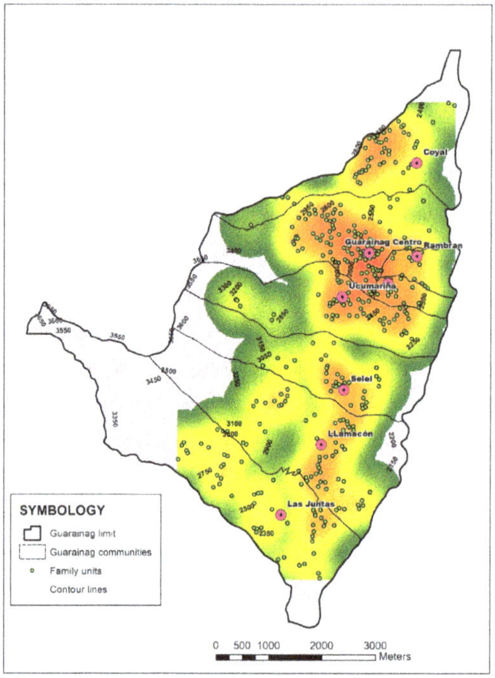

Figure 4. The concentration of family UPAs.

The 95.71% of the land dedicated to family farming is owned; furthermore, 91.69% do not have access to irrigation. Regarding the technology applied to the agricultural area, 86.06% of the family productive units use the animal traction system (plow) and 13.94% use the agricultural tractor; 91.69% practice crop rotation, 93.03% use the sowing technique with associated products, 97.32% use natural fertilizer, and 91.42% use the agricultural calendar.

In relation to access to food, family farming contributes on average to the family diet with 58.57% vegetables, 38.57% legumes, 82.58% cereals, 15.28% tubers, 23.82% fruit, 91.58% dairy, 90.89% chickens, 95.06% guinea pigs, 86.05% sheep, and 87.34% pigs.

According to the analysis carried out in Table 6, it can be seen that 59.79% have high average food sovereignty, 28.95% are rated as average, 5.9% as high, and 5.36% average low, with access to the consumption of healthy food (produced with natural fertilizer), including nutritious (they are important sources of minerals, calcium, potassium, magnesium, iron, iodine, proteins) and diversified products with sustainable environmental and cultural production models (use of animal traction).

Table 6. Ranks of the Food Sovereignty Index.

Ids Ranges	Rating of the Food Sovereignty Index	Ids	Percentage (%)
IDS < 20%	Low	0	0.00
20% < IDS < 40%	Middle-low	20	5.36
40% < IDS < 60%	Half	108	28.95
60% < IDS < 80%,	High average	223	59.79
80% < IDS < 100%	High	22	5.90
	Total	373	100.00

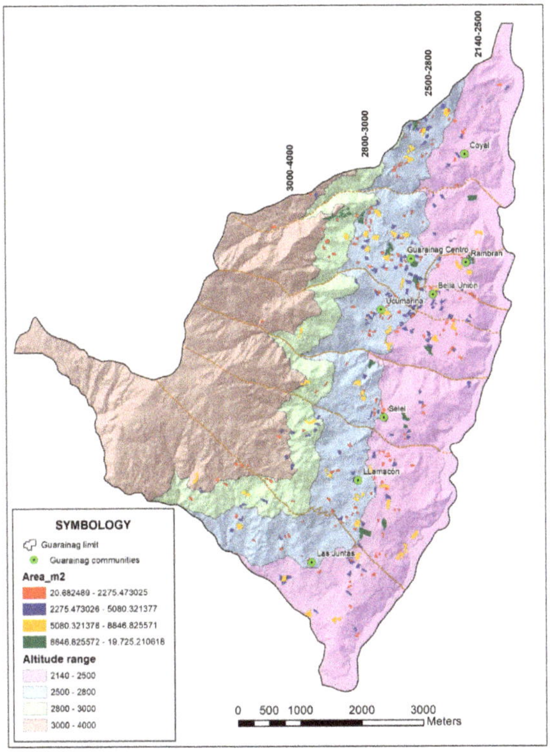

Figure 5. Location of family UPAs by bioclimatic floors.

3.2. *Quantitative Indicators*

This type of indicator analyzes the contribution of family farming to food sovereignty through the "Logit Model", which allows for a binary analysis between independent and dependent variables. As mentioned in the methodological part, food sovereignty has been chosen as the dependent variable. The results of the descriptive statistics of this variable are shown in Table 7.

Table 7. Dependent variable results.

IDS	Frequency	Percentage	Accumulated
0	128	34.32	34.32
1	245	65.68	100.00
Total	373	100.00	

Table 8 shows the results of the binary logit model which was run with the STATA software version 14.0.

The results in Table 8 show that the variables (Number of household members managed by the Productive Unit, Labor used UPA, Group of products that it harvests, Type of animals you raise, and Altitude of 2500–2799) are significant at a 95% confidence level. In addition, the variable (Use of natural fertilizer) is significant at 90% confidence level. The variables (Use of the associated sowing technique, Use of rotation practices, Tenure of the Land, Acquisition of seeds, Altitude of 2800–2999 and Altitude of 3000–4000) were not significant, and therefore, they cannot be interpreted. Hence, we could say that the variables that were found to be significant have a positive influence on food sovereignty.

Table 8. Logit model.

IDS	coef.	Std. Err.	Z	$p > z$	[95% Conf.	Interval]
Number of household members managed by the Productive Unit	0.5025217	0.1671353	3.01	0.003	0.1749424	0.8301009
Labor used UPA	0.9958676	0.377439	2.64	0.008	0.2561008	1.735634
Group of products that it harvests	2.157564	0.2682683	8.04	0.000	1.631768	2.683361
Type of animals you raise	0.7180205	0.2693447	2.67	0.008	0.1901146	1.245926
Technologies used in tillage	−1.015121	0.4436787	−2.29	0.022	−1.884715	−0.1455264
Use of the associated sowing technique	0.5246096	0.6221096	0.84	0.399	−0.6947027	1.743922
Use of rotation practices	0.4377693	0.9790197	0.45	0.655	−1.481074	2.356613
Use of natural fertilizer	0.9456437	0.5628639	1.68	0.093	−0.1575491	2.048837
Tenure of the land	0.3423796	0.6503339	0.53	0.599	−0.9322514	1.617011
Acquisition of seeds	−0.9963864	0.719643	−1.38	0.166	−2.406861	0.4140879
Altitudinal floors						
2. Altitude 2500–2799	0.6919243	0.3196482	2.16	0.030	0.0654253	1.318423
3. Altitude 2800–2999	0.2862071	0.5218403	0.55	0.583	−0.7365811	1.308995
4. Altitude 3000–4000	0.1517798	0.7863475	0.19	0.847	−1.389433	1.692992
_cons	−12.41432	2.139603	−5.80	0.000	−16.60787	−8.220778

3.3. *Indicators of the Fit of the Logit Model*

Table 9 shows the indicators used to measure the model's goodness of fit which are analyzed through Pseudo R^2 and the ROC Curve.

Table 9. Indicators of the fit of the Logit Model.

Indicators	Parameters
Number of obs =	373
LR chi^2 (13) =	171.28
Prob > chi^2 =	0.0000
Pseudo R^2 =	0.3570
log likelihood	−154.24019

The goodness of the logistic model is based on the Pseudo R^2, so the Pseudo R^2 statistic allows us to measure the goodness of a logistic model. It is a statistic used as a proxy of the model with respect to its endogenous variable based on empirical evidence; the following ranges are proposed [28]:

- First, Pseudo R^2 has a value less than 0.2; the model presents a bad fit.
- Second, Pseudo R^2 is in a range of 0.2 and 0.4; the model presents a normal fit.
- Third, Pseudo R^2 has a value greater than 0.4; the model fits the data appropriately.

For the purposes of this research, the calculation carried out has a normal fit because it corresponds to a Pseudo R^2 of 0.357 and the Probability > chi^2 is 0.0000; therefore, it is less than 0.05, concluding that the model is reliable.

The ROC curve is a graphical representation of the sensitivity and specificity of the logistic regression model. The area at the bottom of the ROC curve is an overall measure of the accuracy of a diagnostic test. The area under the curve must be greater than 0.5 for the model to discriminate adequately. In the research carried out, the curve has an ROC of 0.87 which is a good discrimination since it is close to 1; see Figure 6.

Depending on the statistics obtained for each variable, the following considerations fit: the contribution of family farming to food sovereignty is made based on the marginal effects and Odds Ratio, as shown in Table 10.

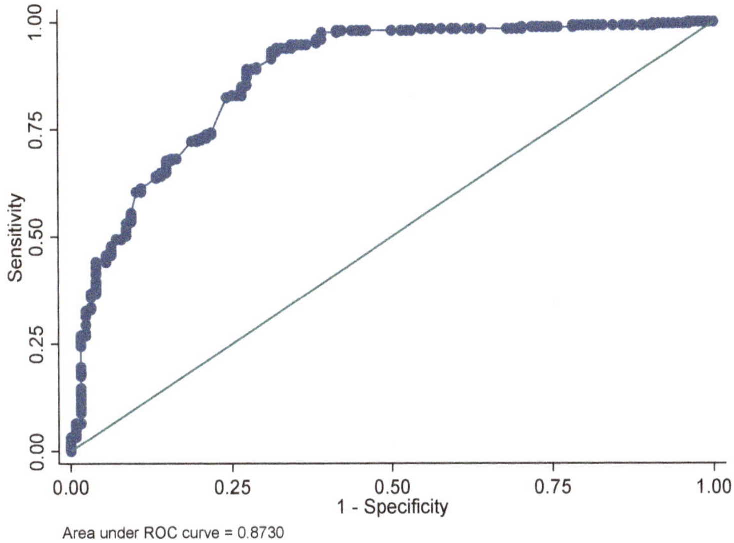

Figure 6. ROC Curve.

Table 10. Marginal Effects and Odds Ratio.

	dy/dx	$p > z$	Odds Ratio
Number of household members managed by the Productive Unit	0.0673586	0.002	1.652884
Labor used UPA	0.1386551	0.008	2.707072
Group of products that it harvests	0.2892024	0.000	8.650045
Type of animals you raise	0.0962443	0.006	2.050371
Technologies used in tillage	−0.1292463	0.013	0.3623587
Use of the associated sowing technique	0.0730374	0.411	1.689799
Use of rotation practices	0.0606742	0.663	1.549247
Use of natural fertilizer	0.1337408	0.099	0.257447
Tenure of the Land	0.0470397	0.606	1.408295
Acquisition of seeds	−0.1414668	0.175	0.3692112
Altitudinal floors			
2. Altitude 2500–2799	0.0916678	0.025	1.997556
3. Altitude 2800–2999	0.0391849	0.577	1.331368
4. Altitude 3000–4000	0.020973	0.845	1.163904

- The variable that refers to the number of members per household of families dedicated to family farming presents a positive coefficient and an Odds Ratio greater than the unit Exp (0.67) = 1.65, which indicates that, for each unit of increase in the number of members in the household, there is 1.65 times more probability of contributing to food sovereignty.
- The variable family labor presents a positive coefficient and the Odds Ratio greater than one unit–Exp (0.13) = 2.70. Thus, it indicates that for each unit of family labor increased, there is 2.70 times the probability of contributing to food sovereignty.
- The variable of the number of groups of products that they cultivate has a positive coefficient and the Odds Ratio is greater than the unit Exp (0.28) = 8.65, indicating that due to an increase of one unit in the diversity of products, it is 8.65 times more likely to contribute to food sovereignty.
- The variable referring to the number of types of animals raised in their production units has a positive coefficient and the Odds Ratio greater than the Exp unit

(0.096) = 2.05, showing that with each increase in the breeding of one type of animal, it is 2.05 times more likely to contribute to food sovereignty.

- The tillage technology variable with the use of the team has a negative coefficient and the Odds Ratio is greater than the Exp unit (−0.091) = 0.36 (inverse 2.77). Thus, it indicates that for each increased unit of use of the team in the UPAs, it is 2.77 times more likely to contribute to food sovereignty.
- The variable related to bioclimatic floors has a positive coefficient and the Odds Ratio is greater than the Exp unit (0.91) = 1.99, indicating that due to the increase of a farm unit located in the bioclimatic floor that goes from 2500 to 2800, it is 1.99 times more likely to contribute to food sovereignty; all of the above mentioned corresponds to the fact that within this bioclimatic floor the soil is loamy and black in color, contains nutrients, moisture, humus, good drainage, and infiltration of water and air which present better characteristics for agricultural cultivation [34].

4. Discussion

The results of the research agree with the criteria of some authors who focus on the direct link between family farming and food sovereignty [33] and consider it as an alternative to capitalist agriculture [28], which is reflected in the Constitution of the Republic of Ecuador 2008, where food sovereignty is proposed as an alternative for people to have a healthy, nutritious diet, and this is achieved with the contribution of family farming [25] with diversified production [6]. Furthermore, food diversity and the use of organic fertilizers are important elements as a contribution to food sovereignty. This coincides with the profile found by García (2023) [35] in a study on self-consumption gardens in the Huelva mountains where farmers committed to the environment grow a range of products with a wide variety and a high commitment to the use of organic fertilizers [36]. In addition, it supports the point of view that food sovereignty is likely to increase when production units have a greater diversity of products [34], and the raising of small and large animals is accomplished with a clean technique [5], the use of ancestral techniques (animal traction), and with mainly family labor [33]. In short, these factors contribute to food sovereignty [20]. Thus, local agrobiodiversity supports food sovereignty [37].

The 59.79% food sovereignty indicator obtained shows that there is still room for improvement in the Guarainag parish case study. Among the aspects that should be worked on is the consumption of fruits, legumes, and tubers. Moreover, a similar study is that of Salgado et al. (2020) [33] who, in their research on vulnerability to food insecurity conducted in Quilombola, found an indicator of food insecurity of 53% of households. In this case, the program applied showed changes in eating habits, creating new habits due to access to a greater diversity of foods.

Food sovereignty occurs when populations have access to healthy, organic, and diverse food. To achieve food sovereignty, family farming plays an important role which becomes a reality when there is a will of families to produce the land with environmentally friendly and ecological techniques and technologies. In this sense, Mann (2014) [36] and Sélingué (2007) [38], also cited by Byaruhanga and Isgren (2023) [39], believe that in order to achieve the above, it is very important to guarantee the rights of people to determine their own food and agricultural systems which guarantee the right to culturally appropriate and healthy food, as well as ecologically sound production.

5. Conclusions

This study concluded that family farming has a positive influence on food sovereignty in the case study of Guarainag parish. The factors that were found to be significant are: Number of household members managed by the Productive Unit, Labor used UPA, Group of products harvested, Type of animals raised, Altitude of 2500–2799 and Use of natural fertilizer.

The food sovereignty variable was obtained from the construction of the Food Sovereignty Index based on ten indicators. The result obtained from its calculation was 59.79%, which places it on a high average scale according to the qualitative classification defined for it. On the other hand, the indicators that most influenced this qualification are: family production units that are characterized by their proximity to population centers, the use of family labor, ancestral practices (animal traction), high diversification of product groups such as vegetables, legumes, cereals, dairy products, fruit, and the raising of various species of animals, and the destination of their production being mainly for family consumption. Nonetheless, it is necessary to point out that one of the limitations of the study is the method of calculating the IDS based on the simple average. Future research will analyze more complex methods of calculation and incorporate other indicators of family farming such as the effect of climate change.

This study is considered a tool of practical importance in local public policy decision-making. In the opinion of the authors, family farming should be treated as a provincial policy, which would have a comprehensive impact on strengthening food sovereignty as a means to achieve zero hunger, as proposed in the 2030 Sustainable Development Goals.

Thus, with the results obtained and as a contribution of the study to the strengthening of family farming in food sovereignty, three lines of action are recommended. They are addressed to the competent and governing bodies of agricultural activity established in the Organic Code of Territorial Organization—COOTAD [36]. These are:

- Incorporate, in the guidelines for updating the land management plans, parameters that establish regulations on permitted and complementary uses which promote family agriculture.
- Since agricultural development is the exclusive responsibility of the Provincial Government and concurrent with the Parochial Governments, comprehensive programs must be promoted to strengthen family farming, and in this way, guarantee the population's food sovereignty.
- Make strategic alliances between the sectional government, families dedicated to family farming, and academia for the transfer of technology adapted to the area and with cultural characteristics that boost the productivity of agricultural production units.

Author Contributions: Conceptualization, G.V., G.C. and Y.C.; methodology, G.V., G.C. and Y.C.; validation, G.V., G.C. and Y.C.; formal analysis, G.C. and Y.C.; investigation, G.V.; data curation, G.V.; visualization, G.V.; original draft preparation, G.V., G.C. and Y.C.; writing—review and editing, G.V., G.C. and Y.C.; supervision, G.C. and Y.C.; final revising, G.C. and Y.C. All authors have read and agreed to the published version of the manuscript.

Funding: This research received no external funding.

Institutional Review Board Statement: Not applicable.

Data Availability Statement: The data in this research are available for appropriate use and may be requested from its author (GV) upon request.

Conflicts of Interest: The authors declare no conflict of interest. The views expressed are strictly those of the authors and do not necessarily represent the positions or policies of their respective institutions.

References

1. Medina, J.; Ortega, L.; Gloriaz, M. Seguridad alimentaria, soberanía alimentaria o derecho a la alimentación. Estado de la cuestión. *Cuad. Desarro. Rural.* **2021**, *18*, 1–19.
2. López, L.; Álvaro, G. Revisión de enfoques de políticas alimentarias: Entre seguridad y la soberanía alimentaria (2000–2013). *Cad. Saude Publica* **2015**, *31*, 1355–1369. [CrossRef] [PubMed]
3. Organización de las Naciones Unidas para la Alimentación y la Agricultura—FAO. *El Estado Mundial de la Agricultura y la Alimentación*; FAO: Rome, Italy, 2015. [CrossRef]
4. FAO. *Reseña de Agricultura Familiar*; FAO: Rome, Italy, 2019; p. 4.
5. Torres, M.; Ramírez, B.; Juárez, J.P.; Aliphat, M.; Ramírez, G. Buen vivir y agricultura familiar en el Totonacapan poblano, México. *Íconos Rev. Cienc. Soc.* **2020**, *68*, 135–154. [CrossRef]
6. Cuevas Reyes, V.; Baca del Moral, J.; Borja Bravo, M.; Grass Ramírez, J.F.; Rivera Martínez, G. Agricultura familiar y tecnología para la elaboración de piloncillo granulado en la comunidad de Aldzulup Poytzén, San Luis Potosí. *Nova Sci.* **2017**, *9*, 481–501. [CrossRef]

7. Barril, F.; Almada, A.Y. *La Agricultura Familiar en los Países del Cono Sur*; IICA: Turrialba, Costa Rica, 2007. Available online: www.iica.int (accessed on 11 April 2023).
8. Domínguez, D. La Soberanía Alimentaria como enfoque crítico y orientación alternativa del sistema agroalimentario global. *Pensam. Am.* **2015**, *8*, 146–175.
9. Arzeno, M.B.; del Pilar Deheza, R.; Muñecas, L.; Zanotti, A.S. Discusiones en torno a las políticas públicas para la soberanía alimentaria y la agricultura familiar en Misiones (Argentina). *Mundo Agrar.* **2015**, *16*. Available online: http://www.mundoagrario.unlp.edu.ar/http://www.mundoagrario.unlp.edu.ar/article/view/MAv16n32a05 (accessed on 7 December 2022).
10. Altieri, M.A.; Funes-Monzote, F.R.; Petersen, P. Agroecologically efficient agricultural systems for smallholder farmers: Contributions to food sovereignty. *Agron. Sustain. Dev.* **2012**, *32*, 1–13. [CrossRef]
11. Grisa, C.; Sabourin, E. *Agri Familiar en ALC de Conceptos a Políticas Públicas en América Latina y el Caribe*; Organización de las Naciones Unidas para la Agricultura y la Alimentación: Santiago de Chile, Chile, 2019; pp. 1–22. Available online: https://hal.archives-ouvertes.fr/hal-02776075 (accessed on 11 April 2023).
12. Cortés, J.; Vieli, L.; Ibarra, J.T. Family farming systems: An index-based approach to the drivers of agroecological principles in the southern Andes. *Ecol. Indic.* **2023**, *154*, 110640. [CrossRef]
13. Ayala, C. La agricultura familiar en El Salvador: ¿Crónica de una muerte anunciada? *Espiral* **2021**, *28*, 231–268.
14. FAO. *Frentes Parlamentarios Contra el Hambre e Iniciativas Legislativas Para el Derecho a Una Alimentación y Nutrición Adecuadas. La Experiencia de América Latina y el Caribe 2009–2016*; FAO: Rome, Italy, 2017.
15. Olivera, G. Políticas neoliberales y agronegocio en Argentina (1991–2002). Vertientes, antiguas y nuevas organizaciones agrarias empresariales, de la agricultura familiar, campesina e indígena. *Trab. Comun.* **2017**, *45*, e033. [CrossRef]
16. do Nascimento, F.S.; Calle, Á.; Muñoz, R. Economía social y solidaria y agroecología en cooperativas de agricultura familiar en Brasil como forma de desarrollo de una agricultura sostenible. *CIRIEC-Esp. Rev. Econ. Publica Soc. Coop.* **2020**, *98*, 189–211. [CrossRef]
17. Pita, X.; Botía, B.; Fonseca, J. Caracterización y tipificación de los atributos ecosistémicos de la Agricultura Familiar Campesina en la Microcuenca del Rio Cormechoque. *RIAA* **2018**, *9*, 49–62. [CrossRef]
18. Rabelo, L.S.; Burte, J.D.P.; Boillot, É.A.S.; Gasmi, H.; Oliveira, W.R.; Fradi, F.; Rios, M.M.; Martins, E.S.P.R. The invisibility of young people in family farming: A gender perspective in the semi-Arid region of Ceará, Brazil. *Cah. Agric.* **2023**, *32*, 6. [CrossRef]
19. Piza, C.; Palacios, L.; Pulido, N.; Dallos, R. Agricultura familiar una alternativa para la seguridad alimentaria. *Conex. Agropecu.* **2016**, *6*, 13–25.
20. Fabron, G.; Castro, M. Agricultura Familiar y soberanía alimentaria en las nacientes de la Quebrada de Humahuaca (Jujuy). *Rev. Esc. Antropol.* **2018**, *24*, 1–20. [CrossRef]
21. Wald, N.; Hill, D.P. 'Rescaling' alternative food systems: From food security to food sovereignty. *Agric. Hum. Values* **2016**, *33*, 203–213. [CrossRef]
22. Vergara, A.; Moreno, A. Soberanía Alimentaria en Ecuador: Fundamentos teóricos y metodológicos para un modelo de medición. *Rev. Científica ECOCIENCIA* **2019**, *6*, 1–18. [CrossRef]
23. Mata, F. Soberanía Alimetnaria. *Kult. Rev. Interdiscip. Sobre Cult. Ciutat* **2018**, *5*, 159–178.
24. Waldmueller, J.; Rodríguez, L. La soberanía alimentaria más allá de la técnica: Una aproximación global hacia su monitoreo intercultural. *Coment. Int. Rev. Cent. Andin. Estud. Int.* **2015**, *15*, 253–286.
25. Asamblea Constituyente del Ecuador. *Constitución de la República del Ecuador 2008*; Registro Oficial; Asamblea Constituyente del Ecuador: Montecristi, Ecuador, 2008; pp. 1–136. Available online: www.lexis.com.ec (accessed on 12 December 2022).
26. Rivas, M.C.P.; Campoverde, C.M.B.; Ortega, Y.C. Modelo de optimización de la cadena de distribución de la agricultura familiar campesina en las parroquias Quingeo y Santa Ana del Cantón Cuenca. *ConcienciaDigital* **2021**, *4*, 114–134. [CrossRef]
27. Vargas, V.H.V.; Pico, J.E.V. Caracterización de Sistemas de Producción Agropecuarios en el proyecto de riego Guargualla-Licto, cantón Riobamba, provincia de Chimborazo. *Cienc. Tecnol.* **2018**, *11*, 42–50. [CrossRef]
28. Samuel, W.A.; William, M.G.; Douglas, L.A. *Estadística Aplicada a los Negocios y la Economía*; McGraw Hill Mexico: Ciudad De Mexico, Mexico, 2014; Volume XVI.
29. Fraser, E.D.G.; Dougill, A.J.; Mabee, W.E.; Reed, M.; McAlpine, P. Bottom up and top down: Analysis of participatory processes for sustainability indicator identification as a pathway to community empowerment and sustainable environmental management. *J. Environ. Manag.* **2006**, *78*, 114–127. [CrossRef] [PubMed]
30. García-Sempere, A.; Morales, H.; Hidalgo, M.; Ferguson, B.G.; Rosset, P.; Nazar-Beutelspacher, A. Food Sovereignty in the city? A methodological proposal for evaluating food sovereignty in urban settings. *Agroecol. Sustain. Food Syst.* **2019**, *43*, 1145–1173. [CrossRef]
31. Mickwitz, P.; Melanen, M.; Rosenström, U.; Seppälä, J. Regional eco-efficiency indicators—A participatory approach. *J. Clean Prod.* **2006**, *14*, 1603–1611. [CrossRef]
32. Yegbemey, R.N.; Yabi, J.A.; Dossa, C.S.G.; Bauer, S. Novel participatory indicators of sustainability reveal weaknesses of maize cropping in Benin. *Agron. Sustain. Dev.* **2014**, *34*, 909–920. [CrossRef]
33. dos Santos Figueiredo Salgado, R.; Dias, M.M.; de Souza, W.J. Family Agriculture, Rural Extension and Sovereignty and Food and Nutrition Security: Delimiting analytical categories in light of the implementation of the Food Acquisition Program in Brazil. *Mundo Agrar.* **2020**, *21*, e137. [CrossRef]

34. Bouzo, C.A. *El Cultivo de la Papa en Argentina*; Cultivos Intensivos II; Facultad de Ciencias Agrarias, Universidad Nacional del Litoral: Santa Fe, Argentina, 2018; pp. 1–42.
35. García, M. La huerta de autoconsumo en la Sierra de Huelva: Calidad, soberanía y apego alimentario. *Cuad. Geográficos* **2023**, *62*, 72–96. [CrossRef]
36. Mann, A. *Global Activism in Food Politics: Power Shift*; Springer: Berlin/Heidelberg, Germany, 2014.
37. Asamblea Nacional Ecuador. *Código Orgánico de Organización Territorial, COOTAD*; Asamblea Nacional Ecuador: Quito, Ecuador, 2019; pp. 1–181. Available online: www.lexis.com.ec (accessed on 13 January 2023).
38. Declaration of Nyéléni. Sélingué, Mali, 2007. Available online: https://nyeleni.org/IMG/pdf/DeclNyeleni-en.pdf (accessed on 17 January 2023).
39. Byaruhanga, R.; Isgren, E. Rethinking the Alternatives: Food Sovereignty as a Prerequisite for Sustainable Food Security. *Food Ethics* **2023**, *8*, 16. [CrossRef]

agriculture

MDPI

Article

An Empirical Investigation of the "Mezzogiorno Trap" in China's Agricultural Economy: Insights from Data Envelopment Analysis (2015–2021)

Xiaotuan Li [1], Panfei Yang [2] and Yuntao Zou [3,4,*]

1 School of Politics and Law, Northeast Normal University, Changchun 130024, China; lixt636@nenu.edu.cn
2 China Electronic Product Reliability and Environmental Testing Research Institute, Guangzhou 510610, China; yangpanfei@ceprei.com
3 School of Computer of Science and Technology, Huazhong University of Science and Technology, Wuhan 430074, China
4 School of Energy and Power Engineering, Huazhong University of Science and Technology, Wuhan 430074, China
* Correspondence: zouyuntao@hust.edu.cn

Abstract: Reducing regional inequality is one of the seventeen Sustainable Development Goals (SDGs) established by the United Nations. However, a persistent regional disparity known as the "Mezzogiorno Trap" presents a significant challenge. The underdeveloped regions that fall into the "Mezzogiorno Trap", even though they can narrow the gap with other regions through substantial support, see the disparity widen again when the level of assistance starts to decline. This paper proposes a methodology for identifying the "Mezzogiorno Trap". By employing this approach and combining panel data on Chinese agriculture from 2015 to 2021, it is discovered that despite the overall development of the Chinese agricultural economy during this period, the "Mezzogiorno Trap" still exists. The paper analyzes the reasons behind the "Mezzogiorno Trap" in the Chinese agricultural economy and presents constructive recommendations based on the research findings. The research process demonstrates that this methodology is better suited for studying regional disparities in specific economic sectors, and the obtained results are more stable and reliable.

Keywords: regional disparity; productivity disparities; agriculture; economy; sustainability; DEA; meta-frontier model

Citation: Li, X.; Yang, P.; Zou, Y. An Empirical Investigation of the "Mezzogiorno Trap" in China's Agricultural Economy: Insights from Data Envelopment Analysis (2015–2021). *Agriculture* **2023**, *13*, 1806. https://doi.org/10.3390/agriculture13091806

Academic Editors: Roberto Henke and Filiberto Altobelli

Received: 14 August 2023
Revised: 7 September 2023
Accepted: 7 September 2023
Published: 13 September 2023

1. Introduction

1.1. The "Mezzogiorno Trap" in Economic Development

"Mezzogiorno" refers to regions in the south of Italy and Sicily, which are often perceived as economically less developed than the cities and regions in the north of the country. This disparity between lagging and developed regions reflects, to some extent, Italy's unbalanced regional development. Since the end of World War II, the economic and social issues in these areas have posed significant challenges to Italy's economic progress. As early as 1970, Watson described the enormous disparity between the north and the south as, "Italy is, in effect, two nations" [1]. Even today, research by Daniele still indicates a productivity gap of up to 30% between southern Italy and the central and northern regions [2]. Additionally, a significant body of research demonstrates that such disparities persist, and they are indeed comprehensive [3–5].

Therefore, the term "Mezzogiorno Trap" is often used by scholars to describe the discrepancy and imbalance in regional economic development. However, its meaning extends beyond mere economic differences between regions. Over the decades, the Italian government has directed substantial aid towards the Mezzogiorno region. When this aid diminishes or disappears, the regional economy experiences significant setbacks, forming a sort of "trap". Research conducted by Iuzzolino et al. also validates this phenomenon.

In the 20 years following World War II, under the impetus of productivity enhancement and structural changes in the south, Italy witnessed the first substantial and sustained convergence between the north and the south. However, this trend abruptly halted in the mid-1970s, resulting in a renewed divergence [6]. Terrasi used the Theil Index to analyze the regional convergence of per capita GDP in Italy from 1953 to 1993 and found that the economic differences between regions in Italy were minimized only during the period from 1960 to 1975 [7]. The characteristics of the "trap" were quite apparent. Research by Torrisi et al. showed that between 1996 and 2008, the transfer payments received by southern Italy from the European Union accounted for 70% to 87% of Italy's total transfer payments. The funds obtained by the south were equivalent to around 11% of total investment and around 40% of public investment, considerably higher than in the central and northern regions [8]. However, in 2006 and 2007, the average income of residents in the richest region (Valle d'Aosta) was 2.6 times that of residents in the poorest region, indicating a significant disparity.

Hence, a succinct definition of the "Mezzogiorno Trap" can be posited: it is a distinct phenomenon of regional economic disparity characterized by the extensive governmental support enjoyed by lagging regions. The peculiarity lies in the fact that once these regions are devoid of such support, the disparities re-emerge and widen.

Economically, the "Mezzogiorno Trap" implies that the region's economic development heavily relies on external aid rather than internal economic activities. This may stem from the region's underdeveloped infrastructure, insufficient industrial development, and lower-quality human resources, which result in a weak economic development capability. Salvati and others conducted an exploratory analysis on 133 indicators across 7 thematic areas (population/housing, labor market, economic structure, quality of life, agriculture/rural development, landscape/water, environmental/soil resources), with results suggesting that latitude, altitude, and urban gradient dictate the complex spatial pattern of socio-economic and environmental variables in Italy [3]. Research by Daniele indicates a significant positive correlation between relative poverty levels and students' mathematics scores [2].

Socially, the "Mezzogiorno Trap" may reflect issues of regional social injustice. For instance, social resources such as education and healthcare might be unevenly distributed across regions, leading to a significantly lower quality of life for residents in certain areas. This social injustice can impact social mobility, thereby further exacerbating economic inequality [4]. In their study of the dynamics of poverty in Italy, Giarda and colleagues conducted a comparative analysis with the UK and Spain. Utilizing econometric methods, they found that the persistence of poverty in Italy exceeds that in the UK and Spain. Research by Bruzzi and others on the performance of healthcare systems across different Italian regions found that despite considerable support, healthcare performance in most southern regions remains poor [9].

From a policy perspective, the "Mezzogiorno Trap" might reflect the shortsightedness and errors of policy makers in their development strategies. For example, if policy makers overly rely on external aid to stimulate regional economic development without adequately considering how to enhance the region's self-sufficiency, issues could arise once external aid decreases. Research by Fazio et al. indicates that during the seven-year implementation of new strategic policy interventions from 1997 to 2003, regional economic disparities remained unchanged [10]. This demonstrates that inappropriate policies cannot effectively alleviate the "Mezzogiorno Trap". In their investigation of disparities among European Union (EU) countries, Geppert et al. discovered that European integration policies facilitated the catch-up process of lagging countries. However, concurrently, the force of economic activity agglomeration often expanded the internal gaps among EU member states [11]. This highlights the significant challenges inherent in policy formulation. In their study examining regional disparities in China before and after the abolition of agricultural tax, Ruan et al. [12] found that improper policy selection can lead to a dramatic widening of regional gaps once support is reduced. Additionally, in their investigation of economic disparities among the 28 European Union countries, López-Villuendas et al. [12,13] observed that since the implementation of the NUTS (Nomenclature of Territorial Units for Statistics) classification, economic disparities have been concentrating at the national level within regions categorized under NUTS2. Simultaneously, in areas delineated by the more granular NUTS3 classification, disparities have been progressively widening.

1.2. Examples of "Mezzogiorno Trap" in Other Areas

The concept of the "Mezzogiorno Trap" is indeed prominent in the southern regions of Italy, but in fact, similar economic disparities and developmental challenges exist in numerous other regions and countries. Here are some examples:

The United States' "Rust Belt": This region was once the industrial hub of the US. However, due to globalization and industrial transformation, many industries in these areas have declined. Lacking sufficient investment and support, these areas may also face analogous developmental predicaments. Research conducted by Harrison et al. found that the community housing vacancy rate in the Rust Belt was significantly higher than in the Sun Belt from 2012 to 2019, persistently remaining elevated [14]. Hegerty demonstrated that Detroit, the most representative city of the "Rust Belt", has fallen into uniformly poor conditions with a certain degree of contagion, in stark contrast to the cities in the southern and western "Sunbelt" [15].

Eastern Germany: Since German reunification, an economic disparity has existed between the East and West. At the time of reunification, the economic development level of the East (former East Germany) was far behind that of the West (former West Germany). Although the West provided substantial fiscal aid to the East, undertook massive infrastructure construction, and implemented various policies to stimulate economic growth, the economic development of the East remains slow. Herrschel found significant regional differences not only between the East and West but also within different states in the East [16]. Berentsen et al. believe that progress has been slow in eliminating regional inequalities in Germany, and these inequalities continue to evolve. However, the research also points out that according to EU standards, the difference between East and West Germany is not large. The East may perform better in certain aspects (such as education and health) than in its economic performance [17]. Dörr et al. still found an increased incidence and mortality rate related to heart failure in East Germany 30 years after reunification, much higher than in West Germany [18].

Rural areas in India: In India, there is a significant economic disparity between rural and urban areas. The economic development of rural areas primarily depends on agriculture and handicrafts, industries often less developed than the modern industries of cities. Therefore, these areas may also face similar developmental dilemmas. Birthal focused on agriculture to study the economic growth of various Indian states, finding absolute differences that require significant improvements in infrastructure and human resources [19]. Jose et al. demonstrated that there is a considerable disparity in socio-economic development between different states in India, evident in basic facilities such as sanitation, banking, road connectivity, clean drinking water, post offices, and telephony, and this disparity continues to increase [20].

1.3. How to Tell If the "Mezzogiorno Trap" Exists

In summary, to determine whether a region has fallen into the "Mezzogiorno Trap", several aspects can be considered:

1. Economic development gap: if a region lags significantly behind other regions or countries in its level of economic development, it may be at risk of the "Mezzogiorno Trap".
2. Dependence on external aid or investment: if a region's economic development heavily relies on external aid or investment, rather than on internal economic activities, then it might be susceptible to the "Mezzogiorno Trap".

3. Internal economic activity level: if a region's internal economic activities, such as industrial production and commercial activities, appear inactive or small-scale compared to the magnitude of external aid or investment, then it could potentially face the risk of the "Mezzogiorno Trap".
4. Changes in external aid or investment: if there is a decrease or disappearance in external aid or investment in a region, and this leads to a significant downturn in the local economy, the region may have already fallen into the "Mezzogiorno Trap".
5. Continuity of policy support: if a region's economic development largely depends on policy support, which may change for various reasons (e.g., regime changes, shifts in economic policy), then it too could be at risk of the "Mezzogiorno Trap".

1.4. Objectives and Contributions of This Paper

China exhibits a typical dualistic economic structure between urban and rural areas. There are significant disparities between cities and villages in terms of economic output, per capita income, social welfare, and other aspects. Despite China's ascent to becoming the world's second-largest economy, driven primarily by rapid urban economic development, agriculture still accounts for a substantial proportion of the population and the economy. Therefore, this paper focuses on the agricultural economy in China to investigate the potential existence of the "Mezzogiorno Trap" and aims to achieve the following objectives and contributions:

1. By studying the research process and findings, a more reasonable approach to assessing the "Mezzogiorno Trap" is summarized, which can be extended to further investigate regional disparities in a wider range of areas and regions.
2. By employing quantitative methods derived from operations research, management science, and economics, an assessment is conducted to determine the presence of the "Mezzogiorno Trap" in China's agricultural economy.
3. Constructive policy proposals and adjustments are put forward to address the "Mezzogiorno Trap". By studying the "Mezzogiorno Trap", the achievements of regional economic disparity research from various countries worldwide can be introduced into the relevant policy research for rural development in China.

2. Literature Review

2.1. The Possible Existence of the "Mezzogiorno Trap" in China

In China, the primary drivers of economic development are located in the eastern and coastal regions, possessing robust industrial bases and international trade networks, whereas the western economy lags significantly. In January 2000, the State Council established the Western Development Leadership Group. On 8 December 2006, the State Council Executive Meeting reviewed and in principle approved the "Eleventh Five-Year Plan for Western Development" [21], aiming to "use the residual economic development capacity of the eastern coastal regions to enhance the economic and social development level of the western regions, and consolidate national defense". A series of supportive policies known as the "Western Development" were subsequently introduced.

Since the implementation of the Western Development policy in 1999, the economic growth rate of the western regions has remained relatively high, with some provinces like Sichuan and Chongqing experiencing particularly rapid economic growth. However, China's eastern coastal regions have very favorable conditions, holding significant advantages over the western regions in terms of geographical location, climate, economy, science and technology, education, and talent. The pace of their development has not slowed, and the economic gap persists, even widening in some instances. This advantage is especially prominent in high-end industries like technology, finance, and services.

When Chen et al. studied the sample data of 815 Chinese listed companies from 1998 to 2004, they found that although China has shifted its development focus from the eastern coast to the inland regions, the difficulty for the government in guiding the economy is increasing. The influence of market mechanisms on the economy far surpasses that of the government, suggesting that reforms need to be further deepened for supportive policies to take effect [22]. Fan et al. also found, through research, that despite the Chinese government's efforts, the gap between the East and West continues to widen [23]. When Zhang et al. studied the factors influencing innovation in China's high-tech industries, they discovered that the central and western regions lag far behind the eastern region in terms of the decisive factor of innovation investment, with considerable gaps in other factors as well. The ultimate result is that the eastern region's technological innovation capability far exceeds that of the West [24].

In summary, there is a possibility that the "Mezzogiorno Trap" indeed exists in the Chinese economy, especially in the western regions. Previous research indicates that despite receiving favorable policies and substantial financial support, the disparities in industrial structure, education, population quality, geographical location, financial environment, and level of marketization in the western regions are challenging to bridge rapidly through simple financial support and preferential policies. Earlier studies demonstrate that in many aspects, the gap between the western regions and the developed areas is still widening.

2.2. A Review of Relevant Studies on China's Agricultural Support Policies and Regional Differences

Since 2000, the Chinese government has implemented a series of agricultural support policies aimed at increasing agricultural productivity, raising farmers' incomes, improving rural infrastructure, and promoting urban–rural economic integration. Below are some significant policy measures:

1. Agricultural subsidy policy: since 2004, the Chinese government has implemented direct agricultural subsidy policies, including grain planting subsidies and agricultural machinery purchase subsidies, aiming to enhance agricultural productivity and safeguard farmers' interests.
2. Abolition of the agricultural tax: the agricultural tax was a levy on farmers' income from planting grains and other agricultural products. In 2006, China completely abolished this tax, significantly reducing farmers' burden, increasing their income, and stimulating the zeal for agricultural production [25].
3. Agricultural insurance system: to mitigate farmers' losses due to natural disasters and other factors, the Chinese government introduced an agricultural insurance system, subsidizing part of the insurance costs for insured farmers.
4. New rural cooperative medical system: this policy, aimed at improving rural medical conditions, provides basic medical security for farmers through government subsidies and social fundraising [26].
5. Rural land system reform: the government relaxed restrictions on the transfer of rural land use rights, allowing farmers to gain income through leasing or transferring land, creating conditions for the modernization and scaling of agriculture.
6. Agricultural technological advancement policy: the government increased support for agricultural scientific research and promotion, to enhance agricultural productivity and yield, including the promotion of quality seeds and agricultural mechanization.
7. Rural infrastructure construction: this involves building rural roads, water supplies, and power supplies to improve rural living conditions and the production environment.
8. Rural poverty alleviation work: this includes offering low-interest loans, vocational training, rural labor transfer, and other poverty alleviation methods to decrease rural poverty.

By 2018, the Chinese government launched a strategy to comprehensively improve the economic, social, and environmental conditions of rural areas. Its core objective is to achieve balanced development between rural and urban areas, enhance the quality of life and work in rural areas, promote agricultural modernization, and strengthen the economic capacity of rural areas. The rural revitalization strategy covers all aspects of rural areas, including industry, talent, culture, ecology, and organization [27].

These intense agricultural support policies have significantly facilitated rapid growth in China's agriculture. However, they have also sparked concern among some scholars and experts. Research by Chan and colleagues indicated that, while rural economies were growing rapidly, disparities among rural regions across different provinces were also widening. The efficiency discrepancy between collective enterprises in the rural areas of the eastern and western provinces was a primary contributor to this divergence [28].

Li and colleagues, through studying the differences reflected by the agriculture, manufacturing, and service sectors from 1995 to 2004, found that the loss of agricultural employment in the central and western regions was not compensated for by growth in other sectors. The speed variance in the transformation from agriculture to secondary and tertiary industries widened the gap between the coastal regions and the rest of the country [29]. Chen and associates analyzed the spatio-temporal changes in arable land use intensity at national and provincial levels and found that developed regions had a lower labor intensity and a higher capital intensity. Less developed regions had a higher labor intensity but a lower capital intensity [30].

Research by Liu and colleagues demonstrated that the overall quality of agricultural development in China was steadily improving, but structural problems were evident. From the perspective of regional differences, a primarily "high in the East, low in the West" pattern was observed, which was mainly caused by interregional differences and showed a gradually declining trend during the selected period [31].

2.3. Possible Problems with the Study of Regional Disparities in the Agricultural Economy

Previous studies have indeed provided a wealth of insight and assistance. However, we believe there are areas for further refinement. In many studies, the regional differences in the overall economy can interfere with the regional differences in a specific field. For example, from an overall perspective, there is a significant difference between eastern and western China, involving multiple aspects such as the economy, society, and policy. The differences are especially profound in high-tech manufacturing and modern financial services. However, agriculture has a long history in all regions of China, and the level of the agricultural economy in a western province may not necessarily be inferior to that of eastern provinces. When many studies target regional differences in the overall economy, they may categorize this western province as less developed due to its geographical location, which could cause bias in the part of the study concerning the agricultural economy.

Moreover, the potential "Mezzogiorno Trap" in regional differences has not received adequate attention. As previously mentioned, the "Mezzogiorno Trap" typically describes a region whose economic development heavily relies on external aid or investment, rather than internal economic activities. When external assistance or investment declines or disappears, the economy of the region may experience a significant downturn. Therefore, examining whether a region has fallen into the "Mezzogiorno Trap" is meaningful for policy adjustment and the self-construction of underdeveloped areas. However, the issue does not receive much attention in studies on China's regional differences. On one hand, the "Mezzogiorno Trap" in regional economic differences may be obscured by inherent disparities between regions. On the other hand, in specific fields such as agriculture, the process of identifying the "Mezzogiorno Trap" can easily be disrupted.

In conclusion, we have decided to prioritize one crucial step in investigating the existence of the "Mezzogiorno Trap" in the Chinese agricultural economy: utilizing quantitative analysis to identify the regions with genuinely low agricultural economic efficiency. We believe that this is an important step for ensuring the credibility and validity of our research findings and will also serve as a significant foundation for future studies on regional economic disparities in specific sectors.

3. Materials and Methods

3.1. Data Sources

This paper uses relevant data from 31 provinces and cities in China from 2015 to 2021 as the research basis. All original data come from the "China Statistical Yearbook" published by the National Bureau of Statistics of China, as well as the "China Rural

Statistical Yearbook" jointly published by the National Bureau of Statistics and the Ministry of Agriculture and Rural Affairs. The period from 2015 to 2021 was chosen for two main reasons. First, the statistical methods used during this period are relatively consistent, and the data are relatively complete. Earlier data contain significant differences due to changes in statistical methods. Second, this period straddles the major "Rural Revitalization Strategy" initiative, enabling effective observation of changes and trends caused by policy.

Data Sources:

- China Statistical Yearbook, 2016–2022 [32]
- China Rural Statistical Yearbook, 2016–2022 [33]

Note: Each annual edition of the China Statistical Yearbook and China Rural Statistical Yearbook publishes statistical data from the preceding year. For instance, the 2016 China Statistical Yearbook provides statistics from the year 2015.

3.2. Research Process

In typical research on regional disparities, subjects are usually categorized into different groups for comparative study. In the context of China, the vast majority of studies directly classify subjects according to geographical variation. This approach is driven by the focus on the disparities between the eastern and western parts of China, where the substantial economic difference is axiomatic from a macroeconomic perspective [23,28,34]. Some research divides Chinese provinces and cities into eastern, central, and western regions, while others compare coastal regions with inland areas in China [29,35].

Our research objective is to determine whether the "Mezzogiorno Trap" exists in China's agricultural economy, considering the possibility that the agricultural economic level of a western province or city might surpass that of an eastern one. Consequently, we argue against the mere reliance on traditional geographical grouping. Instead, we advocate for a quantitative analysis approach to identify regions that lag behind in terms of agricultural economic development for comparison with more advanced regions.

In accordance with the characteristics of the "Mezzogiorno Trap", a simple regional disparity is insufficient for its identification. It is crucial to observe the changes in disparities over a specific period, and to combine this observation with changes in associated aid and investments during the same timeframe, to derive a comprehensive conclusion.

In summary, our research process is as follows, as illustrated in Figure 1.

1. Data collection.
    - Grouping using the super-efficient SBM model.
    - Constructing a meta-frontier SBM model of the agricultural economy in 31 provinces and cities, 2015–2021, calculating intra-group gaps.
    - Calculating financial support for agriculture in China during 2015–2021.
2. Determine if the "Mezzogiorno Trap" exists.
    - Calculating financial support for China's agricultural subgroups during 2015–2021.
    - Constructing an SBM Model of the Agricultural Economy in 31 Provinces and Cities, 2015–2021 We compute the intensity of financial support for agricultural groups in China from 2015 to 2021.
    - Constructing the SBM-Malmquist Model of Agricultural Economy in 31 Provinces and Cities, 2015–2021.
3. Analysis of the factors influencing the "Mezzogiorno Trap".
4. Conclusions and recommendations.

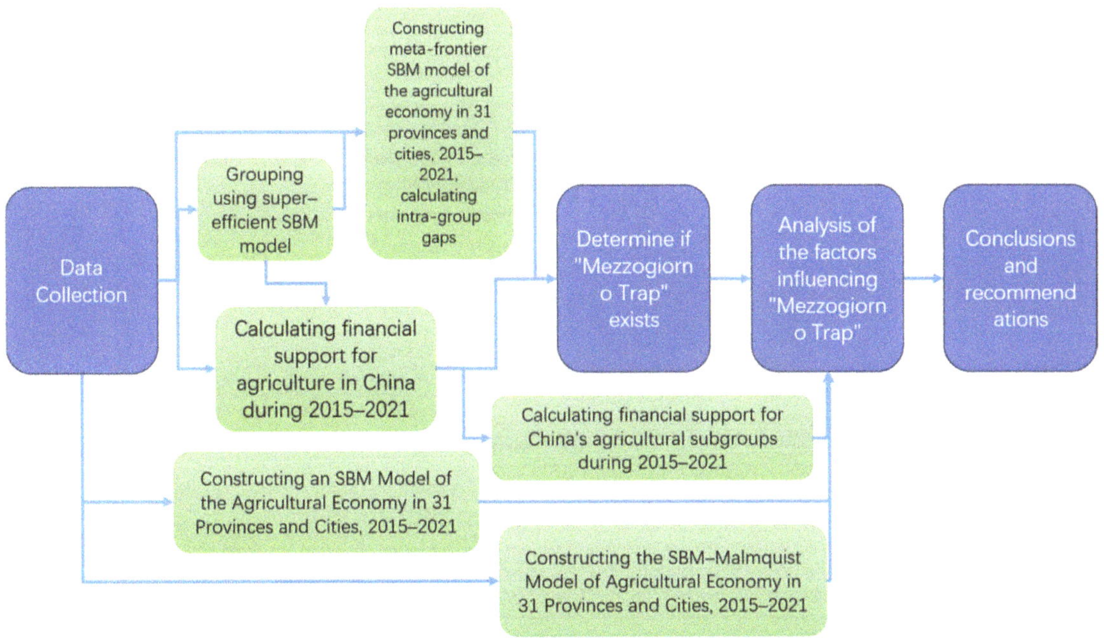

Figure 1. Flowchart of the research process for the "Mezzogiorno Trap" in Chinese agricultural economics.

3.3. Research Methods

The research methodology of this paper is based on efficiency assessment. There is a large heterogeneity in different regions of China, which is very different in terms of the employed population, total agricultural economy, cultivated land area, agricultural production methods, and so on. Efficiency assessments can better eliminate the heterogeneity. Higher production efficiency means more output with less input, and therefore more advanced technology, more efficient management, and less pollution and better sustainability. Efficiency assessment is widely used in the evaluation of economic levels. As early as 1952, Schmookler et al. used efficiency to evaluate the U.S. economy [36]. Bukarica et al. used efficiency evaluation to study energy policy and the level of sustainable development [37]. Bravo-Ureta et al. used efficiency evaluation to study agricultural and resource economics [38]. Paul et al. also used efficiency evaluation in their study of U.S. farm and agricultural economies of scale [39].

The data envelopment analysis (DEA) method, commonly used in efficiency evaluation, is a technique for analyzing relative efficiency through input–output analysis, proposed by Charnes et al. in 1978 [40]. DEA views each evaluated unit as a decision-making unit (DMU), with all DMUs having identical input and output variables. It calculates the production efficiency frontier surface, also known as the envelopment structure, by examining these input and output variables, thereby evaluating the relative efficiency of each DMU. DMUs situated on the frontier surface are considered DEA-efficient, with a comprehensive technical efficiency score of 1. The efficiency scores of other DMUs are determined by their relative position to the frontier surface, specifically ranging between 0 and 1. As a non-parametric method for evaluating relative effectiveness, the advantage of DEA is that it does not require the pre-assignment of weights for input and output, and it can evaluate the relative efficiency of multiple DMUs with multiple inputs and outputs. Hence, it is widely used in operational research, management, econometrics, and other fields.

The basic DEA models include the CCR model (named after its authors A. Charnes, W.W. Cooper, E. Rhodes) [40] and the BCC model (named after its authors Banker, Charnes, and Cooper) [41]. However, as they only consider radial improvement and neglect the slack of input and output variables, their efficiency calculation is not accurate enough and their efficiency improvement suggestions are not scientifically sound. To address this, Tone et al. established the non-radial SBM (slacks-based measure) model based on variable slack measurements by incorporating all slack measurements into the objective function through a scaling method [42]. Compared to traditional DEA models, the SBM model is more reasonable and rigorous. Like traditional DEA models, it decomposes comprehensive technical efficiency into pure technical efficiency and scale efficiency, which is equally convenient when analyzing causes. Chang et al. applied the SBM-DEA model to study the economic and environmental efficiency of 27 global airlines in 2010 [43]. Lin et al. used the SBM-DEA model to study CO2 emissions and the sustainable economy [44].

The methodology of this paper is based on the SBM-DEA model. The equation for SBM can be presented as follows:

In the input-oriented slacks-based measure (SBM) model [45], the objective is to minimize slack variables or equivalently maximize efficiency, subject to constraints on inputs and outputs. To assess the relative efficiency of $DMU_o = (x_o, y_o)$, the following linear programming formulation can be solved. This process is repeated n times for $o = (1, \ldots, n)$.

[SBM-I-C] (Input-oriented SBM under constant returns-to-scale assumption). Formally, the input-oriented SBM model can be articulated as the following linear programming problem:

Objective Function:

$$\rho_I^* = min_{\lambda, s^-, s^+} \ 1 - \frac{1}{m}\sum_{i=1}^{m} \frac{s_i^-}{x_{io}}$$

Subject to:

$$x_{io} = \sum_{j=1}^{n} x_{ij}\lambda_j + s_i^- \quad (i = 1, \ldots, m)$$

$$y_{ro} = \sum_{j=1}^{n} y_{rj}\lambda_j - s_r^+ \quad (r = 1, \ldots, s)$$

$$\lambda_j \geq 0(\forall j), \ \ s_i^- \geq 0(\forall i), \ \ s_r^+ \geq 0(\forall r)$$

ρ_I^* is called SBM-input-efficiency.

The SBM model has certain limitations under specific conditions. For example, in this paper, we take the 31 provinces of China as DMUs to analyze the relative efficiency of agricultural economics, firstly ranking and grouping them based on their overall technical efficiency. If we directly adopt the SBM model, there might be instances where multiple provinces have an efficiency score of 1, making it impossible to group them. Therefore, when ranking and grouping, we use the super efficiency SBM model.

The super efficiency model was proposed by Andersen and Petersen in 1993 with the aim of solving the issue in the DEA model when multiple DMUs are on the efficiency frontier, hence further comparison cannot be made [46]. When evaluating a particular DMU with the super efficiency model, it is excluded from the reference set, meaning it is not allowed to participate in calculating its own efficiency score. If this DMU still lies on the new efficiency frontier (i.e., the super efficiency score is greater than 1), then it can be deemed not only more efficient than the original set of DMUs but also higher in efficiency than other DMUs that are evaluated as efficient. The construction of the super efficiency model can thus solve the issue of ranking and grouping.

With the 31 provinces of China ranked and grouped based on the super-efficiency SBM model, subsequent research can be conducted. As the super-efficiency model might exhibit unboundedness, i.e., the super-efficiency scores for some DMUs might be infinite, we only use the super-efficiency SBM model for ranking and grouping. The subsequent research will be conducted by constructing the meta-frontier SBM model.

The meta-frontier model was first proposed by O'Donnell et al. to study the efficiency differences among grouped DMUs [47]. The technical principle involves, firstly, determining an efficiency frontier within each DMU group, termed the group frontier; thus, each DMU obtains an internal group efficiency score. Secondly, an efficiency frontier termed the meta-frontier is established by considering all DMUs together, from which each DMU obtains an efficiency score relative to the meta-frontier. The ratio of these two efficiency scores is referred to as the technology gap ratio (TGR).

Characteristic of the meta-frontier model is its ability to eliminate the heterogeneity of DMUs, commonly employed for comparative studies among regions and, with adjustments, can be used for comparisons between industries, policy comparisons, etc. O'Donnell et al. used the meta-frontier model to study enterprise efficiency, and empirical application was made using cross-national agricultural sector data [47]. Li et al. combined the meta-frontier model with the Malmquist index model and the Tobit regression model for regional comparative studies of China's high-tech industries [48]. Yu et al. employed the meta-frontier SBM model to study the energy efficiency of Eastern, Central, Western, Northeastern China, and various provinces from 2006 to 2016 [49]. Chen et al. also used the meta-frontier model when researching the agricultural economy at the county level in China [22].

In the meta-frontier analysis, this study employs efficiency scores under the CRS assumption, given that the efficiency score under CRS represents technical efficiency (TE). This score is a product of pure technical efficiency (PTE, which is also the efficiency score under the VRS assumption) and scale efficiency (SE). Consequently, TE can be considered as an overall efficiency that integrates both PTE and SE. Our research aim is to assess the overall efficiency of DMUs in all aspects (both technical and scale), making TE more pertinent. In contrast, the efficiency score obtained under the VRS assumption signifies pure technical efficiency, which holds relatively lesser significance when evaluating the comprehensive level of DMUs.

This is not to suggest that our study entirely overlooks the efficiency scores under the VRS assumption. We have additionally formulated SBM models under both CRS and VRS assumptions. Discussions encompassing the pure technical efficiency scores and scale efficiency scores under the VRS assumption have been conducted, which greatly aid in the analysis of the underlying reasons for the "Mezzogiorno Trap".

We also constructed a Malmquist SBM model for the agricultural economic efficiency of 31 provinces and cities in China from 2015 to 2021 to examine the development trend of China's agricultural economic efficiency during this period. This is because both the SBM model and the super-efficiency SBM model calculate the relative efficiency of DMUs within a specific period, and efficiency scores from different periods cannot be directly compared. To compare efficiencies across different periods, the Malmquist index model is used. The Malmquist productivity index was first proposed by Swedish economist Sten Malmquist in 1953 [50], and was later introduced into data envelopment analysis (DEA) by Färe et al. in 1984 to measure the change in production efficiency of DMUs over different time periods [51].

The basic construction of the Malmquist index is as follows: suppose that in two time periods, t and t + 1, each DMU has a corresponding production possibility set (PPS), which can be described by their input and output vectors. We can calculate the efficiency scores of DMUs in periods t and t + 1 based on the PPS of these two periods. The Malmquist productivity index is defined as the geometric mean of the ratio of the efficiency score in period t to the efficiency score in period t + 1. If EFF denotes the efficiency score, then the Malmquist index can be expressed as:

$$\mathrm{Tfpch} = \mathrm{Effch} \times \mathrm{Techch} = \frac{D^{t+1}\left(X^{t+1}, Y^{t+1}\right)}{D^{t}\left(X^{t}, Y^{t}\right)} \times \sqrt{\frac{D^{t}\left(X^{t+1}, Y^{t+1}\right)}{D^{t+1}\left(X^{t+1}, Y^{t+1}\right)} \times \frac{D^{t}\left(X^{t}, Y^{t}\right)}{D^{t+1}\left(X^{t}, Y^{t}\right)}}$$

If Tfpch (total factor productivity change) is greater than 1, then the efficiency score has improved. If a large proportion of provinces or cities exhibit this trend, it would suggest an overall improvement in the efficiency of China's agricultural economy. A Tfpch value of 1 indicates no change in efficiency, while a value less than 1 signifies a decline in efficiency. The purpose of constructing a Malmquist index model is to explore the potential relationship between the "Mezzogiorno Trap" phenomenon and the efficiency development trend in China's agricultural economy. This is aimed at studying the elusive nature of the "Mezzogiorno Trap": even if the overall economy is continuously progressing, a "Mezzogiorno Trap" may still exist and be less easily detected.

4. Results

4.1. Constructing a Super-Efficient SBM Model of the Agricultural Economy in 31 Provinces and Cities in 2017 and Grouping Them According to Rankings

To construct a DEA (data envelopment analysis) model, the first step is to select input and output variables. In this study, we chose the "Rural Population" across 31 provinces in China as the input variable representing labor. Although publicly available data from the Chinese government include the number of employed persons in rural areas, these figures are verified to only encompass rural enterprise employees who contribute to social insurance. A significant labor force in many rural areas of China does not participate in social insurance schemes [52]. Furthermore, agriculture is often a family-based endeavor involving both the elderly and children [53]. Hence, rural population size serves as a more accurate input variable for labor.

Utilizing rural population numbers as an input variable introduces certain uncontrollabilities. However, China presents a unique scenario. Firstly, China adopts a state-owned land system, which results in a notably weaker connection between farmers and their land compared to other nations. Additionally, the country enforces a household registration system, ensuring clear population registration and statistics. Over past decades, with the Chinese government's robust push for urbanization, the urban population swelled from 191 million in 1980 to 622 million in 2009. By 2011, urban residents constituted 51% of the total population, marking the first instance of surpassing the rural demographic [54]. Subsequent to the launch of the Rural Revitalization Strategy in 2018, a multitude of policies shifted in favor of rural development, causing certain regions to experience a phenomenon of "reverse urbanization" [55]. As such, China's rural population figures and proportions adjust significantly in response to government policy shifts. Therefore, employing rural population counts as an input variable is indeed salient for crafting rural policies.

For the material input variables, we have selected diesel consumption, pesticide consumption, and fertilizer consumption, which constitute the major consumables in Chinese agricultural production. These are not only used in crop cultivation but also in forestry and animal husbandry. The data sources also provide information on the consumption of seeds and plastic films; however, these represent a small proportion and have limited application scope. In consideration of the relationship between input variables and the number of decision making units (DMUs), this study opted for the most representative consumables to ensure the effectiveness of the DEA model.

We chose the total output value of agriculture, forestry, animal husbandry, and fishery as the output variable. Although data sources provide various specific outputs like crop yields, livestock, and aquatic products, these variables exhibit collinearity with the total output value. Moreover, provinces exhibit considerable heterogeneity in terms of specific types of agricultural products. Therefore, monetizing the end results of various types of agricultural production yields the total output value that is most representative.

It is noteworthy that some agricultural studies employ arable land area as an input variable. However, considering the latitude range of 31 provinces in China (3°30′ N to 53°33′ N), there are substantial climatic differences. In northern regions, crops mature once a year, while in the south, they can mature up to three times a year. Consequently, the same arable land area could yield significantly different levels of productivity. To eliminate this heterogeneity among the 31 provinces, arable land area was not utilized as an input variable.

The input and output variables used for assessing agricultural efficiency in the DEA model across China's 31 provinces are summarized in Table 1. For specific numerical values, please refer to Table S1 in the Supplementary Materials.

Table 1. Input and output variables for the DEA model of agricultural efficiency in China's 31 provinces.

	Variables	Data Sources
Input variables	Rural Population	China Statistical Yearbook
	Consumption of Chemical Fertilizers	China Rural Statistical Yearbook
	Consumption of Pesticides	China Rural Statistical Yearbook
	Consumption of Diesel Fuel	China Rural Statistical Yearbook
Output variables	Gross Output Value of Agriculture, Forestry, Animal Husbandry and Fishery	China Rural Statistical Yearbook

To rank and group the 31 provinces and cities in China, we chose the agricultural economic super-efficiency SBM model of 2017 as the basis. In 2017, it was the year before the "Rural Revitalization Strategy" was proposed. We believe that the regional differences in agricultural economy might be large at this time, and grouping at this time point could reflect the gap in agricultural economic levels more realistically. Subsequent calculations also confirmed our thinking: in all the 7-year meta-frontier SBM models, all the meta-frontier comprehensive technological efficiency scores for the 10 provinces and cities representing the relatively advanced Group 1 were 1. This indicates that these 10 provinces have always been at the highest level of agricultural economy. As for the provinces and cities that were grouped into the relatively backward Group 3, their meta-frontier comprehensive technological efficiency scores were almost at the bottom, with only one–two changes, which shows that our grouping method can represent the regional differences in China's agricultural efficiency.

The above variables were imported into the DEARUN software V3.1 edition to calculate the super-efficiency SBM model of agricultural economy for the 31 provinces and cities. According to the efficiency score ranking, the 31 provinces and cities were divided into three groups: Group 1 (rank 1–10), Group 2 (rank 11–20), and Group 3 (rank 21–31). The results are as follows in Table 2.

Additionally, it is worth noting that the choice to categorize the 31 provinces into three groups, rather than two or more than three, is guided by the following considerations: Segmenting the dataset into only two groups would lack a mid-range control group, thereby compromising the robustness of subsequent comparative analyses. On the other hand, dividing into more than three groups would result in each group containing fewer than 10 provinces. Given that the meta-frontier DEA efficiency analysis would then proceed with fewer than 10 DMUs per group and four input–output variables, this scarcity would negatively impact the validity of the DEA model.

Customary guidelines suggest that the number of DMUs should exceed thrice the sum of the input–output variables. In this study, utilizing the meta-frontier SBM model, there are a total of 31 DMUs and five input–output variables. When establishing the common frontier, the DMU count satisfactorily aligns with the general empirical suggestion. However, once segmented into three groups, each group comprises 10–11 DMUs, not meeting the thrice criterion relative to the input–output variable count. Such a shortfall could adversely influence the frontier formation for each group: a limited DMU number might result in more DMUs being adjudged as efficient, subsequently diminishing the discriminative capability among DMUs within a group.

Table 2. Ranking and grouping of agricultural economic efficiency in 31 Provinces in China.

DMU	Efficiency Score	Ranking	Group
Beijing	0.673728	14	2
Tianjin	0.570652	19	2
Hebei	0.443189	27	3
Shanxi	0.320583	31	3
Inner Mongolia	0.80136	11	2
Liaoning	0.915023	9	1
Jilin	0.443526	26	3
Heilongjiang	1.092438	3	1
Shanghai	0.487126	24	3
Jiangsu	1.029901	7	1
Zhejiang	0.606872	17	2
Anhui	0.438897	28	3
Fujian	1.064987	4	1
Jiangxi	0.50359	23	3
Shandong	0.725972	12	2
Henan	0.420875	29	3
Hubei	0.811924	10	1
Hunan	0.548386	20	2
Guangdong	0.522385	22	3
Guangxi	0.611202	16	2
Hainan	1.160764	2	1
Chongqing	0.577371	18	2
Sichuan	0.719591	13	2
Guizhou	1.573564	1	1
Yunnan	0.444038	25	3
Tibet	0.526058	21	3
Shaanxi	1.013779	8	1
Gansu	0.360777	30	3
Qinghai	1.042648	5	1
Ningxia	0.672134	15	2
Xinjiang	1.035293	6	1

Nonetheless, this paper consciously opts for a tripartite division rather than bifurcation for several reasons: Firstly, the negative repercussions predominantly transpire within each group, rendering it "equitable" across groups. Given that the research aim of our meta-frontier SBM model is to discern inter-group disparities, such adverse effects on between-group variations might be mitigated due to this inherent "equity". Secondly, an SBM model without group distinctions has also been formulated in this study to scrutinize variations between DMUs within each subset, serving to counterbalance potential detrimental effects. Moreover, sustaining the triad, inclusive of a median group for comparison, proves pivotal for inter-group comparative analysis. Lastly, while curtailing the number of input–output variables could ostensibly conform to general guidelines, it would profoundly compromise the model's precision, reliability, and robustness, deficits that are challenging to counteract.

The final distribution of the three groups is shown in Table 3.

Table 3. Grouping table of 31 provinces and cities in China.

Group 1	Group 2	Group 3
Liaoning	Beijing	Hebei
Heilongjiang	Tianjin	Shanxi
Jiangsu	Inner Mongolia	Jilin
Fujian	Zhejiang	Shanghai
Hubei	Shandong	Anhui
Hainan	Hunan	Jiangxi
Guizhou	Guangxi	Henan
Shaanxi	Chongqing	Guangdong
Qinghai	Sichuan	Yunnan
Xinjiang	Ningxia	Tibet
		Gansu

4.2. Constructing the Meta-Frontier SBM-DEA Efficiency Model for the Agricultural Economy of 31 Provinces and Cities in China from 2015 to 2021

In constructing the meta-frontier SBM-DEA (slack-based measure—data envelopment analysis) efficiency model for the agricultural economy of 31 provinces in China for the years 2015–2021, the input and output variables employed were identical to those detailed in Table 1. For specific numerical values, please refer to Table S1 in the Supplementary Materials. We incorporated the ranking and group data from Table 2 into the panel data of 31 provinces spanning from 2015 to 2021. The meta-frontier SBM models for each year were formulated using the DEARUN software. Subsequently, we computed the average TGR (technology gap ratio) values for DMUs (decision making units) within each group, which are organized and presented in Table 4.

Table 4. Average TGR values of meta-frontier SBM Model for Chinese agricultural economy efficiency from 2015 to 2021.

	Group 1	Group 2	Group 3
2015	1	0.839242	0.597488
2016	1	0.809165	0.572281
2017	1	0.700432	0.488344
2018	1	0.71545	0.505777
2019	1	0.640931	0.560976
2020	1	0.723399	0.577238
2021	1	0.637693	0.569133

4.3. Constructing the 2015–2021 SBM-Malmquist Index Model for Agricultural Economics across the 31 Provinces in China

When constructing the SBM-Malmquist index model with DEARUN software, the input and output variables selected are as shown in Table 1. The purpose of this analysis is to test the development trend of China's agricultural economic efficiency. The panel data input has not been ranked or grouped to minimize interference. The model calculated is shown in Table 5, after tidying up.

Table 5. Summary table of SBM-Malmquist index model results for Chinese agricultural economy efficiency from 2015 to 2021.

	Number of Provinces with Effch > 1	Number of Provinces with Techch > 1	Number of Provinces with Tfpch > 1
2015–2016	7	27	28
2016–2017	3	29	19
2017–2018	7	31	31
2018–2019	14	31	31
2019–2020	13	31	30
2020–2021	6	31	31

5. Discussion

5.1. Surprising Grouping

In Figure 2, provinces are annotated on the map of China according to their respective groupings as delineated in Table 6. A summary of the gross regional product, per capita gross regional product, and the corresponding rankings among the 31 provinces for the year 2017 is compiled in Table 6. A counterintuitive observation emerges from these graphical and tabular representations: there is a conspicuous lack of a direct correlation between the efficiency of agricultural economics in Chinese regions and their overall economic performance or geographical location. Intriguingly, Group 3, identified as having the lowest agricultural economic efficiency, includes some of China's most economically advanced provinces such as Shanghai and Guangdong. These provinces excel in metrics like gross regional product and per capita gross regional product and are located along the southeastern coast. In various Chinese regional economic studies, including those focusing on agricultural economic disparities, these provinces are frequently classified among the most developed [23,29,31]. Conversely, Group 3 also comprises economically underdeveloped regions like Tibet, Gansu, and Jilin, which lag in economic indicators and are located in western and northeastern China. Moreover, provinces with medium-level economic performance, such as Jiangxi, Anhui, and Henan, are also incorporated in this group and geographically situated in central China.

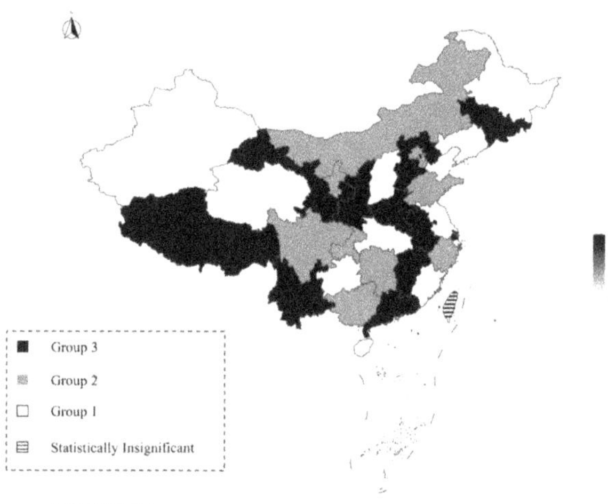

Figure 2. Geographical distribution map of agricultural economic efficiency rankings among China's 31 provinces and municipalities.

Table 6. Summary of gross regional product and ranking, per capita gross regional product and ranking of provinces in Group 3.

	Gross Regional Product (CNY 100 Millions)	Ranking of Gross Regional Product	Per capita Gross Regional Product (CNY)	Ranking of Per Capita Gross Regional Product
Hebei	40,391.27	12	54,231.03	27
Shanxi	22,590.16	20	64,914.26	17
Jilin	13,235.52	26	55,728.49	26
Shanghai	43,214.85	10	173,623.3	2
Anhui	42,959.18	11	70,275.11	13
Jiangxi	29,619.67	15	65,573.77	15
Henan	58,887.41	5	59,584.54	22
Guangdong	124,369.7	1	98,052.41	7
Yunnan	27,146.76	18	57,882.23	23
Tibet	2080.173	31	56,835.32	24
Gansu	10,243.31	27	41,137.77	31

From an agricultural economic perspective, this grouping is justified. Numerous studies have indicated that Shanghai and Guangdong are among the most developed and fastest-growing regions in China [56–58]. However, there may be significant issues regarding agricultural economic efficiency. In economically prosperous areas, there is a prevalent issue of excessively high land-use costs. Research by Gao et al. [59] revealed that urbanization in China is progressing rapidly, leading to over-expropriation of rural land in Shanghai, resulting in land idleness. The ambiguity in China's unique rural land ownership system has made it challenging to utilize this idle land. Studies by Liu et al. [60] show that land rents have heavily impacted areas surrounding Shanghai. After investigating

160 villages in the Qingpu district of Shanghai, Gu et al. [61] found that rural areas on the outskirts of Shanghai are evolving multifunctionally, and solely focusing on agricultural production is no longer the most viable livelihood option in Shanghai's rural areas.

The rural area in Guangdong Province is relatively vast, and the issues therein are more intricate. Rural areas close to the Greater Bay area of the Pearl River delta have been rapidly urbanized over the past few decades. Research by Choy et al. [62] indicates that Shenzhen, situated near Hong Kong, was an agricultural county with an urban built-up area of merely 3 square kilometers in 1980. By 2010, it had transformed into a metropolis with an urban built-up area of 703 square kilometers, with most of its land transitioning from agricultural to industrial use. In contrast, areas far from the Greater Bay Area, due to labor shortages and high transportation costs, exhibit significant land abandonment phenomena. Su et al. developed an algorithm based on phenology and time series and, after analyzing satellite imagery from Google Earth Engine, highlighted that abandoned lands in Guangdong Province have consistently measured around 500,000 hectares. Post-2000, due to the rapid urbanization of Guangdong, the rate of land abandonment has been increasing yearly [63]. In 2021, the Guangdong provincial government announced a cultivated land area of 28,480,000 mu, equivalent to 1,898,667 hectares, with a land abandonment rate reaching 26%. Studies by Hou et al. [64] reveal that land abandonment exists in other areas and that the rate of abandonment is directly proportional to the distance from urban settlements.

Loss of young labor, high land-use costs in areas close to cities, and elevated land abandonment rates in areas distant from cities have profoundly impacted the agricultural economic efficiency of developed provinces and cities like Shanghai and Guangdong. This further substantiates the hypothesis of this paper: when undertaking regional economic research in specific areas, one should not solely rely on empirical judgments. Identifying regions that are advanced and those that are lagging, and distinguishing between them through judicious assessment methods, is a prudent approach.

5.2. China's Agricultural Economy Does Have a "Mezzogiorno Trap" for the Period 2015–2021

Based on the data displayed in Table 4, we take the mean TGR value of Group 1, representing the more advanced agricultural economy, and subtract the mean TGR value of Group 3, representing the less advanced agricultural economy. The resulting difference, which we term "Performance Gap Difference" (PGD), stands for the gap between the advanced and less advanced groups in China's agricultural economy. The larger the PGD value, the larger the gap. The change in the PGD of China's agricultural economic efficiency from 2015 to 2021 is depicted in Figure 3.

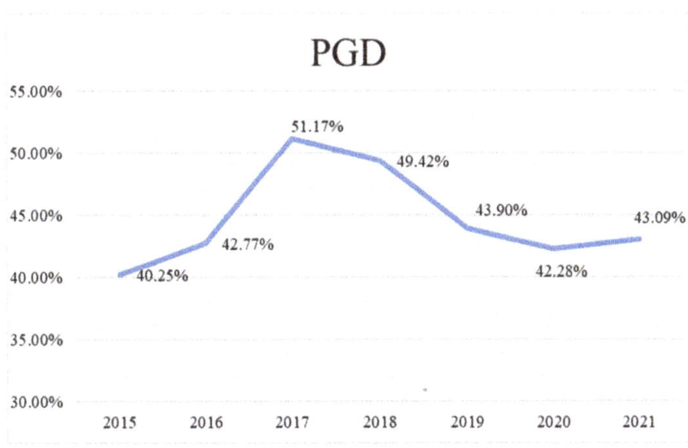

Figure 3. Trend Graph of Chinese Agricultural Economy PGD from 2015 to 2021.

As depicted in Figure 3, the PGD rapidly increased from 2015 to 2017, a period just prior to the introduction of the "Rural Revitalization Strategy". This widening gap was between provinces with high agricultural efficiency and those with lower agricultural efficiency. After the Chinese government proposed the comprehensive "Rural Revitalization Strategy" in 2018, the PGD markedly decreased by 2020. However, by 2021, the PGD started to increase again.

The changes in PGD alone are not sufficient to substantiate the presence of the "Mezzogiorno Trap" in China's agricultural economy. As we previously discussed, the changing circumstances of external aid or investments typically provide compelling evidence for the "Mezzogiorno Trap". We chose to compare the changes in support from the government with the trends in PGD.

For this purpose, we specifically analyzed the government's fiscal support for the agricultural economy from 2015 to 2021. We consider the proportion of expenditure on agriculture, forestry, and water by the governments of each province in their general public budget expenditure during the same period as the degree of agricultural financial support (DAFS). Its expression is:

$$\text{DAFS} = \frac{\text{Expenditure for Agriculture, Forestry and Water Conservancy}}{\text{General Public Budget Expenditure}}$$

The changes in DAFS from 2015 to 2021 are depicted in Figure 4.

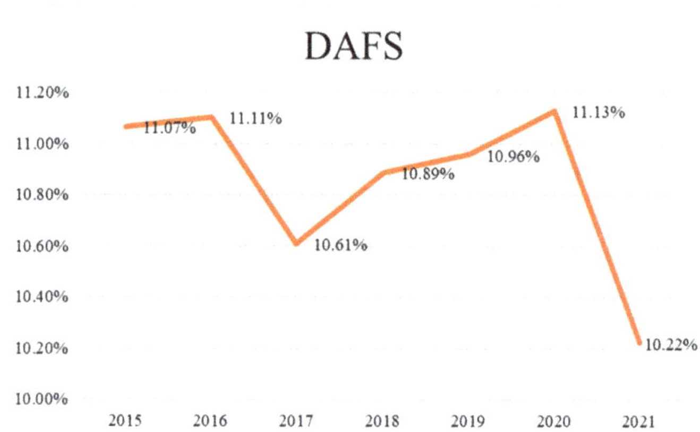

Figure 4. Trend Graph of Chinese Agricultural Economy DAFS from 2015 to 2021.

For easy comparison, we multiplied the DAFS for the years 2015–2021 by 5, and displayed it together with the PGD for the same period in Figure 5. It can be observed that whenever DAFS peaks, the PGD is at a lower level, as evidenced in 2015 and 2020. Conversely, when DAFS dips, PGD rises, as shown in 2017 and 2021. In particular in 2021, we speculate that due to the financial burden brought by COVID-19, DAFS experienced a downturn, leading to an immediate increase in PGD from its continual decline. Such fluctuations reveal a very typical "Mezzogiorno Trap".

In summary, based on our assessment, during the period 2015–2021, the agricultural economic efficiency in China exhibited a clear "Mezzogiorno Trap" for the following reasons:

1. Significant disparities exist between agriculturally lagging regions and agriculturally advanced regions.
2. This disparity remains relatively stable. While the magnitude of the gap may fluctuate, the composition of the lagging and advanced areas remains largely unchanged.
3. This gap correlates with variations in governmental support. When support intensifies, the disparity narrows. Conversely, as support diminishes, the gap widens.

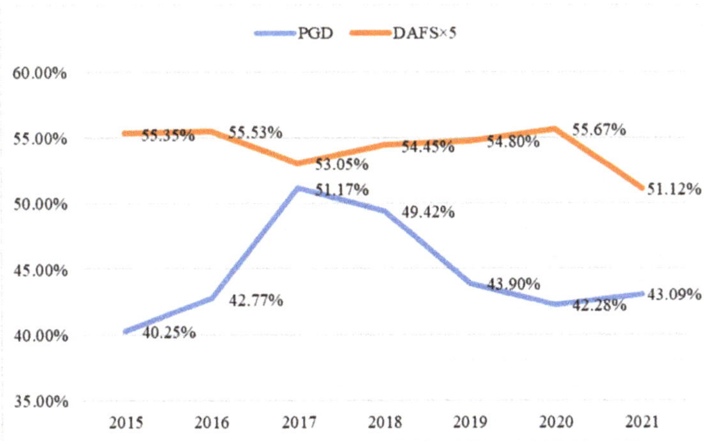

Figure 5. Combined trend graph of PGD and DAFS in Chinese agricultural economy from 2015 to 2021.

This dependence on fiscal support represents the most problematic aspect of the "Mezzogiorno Trap", largely due to the myopic perspective of certain policymakers. These administrators often formulate policies based on simplistic cause-and-effect relationships: subsidies are provided because of poverty, and once these subsidies are dispensed, various metrics immediately improve, leading them to believe that the issue has been resolved. In reality, structural disparities persist, and irrational or excessive subsidies can have severe repercussions, potentially exacerbating the "Mezzogiorno Trap".

McRae's research suggests that subsidies directed towards lagging areas often struggle to be effective due to poor infrastructure [65]. Dvouletý et al. discovered in their study of the Czech food industry that while public subsidies did indeed enhance firms' productivity in the short term, they had a negative impact on total factor productivity (TFP) [66]. Šipikal et al., in their examination of the European Union's regional policies, found that 35% of public subsidies constituted "deadweight" [67]. Research by Tsiouni et al. into Greece's livestock industry revealed that Greek goat farms have developed a significant dependency on government subsidies, with profitability becoming virtually non-existent in the absence of such aid [68].

5.3. Analysis of the Causes of "Mezzogiorno Trap"

As previously analyzed, there is a certain correlation between PGD and DAFS. However, DAFS is not the sole cause of PGD. We grouped the DAFS of the 31 provinces and compiled them into Table 7 and Figure 6. Figure 6 distinctly illustrates that the DAFS of provinces and cities in Group 3 is not the lowest. Throughout all the years, the gray bars representing the mean DAFS for Group 3 are consistently higher than the orange bars for Group 2.

Table 7. Average values table of DAFS groupings in China from 2015 to 2021.

	DAFS of Group 1	DAFS of Group 2	DAFS of Group 3
2015	0.126519343	0.110671251	0.116122018
2016	0.132630976	0.110073457	0.119074741
2017	0.122422637	0.110352681	0.114762524
2018	0.124583928	0.111795782	0.121278352
2019	0.131714679	0.107610036	0.12401321
2020	0.130910846	0.109419802	0.12523234
2021	0.116297515	0.104905896	0.116010143

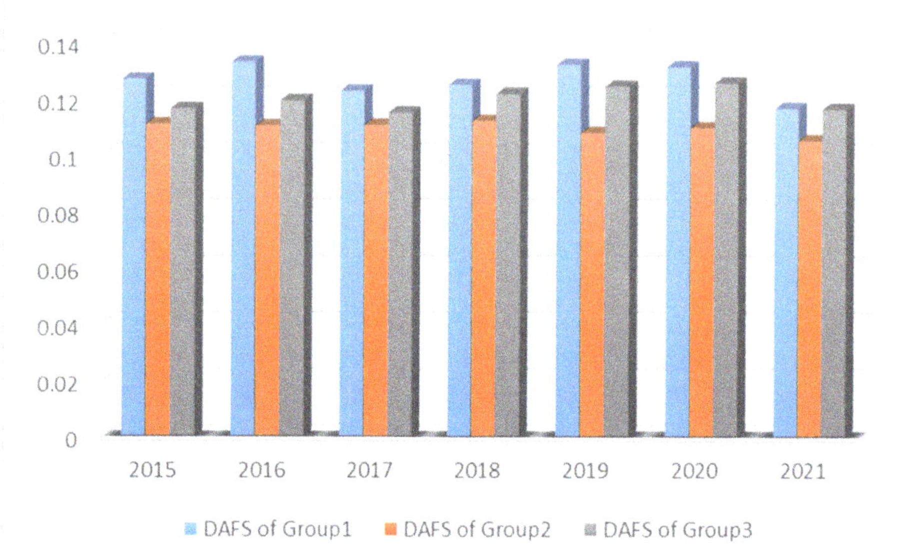

Figure 6. Mean DAFS values by group for the period 2015–2021.

We have reorganized the data from Table 4 into Figure 7. A comparison between Figures 6 and 7 provides a clearer visualization: throughout all years, the mean TGR of Group 3 is consistently lower than that of Group 2, while the DAFS values are consistently higher. This suggests that despite receiving more substantial fiscal support, the agricultural economic efficiency of the provinces and cities in the relatively lagging Group 3 remains inferior to that of Group 2. Fiscal support intensity is not the sole reason for the lower agricultural economic efficiency observed in Group 3.

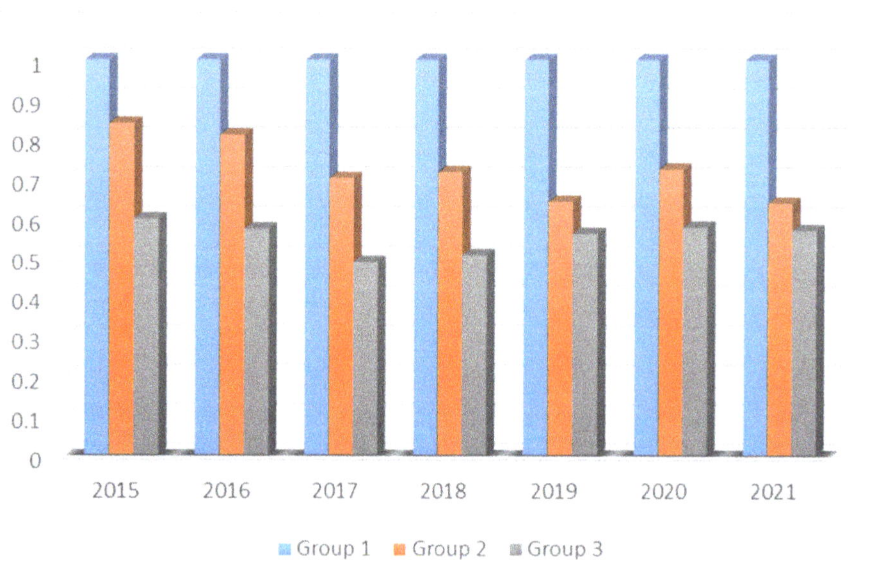

Figure 7. Mean PGD values by group for the period 2015–2021.

Therefore, to further probe into the causes, we continue employing the DEA method. Given that the meta-frontier SBM model cannot be directly decomposed into pure technical efficiency and scale efficiency, we additionally constructed an SBM-DEA model for the agricultural economic efficiency of the 31 provinces from 2015 to 2021 and organized the results as shown in Table 8.

Table 8. Summary table of SBM Model for Chinese agricultural economy from 2015 to 2021.

	Number of Provinces with Crste Values of 1	Number of Provinces with Vrste Values of 1	Number of Provinces with Scale Values of 1	Number of Provinces with DRS	Number of Provinces with IRS
2015	8	17	8	10	13
2016	10	17	10	11	10
2017	8	16	9	12	10
2018	8	16	8	12	10
2019	6	15	6	12	12
2020	8	15	7	14	9
2021	5	14	5	15	11

Table 8 reveals a noticeably larger number of provinces reaching DEA effectiveness in pure technical efficiency than in scale efficiency. This suggests that scale inefficiency is the primary reason behind insufficient agricultural economic efficiency. In light of our grouping of the 31 provinces, we organize the pure technical efficiency and scale efficiency of each group into Table 9. In Group 2, most provinces reach DEA effectiveness in pure technical efficiency. For the more lagging provinces in Group 3, nearly all fail to reach DEA effectiveness in both pure technical efficiency and scale efficiency.

Table 9. Summary table of pure technical efficiency and scale efficiency groupings of SBM Model for Chinese agricultural economy from 2015 to 2021.

	Number of Scale Efficiency Scores of 1 in Group 1	Number of Scale Efficiency Scores of 1 in Group 2	Number of Scale Efficiency Scores of 1 in Group 3	Number of Pure Technical Efficiency Scores of 1 in Group 1	Number of Pure Technical Efficiency Scores of 1 in Group 2	Number of Integrated Technical Efficiency Scores of 1 in Group 3
2015	6	2	0	9	7	1
2016	8	2	0	9	7	1
2017	8	0	0	9	6	1
2018	7	1	0	9	6	1
2019	6	0	0	9	5	1
2020	7	1	0	8	6	1
2021	5	0	0	8	8	1

From Table 9, it can be inferred that the primary reason for the "Mezzogiorno Trap" in China's agricultural economy is the lower scale efficiency, a common issue for provinces and cities in both Group 2 and Group 3. A defining characteristic of China's agricultural production is the fragmentation of farmlands and the dominance of small-scale subsistence farms. This emerged as a consequence of transitioning from the People's Communes to the Household Responsibility System, leading to significant structural issues in agricultural production, as substantiated by several studies [69,70]. Additionally, as previously analyzed, provinces and cities in Group 3, such as Shanghai and Guangdong, are burdened with challenges including labor shortages, high labor costs, elevated land use costs, serious land fallow issues, and high capital costs [71,72]. On the other hand, provinces with generally lagging economies like Tibet and Gansu are confronted with harsh natural environments, outdated infrastructure, and significant labor outflows [73,74]. Hence, for these provinces, enhancing efficiency through scaling proves to be a significant challenge.

Based on the SBM model from 2015 to 2021, provinces and cities in Group 3 with increasing returns to scale [20] and decreasing returns to scale (DRS) are compiled into Table 10. From Table 10, it is evident that the returns to scale status of most provinces and cities remain relatively stable. Shanxi, Shanghai, Tibet, and Gansu consistently exhibit decreasing returns to scale, whereas Hebei, Anhui, Henan, Guangdong, and Yunnan consistently demonstrate increasing returns to scale. Only Jilin and Jiangxi have shown some fluctuations over the period. This indicates that the deficiencies in scale efficiency for provinces and cities in Group 3 are persistent and relatively consistent.

Table 10. Summary table of scale returns groupings of SBM model for Chinese agricultural economy from 2015 to 2021.

	2015	2016	2017	2018	2019	2020	2021
Hebei	DRS	DRS	DRS	DRS	DRS	DRS	DRS
Shanxi	IRS	IRS	IRS	IRS	IRS	IRS	IRS
Jilin	IRS	IRS	IRS	IRS	IRS	DRS	DRS
Shanghai	IRS	IRS	IRS	IRS	IRS	IRS	IRS
Anhui	DRS	DRS	DRS	DRS	DRS	DRS	DRS
Jiangxi	DRS	DRS	CRS	CRS	IRS	IRS	IRS
Henan	DRS	DRS	DRS	DRS	DRS	DRS	DRS
Guangdong	DRS	DRS	DRS	DRS	DRS	DRS	DRS
Yunnan	DRS	DRS	DRS	DRS	DRS	DRS	DRS
Tibet	IRS	IRS	IRS	IRS	IRS	IRS	IRS
Gansu	IRS	IRS	IRS	IRS	IRS	IRS	IRS

Furthermore, Table 9 also indicates a significant deficiency in pure technical efficiency among the provinces and cities in Group 3, with only Tibet achieving DEA efficiency. The lack of pure technical efficiency is the primary reason for the gap between Group 3 and Group 2. Pure technical efficiency reflects factors in agricultural production beyond scale, including management expertise, agricultural science and technology, capital efficiency, sales, deep processing of agricultural products, and so on. Only by achieving a high level in these areas can inputs be efficiently transformed into outputs. The discrepancy in pure technical efficiency also elucidates why Group 2 provinces and cities have lower DAFS than Group 3; however, their overall agricultural economic efficiency is higher than Group 3.

In conclusion, the "Mezzogiorno Trap" in China's agricultural economy has multi-faceted causes. In the short term, improvements in areas such as agricultural science and technology and management levels might yield noticeable results, narrowing the gap with Group 2. However, a fundamental resolution to the issue will likely necessitate challenging adjustments in scale.

5.4. Trend of Overall Agricultural Economic Efficiency Development in China from 2015 to 2021

In our prior analysis on the existence of the "Mezzogiorno Trap" in Chinese agriculture, we constructed SBM models for 31 provinces over multiple years and performed a thorough evaluation. From Table 7, it appears that the overall technical efficiency of Chinese agricultural economics is declining, as the number of provinces achieving DEA efficiency decreases each year. However, SBM models provide a relative description of the agricultural economic efficiency of 31 provinces at a certain time, and models from different periods cannot be directly compared. Utilizing the same data and input–output variables, we constructed an SBM Malmquist index model for 31 provinces from 2015 to 2021, adopting an adjacent benchmarking pattern. We obtained results for six periods, which are consolidated and presented in Table 11.

Table 11. Summary table of SBM-Malmquist index model for China's agricultural economy from 2015 to 2021.

	Number of Provinces with Effch > 1	Number of Provinces with Techch > 1	Number of Provinces with Tfpch > 1
2015–2016	7	27	28
2016–2017	3	29	19
2017–2018	7	31	31
2018–2019	14	31	31
2019–2020	13	31	30
2020–2021	6	31	31

A Tfpch (total factor productivity change index) greater than 1 indicates an enhancement in economic efficiency for the given period. Based on Table 11, when Tfpch values are summarized in Figure 8, it becomes evident that, during various time intervals, the number of provinces and cities with a Tfpch greater than 1 only reached its lowest in 2016–2017 with a figure of 19. This suggests that during that year, only 19 provinces and cities achieved progress in total factor productivity. In other periods, the number of provinces and cities with a Tfpch greater than 1 ranged from 28 to 31. Notably, in 2019 and 2021, the agricultural economy of China might have been severely impacted by the COVID-19 pandemic. The SBM static model, as depicted in Table 7, shows that only six and five provinces and cities, respectively, achieved DEA efficiency during these years. However, Figure 8 demonstrates that 30–31 provinces and cities had a Tfpch exceeding 1 during these intervals.

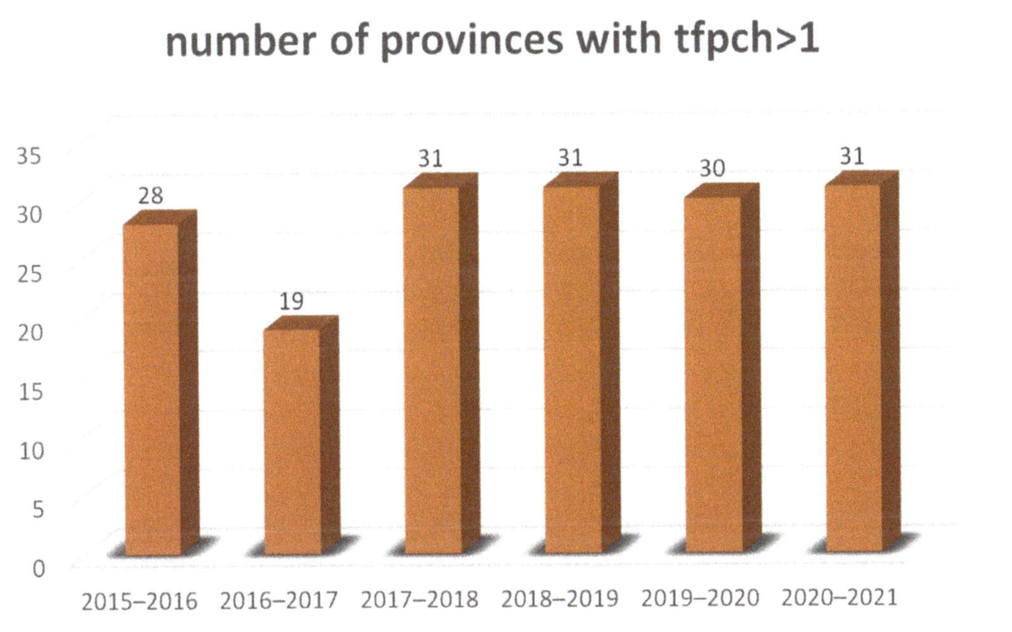

Figure 8. Number of provinces in China with an agricultural Tfpch > 1 from 2015 to 2021.

In fact, the SBM model and the SBM-Malmquist model are not contradictory; they have different reference sets. The SBM model calculates the relative efficiency for a specific year using cross-sectional data from that year as its reference set. In contrast, the Malmquist index model computes the productivity index based on panel data from various periods.

When assessing the trend of production efficiency, the SBM-Malmquist index model should be used as the benchmark.

From the perspective of the SBM-Malmquist index model, the overall development trend of China's agricultural economic efficiency is consistently improving. This trend does not contradict the existence of the "Mezzogiorno Trap": even if the agricultural economic efficiency of less developed provinces is improving, if the rate of improvement lags behind that of more advanced provinces, the gap will continue to widen. This scenario further underscores the subtle nature of the "Mezzogiorno Trap". Even if the overall level of the agricultural economy is on the rise, the trap can still persist or even expand in a concealed manner. Managers and policymakers might overlook the existence of the "Mezzogiorno Trap" due to the overall improvement in agricultural economic efficiency.

Tfpch is the product of Effch and Techch. Summarizing the Effch and Techch from Table 11 into Figure 9, it is evident that most provinces, including those from Group 2 and Group 3, exhibit a significant shortfall in Effch (represented by the green bars). This is a primary factor contributing to the lower values of Tfpch. This aligns with the findings from the SBM model decomposition: the less developed provinces in China's agricultural economy generally suffer from structural issues and require profound industrial structural reforms to enhance scale efficiency.

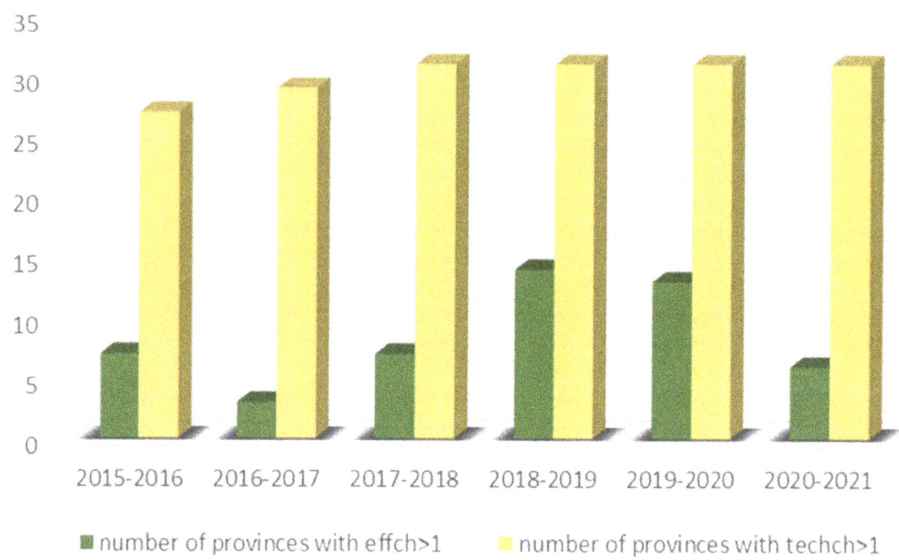

Figure 9. Number of Chinese provinces with agricultural Effch > 1 and Techch > 1 from 2015 to 2021.

The continuous improvement in China's agricultural economic efficiency may render the "Mezzogiorno Trap" more concealed. Some policymakers might be contented with the present achievements, overlooking the structural adjustments with less evident short-term results, thereby failing to address the core of the "Mezzogiorno Trap". However, from an optimistic perspective, the ever-increasing agricultural economic efficiency will eventually aid in ameliorating the "Mezzogiorno Trap": if challenges are identified in a timely manner, the upward development trend provides a broader array of solutions to address the issue.

6. Conclusions, Recommendations, and Shortcomings

6.1. Conclusions

The primary objective of this study is to investigate the existence of the "Mezzogiorno Trap" in China, focusing specifically on the agricultural sector. Based on the results and discussions presented in the preceding sections, we have formulated several conclusions:

1. This paper introduces a methodology tailored for investigating the "Mezzogiorno Trap", particularly within specific industries. Initially, a quantitative analysis is employed to identify underperforming regions. As an example, the super-efficiency SBM model is adopted in this paper to rank and categorize the subjects under study. Subsequently, the disparity between lagging and advanced regions is examined, exemplified in this research by the deployment of a meta-frontier SBM model to compute PGD values. Factors influencing these disparities are then scrutinized to ascertain the presence of the "Mezzogiorno Trap". In this context, a comparative analysis between PGD and DAFS values is utilized to discern additional characteristics of the trap.

Compared to other methodologies, this approach offers an enhanced lens to study the "Mezzogiorno Trap" within specific sectors. Such sectors' regional variations can often be overshadowed by overarching economic differences, thereby inducing biases for researchers. Take, for instance, the agricultural economy focused on in this paper. There exists a pronounced discrepancy between agricultural efficiency and the overall economic standing across Chinese provinces and cities. Some of the economically flourishing provinces paradoxically rank low in terms of agricultural efficiency. Sole reliance on regional economic performance or geographical location for classifying regions as advanced or lagging can lead to substantial inaccuracies, which in turn can significantly impact policy formulation.

Common pitfalls may include excessive enhancement of support levels, engendering a dependency in regions ensnared by the "Mezzogiorno Trap", which is manifestly counterproductive for fully addressing the underlying issues of the trap. Research by Tsiouni et al. [68] illustrates this phenomenon, demonstrating that goat farms in Greece become unsustainable in the absence of governmental subsidies. Furthermore, this dependency is not limited to a single sector but is pervasive across various fields. For instance, Wang et al. investigated China's new energy vehicle industry [75] and discovered a significant negative impact of subsidies on firms' financial performance during the period 2009–2018, requiring structural adjustments for mitigation.

2. Utilizing this approach, we identified the presence of the "Mezzogiorno Trap" in China's agricultural economy from 2015 to 2021. Even during periods characterized by consistently rising overall economic levels, this methodology effectively detected the existence of the "Mezzogiorno Trap". Integrating findings from the SBM-Malmquist index model for the same years, we further corroborated the covert nature of the "Mezzogiorno Trap": even when the overall economic efficiency is on an uptrend, the trap persists and is easily overlooked.
3. Through the decomposable DEA-SBM model, integrated with the unique characteristics of agricultural production, we posit that the primary reason for the "Mezzogiorno Trap" in China's agricultural economy is the insufficiency in scale efficiency. This lack of scale efficiency is not only evident in provinces and cities with relatively lagging agricultural economic efficiency but is also prevalent among those with a moderate performance. Additionally, provinces and cities with a lagging agricultural economic efficiency exhibit deficiencies in pure technical efficiency, marking a significant difference from other regions. The issue of low scale efficiency is a common challenge faced by developing countries, often attributable to an imbalanced industrial structure, as demonstrated by the research conducted by Karimov et al. [76].
4. Fundamentally, regions mired in the "Mezzogiorno Trap" suffer from outdated industrial structures, inferior infrastructure, subpar technological standards, inefficiencies

in capital utilization, and talent deficiencies, among others. The gaps present in these areas cannot be fully bridged solely through basic support policies such as financial subsidies. When such support wanes, the disparities swiftly widen once more.

6.2. Recommendations

1. Maintaining support strength, including fiscal support, is crucial for resolving the "Mezzogiorno Trap". Data analysis reveals a certain correlation between strong support and the "Mezzogiorno Trap", and prematurely weakening support could widen the gap. Several studies have likewise highlighted the significance of government support, including Mutlu's study on Japan's regional economic differentiation [77], Das et al.'s research on the Indian regional economy [78], and Chen et al.'s studies on regional differences in China and Henan Province's agricultural economy [22,79]. These studies show that government support is a material basis and necessary condition for resolving regional differences.
2. However, the essence of addressing the "Mezzogiorno Trap" hinges on structural adjustments tailored to the realities of underdeveloped regions. Fiscal subsidies from the government must be dispensed judiciously; sheer direct capital allocations may inadvertently yield adverse consequences. Integrating the findings from this paper, the primary strategy for China to rectify its agricultural economic "Mezzogiorno Trap" centers on enhancing scale efficiency. Predicated on the characteristics of agricultural production, the emphasis on boosting scale efficiency necessitates a prudent reshaping of the industrial structure, specifically in Group 3 and Group 2 provinces and cities, representing immediate challenges to address.

For provinces evidently mired in the "Mezzogiorno Trap", a rational approach would involve conducting research and analyses tailored to the specific circumstances of these underdeveloped regions, thereby formulating long-term, detailed, and targeted policies.

For Group 3 provinces in the western region where educational levels are low, it is advisable to allocate a portion of financial subsidies towards the attraction and cultivation of agricultural talent, as demonstrated in studies by Démurger et al. [80] and Hitka and Ližbetinová [81]. For provinces with insufficient infrastructure, a portion of the funding could be allocated towards less immediately impactful infrastructure projects, as illustrated by Bhatia [82].

For rural areas surrounding Shanghai, enhancing the role of value-added agricultural processing industries through clustering could be more effective. In Guangdong, the initial focus might be on how to efficiently utilize the currently fallow land.

In the long run, these targeted interventions are likely to yield better outcomes than simply disbursing subsidies, serving to fundamentally address the "Mezzogiorno Trap".

In summation, while sustaining supportive measures, a gradual transition of some direct monetary subsidies to funds dedicated to industrial structural adjustments can act as a catalyst for the industrial evolution in lagging regions. Policymakers need not be overly apprehensive about diminishing subsidies' impact on economic efficiency. As found in the study by Yang et al. examining Jilin province's corn procurement policy shift [83], post-2016, the government annulled the protective purchasing subsidies for corn, ushering in a marked escalation in its marketization level. The agricultural economy exhibited robust performance, manifesting commendable adaptive resilience within the new policy milieu.

3. Our preceding analysis indicates that enhancing pure technical efficiency is crucial for provinces entrapped in the "Mezzogiorno Trap". Pure technical efficiency can be understood as the exclusion of scale-related factors, capturing elements such as technological advancement, managerial improvement, and increased capital efficiency. These distinctions underscore the fundamental differences between modern and traditional agriculture, further highlighting the urgency for industrialization within China's agricultural sector.

It is noteworthy that, compared to Group 2 provinces which have higher TGR averages, the financial support towards agriculture is more pronounced in Group 3 provinces. This suggests that elevating the pure technical efficiency of agriculture in Group 3 provinces to a DEA-efficient level would significantly alleviate the fiscal pressure engendered by agricultural support measures. Such an approach would be highly beneficial in addressing the "Mezzogiorno Trap" faced by these provinces.

4. Advanced provinces should take measures to assist less-developed provinces. Provinces in Group 3 should look to their counterparts in Group 1 for the adoption of advanced agricultural technologies and more efficient policies for agricultural industrialization. This would substantially contribute to the improvement of both pure technical efficiency and scale efficiency. In fact, enhancements in pure technical efficiency and scale efficiency are not mutually exclusive. Due to constraints such as land, climate, and water resources, agricultural production cannot simply optimize through arbitrary expansion or contraction of its scale. Advanced provinces often operate within more efficient production cycles, characterized by robust technological innovation, comprehensive policy formulation, and timely evaluation systems. These best practices offer valuable lessons for provinces that are lagging behind.
5. Provinces in Group 3 should pay particular attention to the heterogeneity of agricultural economies across different regions when formulating localized policies. This notion of diversification has been emphasized in earlier sections. For economically advanced regions like Shanghai and Guangdong, the focus should be on precision agriculture and the corporatization of agriculture. In contrast, less developed regions such as Gansu and Tibet should explore additional revenue streams. For instance, Gansu, which is predominantly characterized by desert and barren landscapes, has made notable strides in the development of solar and wind energy as well as agro-photovoltaic complementation. Recent studies indicate that these initiatives offer opportunities for ameliorating regional disparities [84,85].

6.3. Shortcomings and Improvements

The "Mezzogiorno Trap" reflects an unbridgeable disparity between advanced and less developed regions. However, the data envelopment analysis (DEA) approach assesses relative efficiency, implying perpetual disparities in the DEA model, which can potentially hinder the judgement of the "Mezzogiorno Trap". When determining the "Mezzogiorno Trap" based on the DEA method, we introduced additional supporting evidence, making the assessment process somewhat complex.

There are certainly simpler alternatives. For instance, ranking the overall efficiency of the agricultural economy across different years and comparing the ranking results. If the rankings vary greatly across different years, it suggests that the relative efficiency of each region is constantly changing, with no clear distinction between advanced and backward regions. This could suggest an apparent absence of the "Mezzogiorno Trap".

However, this method is only suitable for eliminating the possibility of the "Mezzogiorno Trap", but insufficient to affirm its existence. Additionally, it is not conducive to subsequent quantitative analyses and examinations of influencing factors. Nevertheless, as part of other agricultural economic research, it appears more succinct.

In terms of studying the influencing factors of the "Mezzogiorno Trap", this paper primarily conducted a comparative analysis with the disparities in fiscal support. We are seeking better ways to establish a more comprehensive "Mezzogiorno Trap" model and introducing more extensive correlation analyses, thereby proposing a broader range of feasible recommendations.

Supplementary Materials: The following supporting information can be downloaded at: https://www.mdpi.com/article/10.3390/agriculture13091806/s1, Table S1: Data Source: Panel Data on Agricultural Economy across 31 Provinces in China, 2015–2021.

Author Contributions: Conceptualization, X.L. and Y.Z.; methodology, P.Y. and Y.Z.; software, P.Y.; validation, X.L. and Y.Z.; formal analysis, X.L.; investigation, P.Y.; resources, P.Y.; data curation, P.Y.; writing—original draft preparation, X.L.; writing—review and editing, Y.Z. All authors have read and agreed to the published version of the manuscript.

Funding: This work was supported by the major project under the National Social Science Fund of China (No. 21AFX001).

Institutional Review Board Statement: Not applicable.

Informed Consent Statement: Not applicable.

Data Availability Statement: The data used in this paper are from the China National Bureau of Statistics, the China Statistical Yearbook and the China Ministry of Agriculture and Rural Affairs, the China Rural Statistical Yearbook.

Conflicts of Interest: The authors declare no conflict of interest.

References

1. Watson, M.M. *Regional Development Policy and Administration in Italy*; Longman Group Ltd.: London, UK, 1970.
2. Daniele, V. Territorial disparities in labour productivity, wages and prices in Italy: What does the data show? *Eur. Urban Reg. Stud.* **2021**, *28*, 431–449. [CrossRef]
3. Salvati, L.; Zitti, M.; Carlucci, M. In-between regional disparities and spatial heterogeneity: A multivariate analysis of territorial divides in Italy. *J. Environ. Plan. Manag.* **2017**, *60*, 997–1015. [CrossRef]
4. Giarda, E.; Moroni, G. The Degree of Poverty Persistence and the Role of Regional Disparities in Italy in Comparison with France, Spain and the UK. *Soc. Indic. Res.* **2018**, *136*, 163–202. [CrossRef]
5. Castelnovo, P.; Morretta, V.; Vecchi, M. Regional disparities and industrial structure: Territorial capital and productivity in Italian firms. *Reg. Stud.* **2020**, *54*, 1709–1723. [CrossRef]
6. Iuzzolino, G.; Pellegrini, G.; Viesti, G. Convergence among Italian regions, 1861–2011. In *Bank of Italy Economic History Working Paper*; Bank of Italy Economic Research Department: Rome, Italy, 2011.
7. Terrasi, M. Convergence and divergence across Italian regions. *Ann. Reg. Sci.* **1999**, *33*, 491–510. [CrossRef]
8. Torrisi, G.; Pike, A.; Tomaney, J.; Tselios, V. (Re-)exploring the link between decentralization and regional disparities in Italy. *Reg. Stud. Reg. Sci.* **2015**, *2*, 123–140. [CrossRef]
9. Bruzzi, S.; Ivaldi, E.; Santagata, M. Measuring Regional Performance in the Italian NHS: Are Disparities Decreasing? *Soc. Indic. Res.* **2022**, *159*, 1057–1084. [CrossRef]
10. Fazio, G.; Piacentino, D.; Vassallo, E. Regional Disparities and Public Policies in Italy: Some considerations in light of a performance analysis. In Proceedings of the ERSA 2006—46th Congress of the European Regional Science Association, Volos, Greece, 30 August–3 September 2006.
11. Geppert, K.; Stephan, A. Regional disparities in the European Union: Convergence and agglomeration. *Pap. Reg. Sci.* **2008**, *87*, 193–217. [CrossRef]
12. Ruan, M.; Zhao, X. Fiscal pressure, policy choices and regional economic disparity: Evidence from a natural experiment in China. *Sustainability* **2022**, *14*, 12402. [CrossRef]
13. López-Villuendas, A.M.; del Campo, C. Regional Economic Disparities in Europe: Time-Series Clustering of NUTS 3 Regions. *Int. Reg. Sci. Rev.* **2023**, *46*, 265–298. [CrossRef]
14. Harrison, A.; Immergluck, D. *The Battle of the Belts: Comparing Housing Vacancy in Larger Metros in the Sun Belt and the Rust Belt since the Mortgage Crisis, 2012 to 2019*; Andrew Young School of Policy Studies, Georgia State University: Atlanta, GA, USA, 2020.
15. Hegerty, S.W. The rust belt, the sunbelt, and the concentration of poverty within large US cities. *Rev. Reg. Stud.* **2019**, *49*, 474–494.
16. Herrschel, T. Economic transformation, locality and policy in eastern Germany: A comparison of two Länder. *Appl. Geogr.* **1997**, *17*, 267–281. [CrossRef]
17. Berentsen, W.H. Changing Regional Inequalities in United Germany. *Eurasian Geogr. Econ.* **2006**, *47*, 462–477. [CrossRef]
18. Dörr, M.; Riemer, U.; Christ, M.; Bauersachs, J.; Bosch, R.; Laufs, U.; Neumann, A.; Scherer, M.; Störk, S.; Wachter, R. Hospitalizations for heart failure: Still major differences between East and West Germany 30 years after reunification. *Wiley Online Libr.* **2021**, *8*, 2546–2555. [CrossRef]
19. Birthal, P.S.; Singh, H.; Kumar, S. Agriculture, economic growth and regional disparities in India. *J. Int. Dev.* **2011**, *23*, 119–131. [CrossRef]
20. Jose, A. India's regional disparity and its policy responses. *J. Public Aff.* **2019**, *19*, e1933. [CrossRef]
21. State Council of China. The Eleventh Five-Year Plan for National Economic and Social Development of the People's Republic of China. Available online: https://www.gov.cn/gongbao/content/2006/content_268766.htm (accessed on 15 July 2023).
22. Chen, Z.; Song, S. Efficiency and technology gap in China's agriculture: A regional meta-frontier analysis. *China Econ. Rev.* **2008**, *19*, 287–296. [CrossRef]

23. Fan, S.; Kanbur, R.; Zhang, X. China's regional disparities: Experience and policy. *Rev. Dev. Financ.* **2011**, *1*, 47–56. [CrossRef]
24. Zhang, Y. The regional disparity of influencing factors of technological innovation in China: Evidence from high-tech industry. *Technol. Econ. Dev. Econ.* **2021**, *27*, 811–832. [CrossRef]
25. State Council of China. Circular of the State Council on Deepening the Pilot Reform of Rural Taxes and Fees in 2005. Available online: https://www.gov.cn/gongbao/content/2005/content_64331.htm (accessed on 1 August 2023).
26. State Council of China. Approval Reply of the State Council on Agreeing to Establish the New Rural Cooperative Medical Treatment Approval of the Inter-Ministerial Joint Conference System. Available online: https://www.gov.cn/gongbao/content/2003/content_62401.htm (accessed on 1 August 2023).
27. The Central Committee of the Communist Party of China and the State Council. The Central Committee of the Communist Party of China and the State Council Opinions on Implementing the Rural Revitalization Strategy. Available online: https://www.gov.cn/gongbao/content/2018/content_5266232.htm (accessed on 1 September 2023).
28. Chan, H.L.; Chan, K.T. The analysis of rural regional disparity in China. *Asian Econ. J.* **2000**, *14*, 23–38. [CrossRef]
29. Li, H.; Haynes, K.E. Economic structure and regional disparity in China: Beyond the Kuznets transition. *Int. Reg. Sci. Rev.* **2011**, *34*, 157–190. [CrossRef]
30. Chen, Y.; Li, X.; Tian, Y.; Tan, M. Structural change of agricultural land use intensity and its regional disparity in China. *J. Geogr. Sci.* **2009**, *19*, 545–556. [CrossRef]
31. Liu, X.; Wang, X.; Xin, X. Did agricultural technological changes affect China's regional disparity? *China Agric. Econ. Rev.* **2012**, *4*, 440–449. [CrossRef]
32. National Bureau of Statistics of China. China Statistical Yearbook. Available online: http://www.stats.gov.cn/english/Statisticaldata/yearbook/ (accessed on 1 August 2023).
33. Ministry of Agriculture and Rural Development of China. China Rural Statistical Yearbook. Available online: https://navi.cnki.net/knavi/yearbooks/YMCTJ/detail?uniplatform=NZKPT&language=chs (accessed on 1 August 2023).
34. Chen, F.; Yang, K.; Wang, L.; Zhang, Y.; Liu, H. Comprehensive evaluation of power transmission and transformation project based on electric power big data. *Energy Rep.* **2022**, *8*, 733–741. [CrossRef]
35. Chen, M.; Zheng, Y. China's regional disparity and its policy responses. *China World Econ.* **2008**, *16*, 16–32. [CrossRef]
36. Schmookler, J. The changing efficiency of the American economy, 1869–1938. *Rev. Econ. Stat.* **1952**, *34*, 214–231. [CrossRef]
37. Bukarica, V.; Tomšić, Ž. Energy efficiency policy evaluation by moving from techno-economic towards whole society perspective on energy efficiency market. *Renew. Sustain. Energy Rev.* **2017**, *70*, 968–975. [CrossRef]
38. Bravo-Ureta, B.E.; Pinheiro, A.E. Efficiency analysis of developing country agriculture: A review of the frontier function literature. *Agric. Resour. Econ. Rev.* **1993**, *22*, 88–101. [CrossRef]
39. Paul, C.; Nehring, R.; Banker, D.; Somwaru, A. Scale economies and efficiency in US agriculture: Are traditional farms history? *J. Product. Anal.* **2004**, *22*, 185–205. [CrossRef]
40. Charnes, A.; Cooper, W.W.; Rhodes, E. Measuring the efficiency of decision making units. *Eur. J. Oper. Res.* **1978**, *2*, 429–444. [CrossRef]
41. Banker, R.D.; Charnes, A.; Cooper, W.W. Some models for estimating technical and scale inefficiencies in data envelopment analysis. *Manag. Sci.* **1984**, *30*, 1078–1092. [CrossRef]
42. Tone, K. A slacks-based measure of efficiency in data envelopment analysis. *Eur. J. Oper. Res.* **2001**, *130*, 498–509. [CrossRef]
43. Chang, Y.-T.; Park, H.-s.; Jeong, J.-b.; Lee, J.-w. Evaluating economic and environmental efficiency of global airlines: A SBM-DEA approach. *Transp. Res. Part D Transp. Environ.* **2014**, *27*, 46–50. [CrossRef]
44. Lin, X.; Zhu, X.; Han, Y.; Geng, Z.; Liu, L. Economy and carbon dioxide emissions effects of energy structures in the world: Evidence based on SBM-DEA model. *Sci. Total Environ.* **2020**, *729*, 138947. [CrossRef] [PubMed]
45. Tone, K. Slacks-based measure of efficiency. In *Handbook on Data Envelopment Analysis*; Springer: Boston, MA, USA, 2011; pp. 195–209.
46. Andersen, P.; Petersen, N.C. A procedure for ranking efficient units in data envelopment analysis. *Manag. Sci.* **1993**, *39*, 1261–1264. [CrossRef]
47. O'Donnell, C.J.; Rao, D.P.; Battese, G.E.J.E.E. Metafrontier frameworks for the study of firm-level efficiencies and technology ratios. *Empir. Econ.* **2008**, *34*, 231–255. [CrossRef]
48. Li, L.-B.; Liu, B.-L.; Liu, W.-L.; Chiu, Y.-H.J.S.-E.P.S. Efficiency evaluation of the regional high-tech industry in China: A new framework based on meta-frontier dynamic DEA analysis. *Econ. Plan. Sci.* **2017**, *60*, 24–33. [CrossRef]
49. Yu, Y.; Huang, J.; Zhang, N. Industrial eco-efficiency, regional disparity, and spatial convergence of China's regions. *J. Clean. Prod.* **2018**, *204*, 872–887. [CrossRef]
50. Malmquist, S. Index numbers and indifference surfaces. *Trab. Estadística* **1953**, *4*, 209–242. [CrossRef]
51. Färe, R.; Grosskopf, S.; Lindgren, B.; Roos, P.J.J.o.p.A. Productivity changes in Swedish pharamacies 1980–1989: A non-parametric Malmquist approach. *J. Product. Anal.* **1992**, *3*, 85–101. [CrossRef]
52. National Bureau of Statistics of China. Statistical Bulletin of the People's Republic of China on National Economic and Social Development. 2022. Available online: https://www.gov.cn/xinwen/2023-02/28/content_5743623.htm (accessed on 1 August 2023).
53. Tang, C.; Zhao, L.; Zhao, Z. Child labor in China. *China Econ. Rev.* **2018**, *51*, 149–166. [CrossRef]

54. Gong, P.; Liang, S.; Carlton, E.J.; Jiang, Q.; Wu, J.; Wang, L.; Remais, J.V. Urbanisation and health in China. *Lancet* **2012**, *379*, 843–852. [CrossRef] [PubMed]
55. Zhao, L. Massive return migration emerging in China. *East Asian Policy* **2018**, *10*, 75–86. [CrossRef]
56. Fan, C.C. Economic opportunities and internal migration: A case study of Guangdong Province, China. *Prof. Geogr.* **1996**, *48*, 28–45. [CrossRef]
57. Zhang, L.-Y. Economic development in Shanghai and the role of the state. *Urban Stud.* **2003**, *40*, 1549–1572. [CrossRef]
58. Lo, C.W.-H.; Tang, S.-Y. Institutional reform, economic changes, and local environmental management in China: The case of Guangdong province. In *Environmental Governance in China*; Routledge: Abingdon, UK, 2013; pp. 42–62.
59. Gao, Y.; Ma, Y. What is absent from the current monitoring: Idleness of rural industrial land in suburban Shanghai. *Habitat Int.* **2015**, *49*, 138–147. [CrossRef]
60. Liu, Z.; Liu, L. Characteristics and driving factors of rural livelihood transition in the east coastal region of China: A case study of suburban Shanghai. *J. Rural. Stud.* **2016**, *43*, 145–158. [CrossRef]
61. Gu, X.; Xie, B.; Zhang, Z.; Guo, H. Rural multifunction in Shanghai suburbs: Evaluation and spatial characteristics based on villages. *Habitat Int.* **2019**, *92*, 102041. [CrossRef]
62. Choy, L.H.T.; Lai, Y.; Lok, W. Economic performance of industrial development on collective land in the urbanization process in China: Empirical evidence from Shenzhen. *Habitat Int.* **2013**, *40*, 184–193. [CrossRef]
63. Su, Y.; Wu, S.; Kang, S.; Xu, H.; Liu, G.; Qiao, Z.; Liu, L. Monitoring Cropland Abandonment in Southern China from 1992 to 2020 Based on the Combination of Phenological and Time-Series Algorithm Using Landsat Imagery and Google Earth Engine. *Remote Sens.* **2023**, *15*, 669. [CrossRef]
64. Hou, D.; Meng, F.; Prishchepov, A.V. How is urbanization shaping agricultural land-use? Unraveling the nexus between farmland abandonment and urbanization in China. *Landsc. Urban Plan.* **2021**, *214*, 104170. [CrossRef]
65. McRae, S. Infrastructure quality and the subsidy trap. *Am. Econ. Rev.* **2015**, *105*, 35–66. [CrossRef]
66. Dvouletý, O.; Blažková, I. The impact of public grants on firm-level productivity: Findings from the Czech food industry. *Sustainability* **2019**, *11*, 552. [CrossRef]
67. Šipikal, M.; Pisár, P.; Labudová, V. Are subsidies reall y needed? The case of eu regional policy in the czech and slovak republics. In *E+ M Ekonomie a Management/E+ M Economics Management Science*; EBSCO Industries, Inc.: Alpharetta, GA, USA, 2013; ,
68. Tsiouni, M.; Aggelopoulos, S.; Pavloudi, A.; Siggia, D. Economic and financial sustainability dependency on subsidies: The case of goat farms in greece. *Sustainability* **2021**, *13*, 7441. [CrossRef]
69. Wang, H.; Tong, J.; Su, F.; Wei, G.; Tao, R. To reallocate or not: Reconsidering the dilemma in China's agricultural land tenure policy. *Land Use Policy* **2011**, *28*, 805–814. [CrossRef]
70. Kendall, H.; Clark, B.; Li, W.; Jin, S.; Jones, G.D.; Chen, J.; Taylor, J.; Li, Z.; Frewer, L.J. Precision agriculture technology adoption: A qualitative study of small-scale commercial "family farms" located in the North China Plain. *Precis. Agric.* **2022**, *23*, 319–351. [CrossRef]
71. Liu, H.; Li, G.; Wang, K. Homestead reduction, economic agglomeration and rural economic development: Evidence from Shanghai, China. *China Agric. Econ. Rev.* **2022**, *14*, 274–293. [CrossRef]
72. Shi, Y.; Yang, Q.; Zhou, L.; Shi, S. Can Moderate Agricultural Scale Operations Be Developed against the Background of Plot Fragmentation and Land Dispersion? Evidence from the Suburbs of Shanghai. *Sustainability* **2022**, *14*, 8697.
73. Li, S.; Yao, Z.; Liu, Z.; Wang, R.; Liu, M.; Adam, J.C. The spatio-temporal characteristics of drought across Tibet, China: Derived from meteorological and agricultural drought indexes. *Theor. Appl. Climatol.* **2019**, *137*, 2409–2424. [CrossRef]
74. Ma, L.; Chen, M.; Fang, F.; Che, X. Research on the spatiotemporal variation of rural-urban transformation and its driving mechanisms in underdeveloped regions: Gansu Province in western China as an example. *Sustain. Cities Soc.* **2019**, *50*, 101675. [CrossRef]
75. Wang, X.; Li, Z.; Shaikh, R.; Ranjha, A.R.; Batala, L.K. Do government subsidies promote financial performance? Fresh evidence from China's new energy vehicle industry. *Sustain. Prod. Consum.* **2021**, *28*, 142–153. [CrossRef]
76. Karimov, A.; Amoke Awotide, B.; Timothy Amos, T. Production and scale efficiency of maize farming households in South-Western Nigeria. *Int. J. Soc. Econ.* **2014**, *41*, 1087–1100. [CrossRef]
77. Mutlu, S. Regional disparities, industry and government policy in Japan. *Dev. Change* **1991**, *22*, 547–586. [CrossRef]
78. Das, K. Regional disparity in India. *ACADEMICIA Int. Multidiscip. Res. J.* **2017**, *7*, 53–69. [CrossRef]
79. Chen, S.; Yang, J.; Kang, X. Effect of Fiscal Expenditure for Supporting Agriculture on Agricultural Economic Efficiency in Central China—A Case Study of Henan Province. *Agriculture* **2023**, *13*, 822. [CrossRef]
80. Démurger, S. Infrastructure Development and Economic Growth: An Explanation for Regional Disparities in China? *J. Comp. Econ.* **2001**, *29*, 95–117. [CrossRef]
81. Hitka, M.; Ližbetinová, L. Human potential of talents in agriculture and forestry in context of Agriculture 4. *0. Agric. Econ.* **2023**, *69*, 223–233. [CrossRef]
82. Bhatia, M. Rural infrastructure and growth in agriculture. *Econ. Political Wkly.* **1999**, *34*, A43–A48.
83. Yang, Q.; Zhang, P.; Li, Y.; Ning, J.; Chu, N. Does the Policy of Decoupled Subsidies Improve the Agricultural Economic Resilience?—Evidence from China's Main Corn Producing Regions. *Sustainability* **2023**, *15*, 10164.

84. Zhang, Y.; Ren, J.; Pu, Y.; Wang, P. Solar energy potential assessment: A framework to integrate geographic, technological, and economic indices for a potential analysis. *Renew. Energy* **2020**, *149*, 577–586. [CrossRef]
85. Sun, Y.; Li, Y.; Wang, R.; Ma, R. Measuring dynamics of solar energy resource quality: Methodology and policy implications for reducing regional energy inequality. *Renew. Energy* **2022**, *197*, 138–150. [CrossRef]

agriculture

MDPI

Article

An Analysis of Preference Weights and Setting Priorities by Irrigation Advisory Services Users Based on the Analytic Hierarchy Process

Itzel Inti Maria Donati [1], Davide Viaggi [2], Zorica Srdjevic [3], Bojan Srdjevic [3], Antonella Di Fonzo [4,*], Teresa Del Giudice [5], Orlando Cimino [4], Andrea Martelli [4], Anna Dalla Marta [6], Roberto Henke [4] and Filiberto Altobelli [4,*]

1 Department of Architecture and Design Landscape and Environment Sapienza, University of Rome, Piazza Borghese 9, 00186 Rome, Italy; itzelinti.donati@uniroma1.it
2 Department of Agricultural and Food Sciences, Alma Mater Studiorum Università di Bologna, Viale Fanin 50, 40127 Bologna, Italy; davide.viaggi@unibo.it
3 Department of Water Management, Faculty of Agriculture, University of Novi Sad, Trg D. Obradovica 8, 21000 Novi Sad, Serbia; zorica.srdjevic@polj.uns.ac.rs (Z.S.); bojan.srdjevic@polj.uns.ac.rs (B.S.)
4 Council of Research in Agriculture and Analysis of Agricultural Economics, Research Centre for Agricultural Policies and Bio-Economy, Via Barberini 36, 00187 Rome, Italy; orlando.cimino@crea.gov.it (O.C.); andrea.martelli@crea.gov.it (A.M.); roberto.henke@crea.gov.it (R.H.)
5 Department of Agricultural Sciences, University of Naples Federico II, Reggia di Portici, Via Università 100, 80055 Naples, Italy; teresa.delgiudice@unina.it
6 Department of Agrifood Production and Environmental Sciences, University of Florence, Via delle Cascine 5, 50144 Florence, Italy; anna.dallamarta@unifi.it
* Correspondence: antonella.difonzo@crea.gov.it (A.D.F.); filiberto.altobelli@crea.gov.it (F.A.)

Abstract: *Objective:* Stakeholders—farmers from four different European areas (Campania (IT), Kujawsko-Pomorskie (PL), Limburg (NL), Andalusia (ES))—are asked to share, from the OPERA project, their opinions on five criteria that all aim at improving the use of irrigation advisory services (IASs). Each criterion has different characteristics that affect the way farmers rank it. The present study has two objectives. The first is to individuate the priorities of the preferences expressed by the stakeholders. The second objective is to carry out a ranking of the weights of the criteria by case study, ranking the groups and their associated properties among farmers' profiles. *Methods*: The answers to 120 questionnaires dispensed to the future users of IASs in the four agricultural sites were analyzed in detail, and then the given priorities were evaluated through the analytic hierarchy process (AHP). The AHP methodology was used to determine the relative weights of the five assessment criteria, and finally, to select the one with major value. *Results and conclusions*: The results show that A5 (assuring economic sustainability) was the most important criterion. The contributions provided by this study are twofold: Firstly, it presents an application of a methodology that involves the conversion of a linguistic judgement of farmers in a correspondence weight. Secondly, it tackles decision making regarding improving the use of IASs, evaluating the preferences expressed by the stakeholders. Irrigation advisory services can play a key role in assisting users to adopt new techniques and technologies for more efficient water use and increased production.

Keywords: multicriteria decision analysis; AHP; irrigation advisory services; agricultural decision making; economic sustainability

Citation: Donati, I.I.M.; Viaggi, D.; Srdjevic, Z.; Srdjevic, B.; Di Fonzo, A.; Del Giudice, T.; Cimino, O.; Martelli, A.; Dalla Marta, A.; Henke, R.; et al. An Analysis of Preference Weights and Setting Priorities by Irrigation Advisory Services Users Based on the Analytic Hierarchy Process. *Agriculture* **2023**, *13*, 1545. https://doi.org/10.3390/agriculture13081545

Academic Editor: David Maxwell Freebairn

Received: 1 June 2023
Revised: 24 July 2023
Accepted: 25 July 2023
Published: 2 August 2023

1. Introduction

The interest in promoting a form of agriculture capable of adapting to climate change has made the management of water resources one of the key points in the reform of the CAP 2023–2027 [1]. As a matter of fact, "*Foster sustainable development and efficient management of natural resources such as water, soil, air*" and "*Fostering knowledge, innovation and digitalisation in agriculture"* have been designated as two of the ten new strategic objectives of the new CAP 2023–2027. In order to pursue more efficient and sustainable water use, EU countries are called, among other things, to encourage research and innovation in the sector by the implementation of "smart irrigation" technologies. An efficient use of water for irrigation is a priority driven by the evidence that many areas in the Mediterranean region suffer structural water scarcity, imposed by the periodic droughts and by the expansion of water demands of agriculture and other sectors of society [2]. With the advancement of climate change, higher temperatures, and changing precipitation patterns, the demand for water by the agricultural sector has increased. It has started to affect not only areas where irrigation has always been an essential element of agricultural production (southern Europe) but also areas traditionally considered not irrigated, such as some areas of central and northern Europe. In this context, the EU research project OPERA—"Operationalizing the increase of water use efficiency and resilience in irrigation", http://opendata.waterjpi.eu/dataset/2a2a87e0-5c84-42cd-a9da-ecac0bbb9257/resource/1b07850f-c7e8-4a0d-86c8-180ff3e1bae5/download/d5.1_inception_report_opera.pdf (accessed on 24 July 2023) is a program financed under ERA-NET, which is part of Water JPI. Water JPI aims to tackle the challenge of "achieving sustainable water systems for a sustainable economy in Europe and abroad". Within the context of a sustainable economy, OPERA focuses on the sustainable management of water resources in agriculture and the use of irrigation advisory services (IASs), and thus, intelligent irrigation systems that provide information to a large number of farmers have become useful tools for irrigation programs. The issue is not new, and extensive research and investments have been made to develop more advanced methods and practices to accurately provide water to the crops based on their needs.

Technological advances in IASs continue to increase rapidly [3–9]. Along with it, the behavioral and socio-economic determinants of farmers for the adoption of these efficient irrigation technologies are also evolving. The success of these technologies can be supported by the integration of stakeholders' needs in the design of IASs [10,11].

Nevertheless, the literature on the subject appears to be poor in case studies focused on identifying the needs of farmers, who are the end users of IASs.

There is still a significant disparity between the availability of technologies for efficient water usage and the acceptance of these technologies. One of the reasons is the lack of emphasis on establishing an efficient support system to aid farmers in adopting and effectively operating new techniques and technologies. Through four case studies in the EU context, this paper will address the following questions:

- How can OPERA cope with these issues while taking into consideration the feedback obtained from the stakeholders' answers and making use of the current AHP analysis?
- Are there any spatial differences or correlations among the criterion improvements selected by the stakeholders?
- Which of the criteria seem to have major weight according to the stakeholders?

This study is organized into six sections: Sections 1 and 2 describes the background research, general area of interest, and the topic of focus; Section 3 presents the study areas chosen for the research activity and the data and research methodology; in Sections 4 and 5, the research results and discussions are presented; and, finally, in Section 6, the main conclusions and future research design are reported. Within this framework, the application of AHP demonstrates that a multi-criteria problem can be approached specifically for each case study. Nevertheless, an overall result involving all study areas can be achieved. One of the advantages of the AHP method is to support both individual and group decision-making processes not only with a quantitative but also a qualitative approach. Since the 1990s, there has been a growing number of studies applying the AHP to deal with decision-making problems in agriculture [12–23]. From these works, the utility of the AHP has arisen for understanding heterogeneous farming systems and how farmer behavior is needed for tailoring policy instruments. Against an agricultural water management background, it also helps to share these frontiers for more efficient, equitable, and sustainable outcomes.

2. Background

Irrigation Advisory Services

Several factors determine the quantity of irrigation water employed in agriculture, ranging from the variety of crops and cultivation approaches to soil properties and the irrigation method, among others. Hence, agriculture itself presents prospects for improved water administration and conservation, encompassing both conventional farming practices and innovative agricultural technologies. Among the latter, irrigation decision support systems (DSSs) can assist farmers in making informed decisions, leading to enhanced profitability by optimizing water usage and ensuring maximum crop yield in a particular growing season. These systems are primarily designed to simulate or forecast crop water requirements, presenting a range of choices. Under this scenario, irrigation advisory services (IASs) are considered a useful DSS to help farmers achieve the best efficiency in irrigation water use and to increase their incomes by obtaining the highest possible crop yield. Irrigation advisory services are a set of activities that aim to provide technical and professional support to farmers and agricultural operators in the management of cropland irrigation. In recent decades, the research has focused on investigating new IASs tools, which has contributed to the evolution of the performance capabilities of the services. Nowadays, IASs can be implemented in a broad range of agricultural situations, and they can easily be combined with several software programs. IASs are able to deal with the following:

- Satellite-based irrigation volumes are able to perform a site-specific evaluation of irrigation volumes, integrating remote sensing data with a geographic information system (GIS) [24]. In some cases, the research has been focused on quantifying several irrigation and drainage performance indicators with the support of a GIS.
- Development delivery data from a desktop application to via the web, considering that the graphical user interface is a key element for the successful use of the services (PlanteInfo, WIESE, IRRINET, BEWARE, ISS-ITAP, IrriSAT, IRRISAT) [25–27].
- Biophysical variables, surface soil water content, and canopy water content; for example, some studies have been inquiring about how to estimate separately determine soil evaporation and crop transpiration [28–30].
- In the context of remote-sensing tools, some studies have been carried out as a part of the project DEMETER (Demonstration of Earth observation technologies in routine irrigation advisory services), which deals with the transmission of personalized irrigation scheduling information to the users, related to an extended period of time (e.g., on past, present, and future weather) [31].
- Some studies have investigated the idea of an IAS tailored to the distinct circumstances of farmers. The findings indicate distinct farmers' inclinations, particularly for obtaining weather predictions from the service and for the characteristics associated with water data registration [32,33].

As indicated above, the research has made notable steps forward, progressing in the technical aspects at the basis of DSS programming for irrigation, and has made the use of IASs more and more efficient. It has also made the use of these tools applicable in various agricultural contexts.

The strengthening of the aspects of the research activity mentioned above deserves to be further investigated to understand the judgment of the end users and their needs, with the aim of favoring the implementation of IASs in the management of water resources in the field.

3. Materials and Methods

3.1. Study Areas

The analytic hierarchy process (AHP) is a general theory of measurement [34], and it has been used in the present study to analyze the verbal judgments of IASs end users belonging to four different geographical areas: Italy, Netherlands, Spain, and Poland. The characteristics of these regions are summarized below.

Campania (IT)—For this case study, 40 interviews were collected. The key stakeholders featured in the sample interviews included not only farmers but also representatives from the regional government, land and water reclamation authorities, farmer associations, local policymakers, and legislators. The farm activity is focused on growing silage maize, wheat, tomato, and other vegetables (such as peppers and beans). Additionally, 12% of farmers in the area are engaged in cereal production and livestock rearing. Within the project, the crucial investigation consisted of exploring the potential for adaptation and addressing the needs of the end users, as well as identifying optimal approaches for dynamically forecasting crop water requirements through the assimilation of remote sensing observations and numerical weather predictions in a crop growth model.

In the Kuyavian-Pomeranian province (PL), 53 interviews were recollected. In this case study, two demonstration areas were involved: the first one was located in the Zglowiaczka River catchment in a small productive field, and the second one was located in the Upper Notec River catchment in a big productive field. Vegetables are cultivated as well as irrigated in both fields. Although this is a region of intensive agricultural crop production, in an average year, it suffers concerning rainfall water deficits. The main issues affecting these areas are water shortage in the growing season, the use of clean renewable deep groundwater resources for irrigation, insufficient efficiency of water used for irrigation, and the lack of an irrigation decision support system for farmers based on the current crop water needs and weather forecasts.

Andalucía (ES)—Agriculture in the Mediterranean region is dealing with serious problems related to the present drought and the general scarcity of water resources, resulting in an increasing water demand [35]. These difficulties are expected to worsen due to the future predicted severe water scarcity in the Mediterranean area. Olive cultivation has been chosen as a case study, as it is a crucial economic sector representing 24% of the value of agricultural production in the Andalusia area. It covers an area of approximately 1.5 million hectares (around 17% of the total region's surface, accounting for 60% of the national surface dedicated to olive crops and 30% of the European surface) and contributes to about 40% of global olive oil production and 20% of global table olive production. Additionally, it is a significant source of wealth and employment, supporting over 22 million wages annually. It is also essential for social and territorial cohesion and possesses high environmental value, shaping the Andalusian territory and culture. While some facilities and advice services are available in this area, the current solutions have not been effectively implemented, leading to unsatisfactory results.

Limburg (NL)—Seven interviews were collected in one of the less dry regions of Northern Europe. Among the interviewed farmers, mobile irrigation (overhead sprinkling) systems are in widespread use. Irrigation management is supported by weather forecasts that anticipate crop water availability and by supplementary information through sensors (local or remote). Water availability for irrigation in the Netherlands in previous years was not a frequent concern. However, due to climate change, they encounter more drought periods, and local water boards temporarily proscribe farmers from using surface water and groundwater for irrigation. When this occurs more frequently in the foreseeable future, an enhanced water supply will be necessary.

3.2. Data Collection

3.2.1. Identifying Respondents' Profiles

As shown in Figure 1, this study was organized into a four-phase methodology. The first step of the investigation was identifying the needs and demands of the users. Each case study partner selected the particular stakeholders on the basis of the "Guidelines for analysis and selection of stakeholders". This approach is based on a "snowball sampling design". Information is available in the "D1.1 Assessment of user requirements of the sector" of OPERA at this link: http://opendata.waterjpi.eu/dataset/2a2a87e0-5c84-42cd-a9da-ecac0bbb9257/resource/

09d7444c-c5e2-4473-835b-9c28f27d20d3/download/d1.1_report_stakeholder_opera.pd (accessed on 24 July 2023). Identified and contacted stakeholders were asked to identify further stakeholders, starting with the case study partners. The questionnaires were addressed to a total of 120 farmers (users of IAS), distributed among the following study areas: Campania (IT): 40; Limburg (NL): 7; Kujawsko-Pomorskie (PL): 53; and Andalusia (ES): 20.

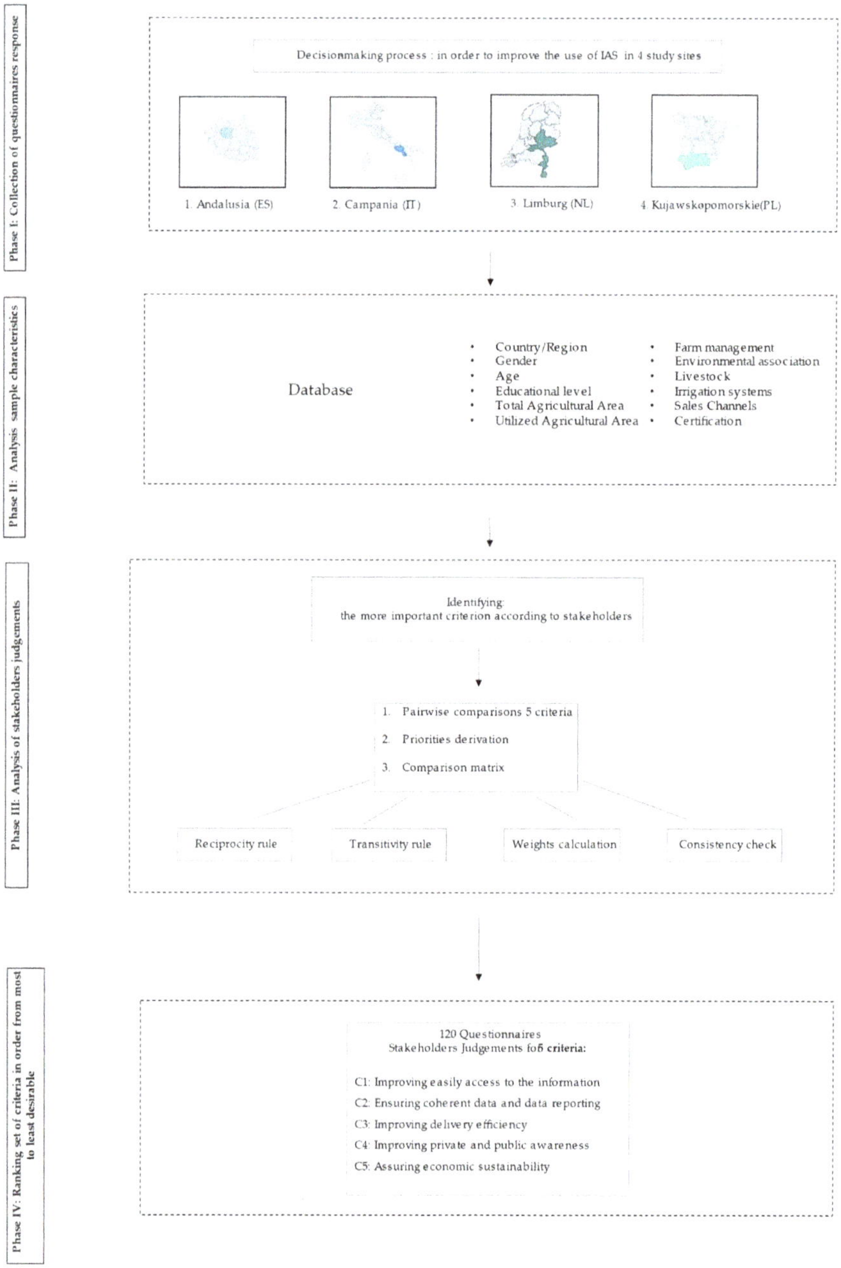

Figure 1. Research methodology.

3.2.2. Questionnaires

The questionnaire was structured into four sections. The first section was focused on general interview information: activity, gender, age, educational level, farm's location, farm surface, farm management, cultivated crop, main irrigation systems, sales channels, etc. Using this information, a database was developed. Phase two of the research (Figure 1) was possible to realize, combining the outcomes of the AHP and the database analysis (Figure 1). The above-mentioned information was important in order to group the weights of criteria evaluated by the stakeholders and to reveal the inter-relations between the technical factors expressed in the proposed questionnaires and the weights resulting from the employment of the AHP.

Section 2 of the questionnaire was named "Improving water use efficiency and the use of advisory services" and was related to the data that were analyzed by the AHP. The farmers, regarded as stakeholders, expressed their preferences among a set of criteria (Table 1), answering the follow question: "Compare criteria C1 and C5 using Saaty's scale 1–9. According to you, is C1 more important than C2, and by how much?" The pairwise comparison at the core of the AHP methodology was inserted into the questionnaires and is discussed below. Section 3 of the questionnaire, named *"Developing policies and strategy suggestions for improving use of irrigation advisory services at farm level: an Italian experience"* included: (i) four questions related to the internal strengths that farmers could come across when adopting IASs; (ii) the weaknesses that farmers may face when adopting IASs; (iii) the opportunities arising for farmers when adopting IASs; and (iv) the threats to companies when adopting such IASs. Finally, Section 4, named *"Eliciting farmers' individual risk attitude"*, provided suggestions about the adoption of innovative tools proposed by the OPERA project. These tools depend on the farmers' subjective attitudes toward taking risks. The questionnaires were translated from English into the native language of each country.

Table 1. Description of AHP criteria.

Type of Criteria	Description
C1. Improving easy access to information	Refers to the ease of access to information for farmers, either through electronic information (SMS, email, etc.), more traditional communication systems, technical operators and journals, newspapers, etc.
C2. Ensuring coherent data and data reporting.	Refers to the ability to implement an IAS based on high-quality data, providing valuable technical information to farmers.
C3. Improving delivery efficiency	Refers to the ability to ensure prompt and constant delivery of information to farmers.
C4. Improving private and public awareness	Refers to improving public awareness and preparedness by informing the public about the risks and consequences in case of excessive use of water for irrigation related to environmental and economic phenomena (e.g., water scarcity, conflict over use of water with others economic sectors).
C5. Assuring economic sustainability	Refers to the cost of IAS, which should be economically justified (i.e., economically affordable).

3.3. Multi-Criteria Decision Analysis—Selection of the Analytical Hierarchy Process (AHP)

In the agricultural sector, the main problems that multi-criteria decision analysis (MCDA) is facing are related to their evolution in terms of technological progress (equipment, fertilizers, pesticides, new plant varieties, irrigation systems).

This use of new production systems has been associated with an increase in the size and degree of specialization of agricultural operations [36]. In the literature, several approaches are proposed in the agricultural sector to assist decision makers, such as farmers and their associations, policy makers, and local and regional authorities, to efficiently explore a range of criterion farm management practices, and thus, identify pathways toward sustainability.

When criteria need to be classified into ordered classes, a sorting method has to be applied, but much less attention has been paid to investigating this kind of problem, especially in the case of multiple decision-makers asked to give subjective scores to different criteria based on qualitative criteria.

The analytic hierarchy process (AHP) is an MCDA developed by Thomas L. Saaty in the 1970s [34].

Considering the number of MCDA methods available (PROMETHEE, MACBETH, ELECTRE, TOPSIS), as suggested by the literature [37–39], there are several methods to choose an appropriate MCDA.

The decision to use the AHP in this work was guided by a series of drivers/reasons, summarized as follows:

- Ratio scale and pairwise comparison: The fundamental process involves the comparison of two stimuli, which are also referred to as alternatives, under a particular criterion or two criteria. The decision maker was asked to determine if they were indifferent towards the two stimuli or if they had a weak, strict, strong, or very strong preference for one of them. Understanding this structure is more intuitive for the respondent and facilitates stakeholder participation. The criteria analyzed in this study were identified within the OPERA project, for which detailed information can be found at the following link: http://opendata.waterjpi.eu/dataset/opera-operationalizing-the-increase-of-water-use-efficiency-and-resilience-in-irrigation (accessed on 24 July 2023).
- *Stakeholders*: The AHP can support complex decisions in which several stakeholders are involved, as in the case of the present study. The construction of the database (areas, farm management, irrigation systems) demonstrates that different interest groups are implicated [40].
- *Software*: The AHP is one of the most popular MCDA methods and is backed by a large variety of software offering diverse data management and representation capabilities [41].

3.4. Application of Analytic Hierarchy Process

The AHP comprises three principal operations, including hierarchy construction, priority examination, and consistency analysis. In the present study, these steps were carried out as shown in Figure 2 [42].

As mentioned above, the objective of the mathematical procedure is to estimate the weights of five criteria from a matrix of pairwise comparisons $A(a_{ij}))$ generated following both the transitivity rule and the reciprocity rule. The reciprocal condition or Axiom 1 defines that given two criteria $(A_i,A_j) \in A \times A$, the intensity of preference of A_i over A_j is inversely related to the intensity of preference of A_j over A_i.

Transitivity rule is:

$$a_{ij} = a_{ik} \cdot a_{kj} \tag{1}$$

Reciprocity rule is:

$$a_{ij} = \frac{1}{a_{ji}} \tag{2}$$

where a_{ij} is the comparison of criteria i and j.

If we suppose that preferences (weights) p_i are known, a perfectly consistent matrix can be constructed because all of the comparisons a_{ij} satisfy equality:

If the preferences (weights) p_i

$$a_{ij} = \frac{p_i}{p_j} \tag{3}$$

where p_i is the priority of the alternative i, and the completely consistent matrix is:

$$A = \begin{bmatrix} {}^{p_1}/{}_{p_1} & \cdots & {}^{p_1}/{}_{p_n} \\ \cdots & 1 & \cdots \\ {}^{p_n}/{}_{p_1} & \cdots & 1 \end{bmatrix} \quad (4)$$

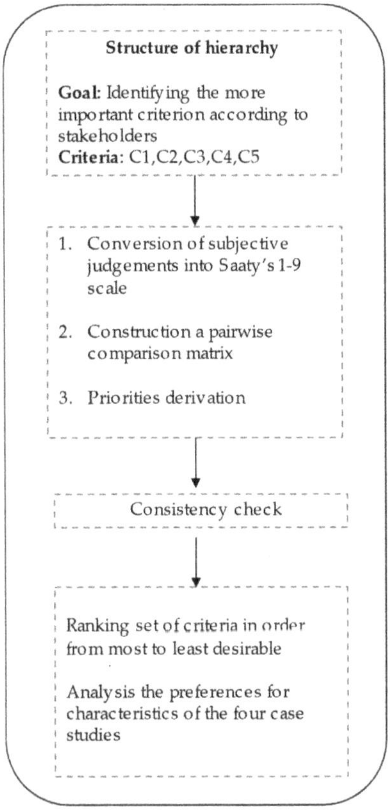

Figure 2. AHP steps in the present study.

We can apply the following formula from the matrix above:

$$\sum_j \frac{p_i}{p_j} \cdot p_j = np_i \quad (5)$$

Thus, the product of row i by the priority vector p gives n times the priority p_i. By multiplying all the elements of the comparison matrix A by the priority vector p, the following equality is obtained:

$$A = \begin{bmatrix} {}^{p_1}/{}_{p_1} & \cdots & {}^{p_1}/{}_{p_n} \\ \cdots & 1 & \cdots \\ {}^{p_n}/{}_{p_1} & \cdots & 1 \end{bmatrix} \begin{bmatrix} p_1 \\ \cdots \\ p_n \end{bmatrix} = n \begin{bmatrix} p_1 \\ \cdots \\ p_n \end{bmatrix} \quad (6)$$

The priorities (weights) of the compared criteria are not known in advance. As priorities only make sense if derived from consistent or near-consistent matrices, a consistency check must be applied. Several methods have been proposed to measure consistency [43,44] Also, normalization plays a key role in obtaining meaningful results from AHP analysis.

The functioning of the model is strictly related to pairwise comparison, involving the level of criteria in the present study. The result of the pairwise comparison is expressed by Saaty's scale. The function in detail is as follows:

First, two criteria—A and B—are compared using the numerical scale ranging from 1 to 1/9 (Saaty's scale), where number 1 means both criteria have the same importance, and they are equal; number 9 means criterion A is 9 times more important than criterion B; 1/9 means that criterion B is nine times more important than criterion A.

Second, it is necessary to compare all elements pairwise with respect to the objective. In the following step, the comparisons are arranged in a matrix. From this matrix, the computed weights for the different criteria are generated.

In 1977, Saaty [45] proposed the consistency ratio (*CR*) to measure the reliability of information contained in a pairwise comparison matrix:

$$CR = \frac{CI}{RI}$$

This is a ratio of the consistency index (*CI*) and random index (*RI*), and it is given by Relation (7)

$$CI = \frac{\lambda_{max} - n}{n - 1}, \tag{7}$$

where λ_{max} is computed as

$$\lambda_{max} = \frac{\left(\sum_{j=1} a_{1j}\, p_j\right)}{p_1}$$

The *CR* provides a way of measuring how many errors were created when providing the judgments; a rule-of-thumb is that if the *CR* is below 0.1, the errors are fairly small and thus, the final estimate can be accepted. The first step for computing the *CR* is determining the eigenvalue (λm), followed by determining the *CI* [46].

4. Results

The AHP was applied to guide a decision-making process, with the ultimate goal of improving the use of IAS among farmers. The first results of the study demonstrate that the most common decision was *Assuring economic sustainability* (C5), as shown in Table 2.

Table 2. Overall results of the four study areas.

	Criteria	Weights of Criteria	Final Ranking
Evaluating Possible Adoption Options of IAS	C1: Improving easy access to information	0.207	3
	C2: Ensuring coherent data and data reporting	0.218	2
	C3: Improving delivery efficiency	0.196	4
	C4: Improving private and public awareness	0.148	5
	C5: Assuring economic sustainability	0.231	1

There was heterogeneity among the farmers' judgments, which involved a clear difference in weights between the most important criterion and the criterion with a lower weight. By looking at the distribution of the priority values (Table 2), the weights vary, with a minimum weight of 0.15, attributed to criterion C4—*Improving private and public awareness*, which refers to improving public awareness and preparedness by informing the public about the risks and consequences in case of the excessive use of water for irrigation related to environmental and economic phenomena e.g., water scarcity, conflict over the use of water with other economic sectors. A maximum weight of 0.23 was attributed to criterion C5—*Assuring economic sustainability*.

In the following stage, the weights were grouped according to the decision makers' profiles. The next step was to relate the preferences expressed by the stakeholders and to analyze the key information provided by the interviews. Subsequently, a mathematical aggregate of the weights of each criterion was calculated using the geometric mean method. While this section is not intended to be an exhaustive account of all results of this work, it aims to provide a broad picture of the most relevant results for each pilot area.

Campania (IT): The results show that the priority rankings of the group are quite "flattened", which may be partly due to the inconsistencies among the elements of the pairwise comparison matrix (hinting at some randomness in the answers). Table 3 shows that C5—*Assuring economic sustainability* and C4—*Improving private and public awareness* are the most preferred options. The proposed grouping procedure can be used to discuss some observations. The criteria weights of the Italian study site were compared to the results obtained from the aggregate weight of the Netherlands pilot area. In both areas, (Figures 3 and 4) farmers rear livestock as one of the main farm activities, which involves growing feed crops. Among these samples, the farmers who grew grassland and ryegrass preferred criterion C5—*Assuring economic sustainability.*

Table 3. Weights and ranks of criteria in the four study sites.

	Andalusia (ES)		Campania (IT)		Kujawsko-Pomorskie (PL)		Limburg (NL)	
Criteria	**Weights**	**Final Ranking**	**Weights**	**Final Ranking**	**Weights**	**Final Ranking**	**Weights**	**Final Ranking**
C1	0.194	3	0.194	3	0.163	5	0.233	1
C2	0.177	4	0.177	4	0.209	3	0.194	3
C3	0.126	5	0.126	5	0.222	1	0.222	2
C4	0.196	2	0.196	2	0.185	4	0.089	5
C5	0.296	1	0.296	1	0.206	2	0.182	4

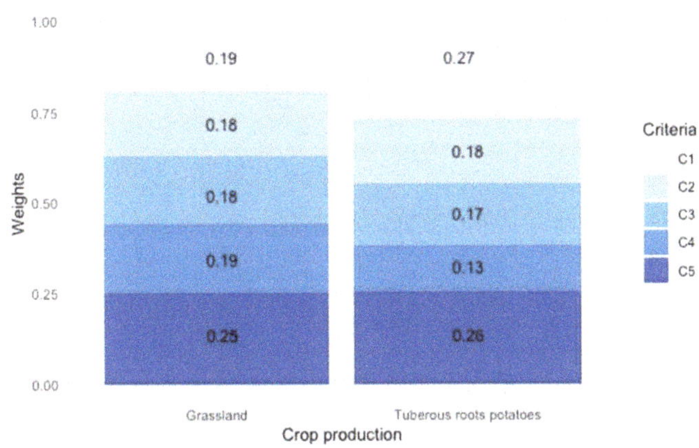

Figure 3. Weights of criteria grouped according to the crop production in the Netherlands. C1—Improve easy access to information; C2—Ensure coherent data and data reporting; C3—Improve delivery efficiency; C4—Improve public and private awareness; C5—Ensure economic sustainability.

Limburg (NL): The preference of the farms that have adopted surface irrigation was C5—*Assuring economic sustainability*. As Walker argues, in his study published by FAO in 1980 [47], one of the advantages of surface irrigation is that these systems are inexpensive to develop at the farm level. The control and regulation structures are simple, durable, and easily constructed with cheap and readily available materials. The survey illustrates the financial aspect, which is an important issue to be considered for designing and developing water management strategies for farmers belonging to this profile. As shown in Figure 5, farms that use sprinkler irrigation identified C1—*Improve easy access to information* as more important. Among the three types of irrigation systems, these are the most sensitive to weather conditions. For example, strong winds can affect the efficiency of the spraying of water from sprinklers.

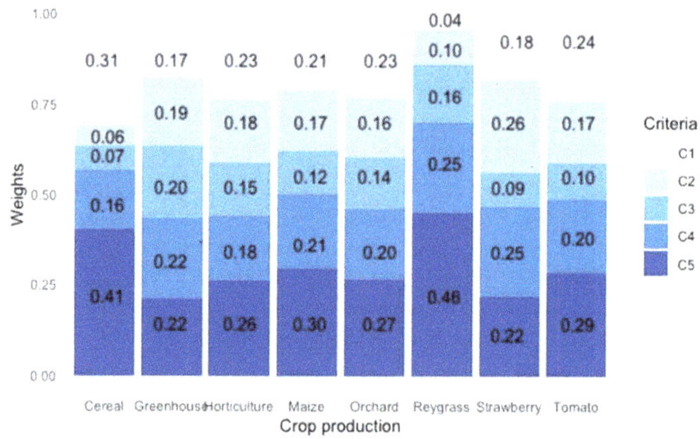

Figure 4. Weights of criteria grouped according to the crop production in Campania (IT). C1—Improve easy access to information; C2—Ensure coherent data and data reporting; C3—Improve delivery efficiency; C4—Improve public and private awareness; C5—Ensure economic sustainability.

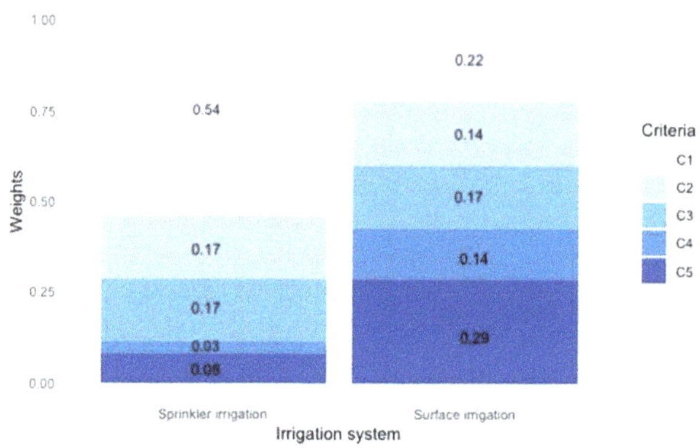

Figure 5. Weights of criteria grouped according to irrigation systems in Limburg (NL). C1—Improve easy access to information; C2—Ensure coherent data and data reporting; C3—Improve delivery efficiency; C4—Improve public and private awareness; C5—Ensure economic sustainability.

It appears evident that farmers who use this type of system are more interested in information concerning remote detection and weather forecasting. According to this profile of respondents, IASs will have to offer these services, which are of fundamental importance to the farmers.

Kujawsko-Pomorskie (PL): Results in Table 3 show that the stakeholders preferred C3—*Improve delivery efficiency*, which refers to the ability to ensure prompt and constant delivery of information to farmers. According to the results generated through the performed surveys, the farmers' expectations in the region are to obtain reliable information on the actual meteorological and soil moisture conditions. They also expect to know when, how much, and which crop should be irrigated. The above-mentioned need should be the main feature of the IAS. However, if we analyze the weights by grouping them according to the irrigation manager, it is evident that the most significant criterion for farmers who adopt drip and sprinkler irrigation is C5—*Assuring economic sustainability*. The long-term viability of drip irrigation also depends on its economic sustainability. Despite the potential increase in water use efficiency and yield, the system also needs to generate higher income to be popular among farmers [4].

Andalucía (ES): The respondents from the Andalucia study site were all olive farmers. There were no farmers with diversification of production in the sample. The final ranking shows that, according to the stakeholders' judgments, C2—*Ensure coherent data and reporting are priorities*, was a criterion with a higher weight (Table 3).

5. Discussion

In the present study, there were a few limitations in the application of the AHP. It is worth noting that these priority rankings based on the collected data are, at times, quite flattened. As previously mentioned, one of the possible causes attributed to the homogeneity of the weights could be the inconsistencies in the matrices that express the farmers' judgments. The AHP has a means for measuring any inconsistencies by a formula called the consistency ratio. A ratio of 0 means a perfect consistency, while any ratio over 0.1 is considered inconsistent [48]. In the present work, only 26% of the subjects had a consistency ratio equal or lower than this limit. In this study, the inconsistency is mainly attributable to two aspects:

1. Method of structuring the model and criteria considered: Ideally, one would structure a complex decision through a hierarchy where factors at any level are comparable. If this condition does not occur during the criteria selection process, the possibility of generating inconsistencies among the elements of the pairwise comparison matrix (hinting at some randomness in the answers from the respondents) increases.
2. Method of administration of the questionnaire: It emerged that the mailed surveys made it difficult for respondents and researchers to interact. The letter was a necessary condition to explain the meaning of the pairwise comparison involved in the multi-criteria AHP analysis and to ensure that the respondents had full awareness and understanding of the criteria that they had to compare. It would have been appropriate to ask the interviewees to re-evaluate their judgments within the matrices, but this was not carried out because it would have been a difficult and time-consuming process.

However, it is evident from the results that there are further chances to improve the application of the AHP model for a better evaluation of the stakeholders' judgments. The literature offers an extremely broad overview of the advantages and disadvantages of the method on the consistency/inconsistency of the answers given by decision makers, and therefore, on the reliability of the model. Forman [49] introduced several comments related to the AHP. The most common reason for inconsistency is the lack of perfect knowledge. For this reason, it becomes essential to support the interviewee during the interview to clarify any doubts and reduce the possibility of error in their answers. In order to increase the use of IASs among farmers, it is necessary that these services acquire characteristics and performances that allow for "*Assuring economic sustainability*" (C5). If an investment is needed to improve irrigation management through the IASs, it will be justified by farmer users only if it is profitable. It would be interesting to extensively discuss the meaning of "*Assuring economic sustainability*" for a farmer, as it is a very complex condition that depends on many factors that are unpredictable, especially in the long term, because they could depend on future economic conditions. In a future project, the AHP could be a useful tool to deeply investigate these unpredictable factors in a qualitative and quantitative framework.

6. Conclusions

This paper addresses a very relevant issue framed in the process of the new CAP reform, which is the efficient and sustainable use of water, especially in the context of progressive scarcity of this resource in the Mediterranean area within and outside of the EU. This issue is a relevant commitment of the agricultural sector in the wider framework of the Agenda 2030 and the fight against the climate change, and there is a busy and rich research agenda ahead on this matter.

This work contributes to the general issue of water use in agriculture by developing a methodological approach based on the analytic hierarchy process to support the decision-making objective of "improving the use of irrigation advisory services". The results highlight that the most important criterion is C5—"Assuring economic sustainability", which means that the cost of IAS should be economically justified (i.e., economically affordable).

In order to enhance our comprehension of this topic, it is necessary to address the subsequent queries: What are the farmers' objective priorities (economic, non-economic, or both)? What are the tools employed by farmers to attain their aims, and what are their genuine objectives? Which factors exert an influence on them? The proposed reflections may be developed in future research activity, departing from the results of the present analysis. Finally, it is important to continue with this type of discussion to ensure that the decision-making process is able to contribute effectively to agricultural development in terms of sustainable irrigation management. Furthermore, using a database (the key information provided by the interviews) made it possible to aggregate the individual priorities in each study area and for each characteristic of the samples by relating the preferences expressed by the stakeholders.

Finally, the findings indicate that the criteria (C1, C2, C3, C4, C5) had varied forms of impact on the end users' judgments, and these attributes play a crucial role in shaping the strategies and scenarios for advancing the implementation of IASs.

Author Contributions: Conceptualization. I.I.M.D., F.A., A.D.M., T.D.G. and O.C.; methodology, I.I.M.D., F.A., A.D.M., T.D.G., O.C., Z.S., B.S. and D.V.; validation, F.A., O.C. and A.D.M.; investigation, F.A., T.D.G., O.C., A.D.M. and I.I.M.D.; data curation, I.I.M.D. and F.A.; writing—original draft preparation, I.I.M.D., F.A. and A.D.F.; writing—review and editing, A.D.F., A.M.; R.H. and I.I.M.D.; visualization, I.I.M.D.; supervision, F.A. and A.D.F.; project administration, F.A. and A.D.M.; funding acquisition, F.A. and A.D.M. All authors have read and agreed to the published version of the manuscript.

Funding: This research received WaterWorks2015 ERA-NET Cofunded Call (JPI Water) funding.

Institutional Review Board Statement: Not applicable.

Data Availability Statement: Data are available on request from the corresponding author.

Acknowledgments: This study was carried out as part of the OPERA project. The authors would like to thank the EU and The Ministry of Economic Affairs (The Netherlands), CDTI (Spain), MINECO (Spain), ANR (France), MIUR (Italy), NCBR (Poland) and WRC (South Africa) for funding, in the frame of the collaborative international consortium OPERA financed under the ERA-NET Cofund WaterWorks2015 Call. This ERA-NET is an integral part of the 2016 Joint Activities developed by the Water Challenges for a Changing World Joint Program Initiative (Water JPI).

Conflicts of Interest: The authors declare no conflict of interest. The partners of the OPERA project were informed and involved in the work, facilitating the access, and the use of data.

References

1. Picone, C.; Henke, R.; Ruberto, M.; Calligaris, E.; Zucaro, R. A Synthetic Indicator for Sustainability Standards of Water Resources in Agriculture. *Sustainability* **2021**, *13*, 8221. [CrossRef]
2. Noto, L.V.; Cipolla, G.; Pumo, D.; Francipane, A. Climate Change in the Mediterranean Basin (Part II): A Review of Challenges and Uncertainties in Climate Change Modeling and Impact Analyses. *Water Resour. Manag.* **2023**, *37*, 2307–2323. [CrossRef]
3. Mirás-Avalos, J.M.; Rubio-Asensio, J.S.; Ramírez-Cuesta, J.M.; Maestre-Valero, J.F.; Intrigliolo, D.S. Irrigation-Advisor—A Decision Support System for Irrigation of Vegetable Crops. *Water* **2019**, *11*, 2245. [CrossRef]
4. Smith, M.; Muñoz, G.; Smith, M.; Muñoz, G. Irrigation Advisory Services and Participatory Extension in Irrigation Management Workshop organised by FAO-ICID. In Proceedings of the Irrigation Advisory Services for Effective Water Use: A Review of Experiences, Montereal, QC, Canada, 24 July 2002.
5. Ortega, J.; de Juan, J.; Tarjuelo, J. Improving water management: The irrigation advisory service of Castilla-La Mancha (Spain). *Agric. Water Manag.* **2005**, *77*, 37–58. [CrossRef]
6. Olalla, F.M.d.S.; Calera, A.; Dominguez, A. Monitoring irrigation water use by combining Irrigation Advisory Service, and remotely sensed data with a geographic information system. *Agric. Water Manag.* **2003**, *61*, 111–124. [CrossRef]

7. Wang, Y.-P.; Chen, C.-T.; Tsai, Y.-C.; Shen, Y. A Sentinel-2 Image-Based Irrigation Advisory Service: Cases for Tea Plantations. *Water* **2021**, *13*, 1305. [CrossRef]
8. Santini, A.; Di Fonzo, A.; Giampietri, E.; Martelli, A.; Cimino, O.; Marta, A.D.; Annosi, M.C.; Blanco-Velázquez, F.J.; Del Giudice, T.; Altobelli, F. A Step toward Water Use Sustainability: Implementing a Business Model Canvas for Irrigation Advisory Services. *Agriculture* **2023**, *13*, 1081. [CrossRef]
9. Malamos, N.; Chalvantzis, C.; Tsirogiannis, I.L. Operation of a Digital Advisory Service for Irrigation Water Management: Case Study for Maize and Alfalfa Crops in Greece. Available online: https://irmasys.com (accessed on 24 July 2023).
10. Allani, M.; Frija, A.; Nemer, R.; Ribbe, L.; Sahli, A. Farmers' Perceptions on an Irrigation Advisory Service: Evidence from Tunisia. *Water* **2022**, *14*, 3638. [CrossRef]
11. Altobelli, F.; Marta, A.D.; Heinen, M.; Jacobs, C.; Giampietri, E.; Mancini, M.; Cimino, O.; Trestini, S.; Kranendonk, R.; Chanzy, A.; et al. Irrigation Advisory Services: Farmers preferences and willingness to pay for innovation. *Outlook Agric.* **2021**, *50*, 277–285. [CrossRef]
12. Barati, A.A.; Azadi, H.; Pour, M.D.; Lebailly, P.; Qafori, M. Determining Key Agricultural Strategic Factors Using AHP-MICMAC. *Sustainability* **2019**, *11*, 3947. [CrossRef]
13. Draginčić, J.; Vranešević, M. AHP-based group decision making approach to supplier selection of irrigation equipment. *Water Resour.* **2014**, *41*, 782–791. [CrossRef]
14. Jha, G.; Nicolas, F.; Schmidt, R.; Suvočarev, K.; Diaz, D.; Kisekka, I.; Scow, K.; Nocco, M.A. Irrigation Decision Support Systems (IDSS) for California's Water–Nutrient–Energy Nexus. *Agronomy* **2022**, *12*, 1962. [CrossRef]
15. Mishra, A.K.; Deep, S.; Choudhary, A. Identification of suitable sites for organic farming using AHP & GIS. *Egypt. J. Remote Sens. Space Sci.* **2015**, *18*, 181–193. [CrossRef]
16. Poveda-Bautista, R.; Roig-Merino, B.; Puerto, H.; Buitrago-Vera, J. Assessment of Irrigation Water Use Efficiency in Citrus Orchards Using AHP. *Int. J. Environ. Res. Public Health* **2021**, *18*, 5667. [CrossRef]
17. Karami, E. Appropriateness of farmers' adoption of irrigation methods: The application of the AHP model. *Agric. Syst.* **2006**, *87*, 101–119. [CrossRef]
18. Karleuša, B.; Ožanić, N.; Deluka-Tibljaš, A. Improving decision making in defining priorities for implementation of irrigation plans using AHP methodology. *Teh. Vjesn. Tech. Gaz.* **2014**, *21*, 673–680.
19. Ávila Mogollón, R.M. El AHP y su Aplicación Para Determinar los Usos de Tierras. El Caso de Brasil, Proyecto Regional Información Sobre Tierras y Aguas Para un Desarrollo Agrícola Sostenible; Proyecto GCP/RLA/126/JPN; Santiago, Chile. 2000. Available online: https://docplayer.es/93452079-Toma-de-decisiones-proceso-analitico-jerarquico-aplicado-a-la-seleccion-de-herramienta-de-educacion-virtual.html (accessed on 31 May 2023).
20. Srdevic, Z.; Blagojevic, B.; Srdevic, B. AHP based group decision making in ranking loan applicants for purchasing irrigation equipment: A case study. *Bulg. J. Agric. Sci.* **2011**, *17*, 531–543.
21. Bozdağ, A. Combining AHP with GIS for assessment of irrigation water quality in Çumra irrigation district (Konya), Central Anatolia, Turkey. *Environ. Earth Sci.* **2015**, *73*, 8217–8236. [CrossRef]
22. Omeka, M.E.; Igwe, O.; Onwuka, O.S.; Nwodo, O.M.; Ugar, S.I.; Undiandeye, P.A.; Anyanwu, I.E. Efficacy of GIS-based AHP and data-driven intelligent machine learning algorithms for irrigation water quality prediction in an agricultural-mine district within the Lower Benue Trough, Nigeria. *Environ. Sci. Pollut. Res.* **2023**, 1–30. [CrossRef]
23. Bonfante, A.; Monaco, E.; Manna, P.; De Mascellis, R.; Basile, A.; Buonanno, M.; Cantilena, G.; Esposito, A.; Tedeschi, A.; De Michele, C.; et al. LCIS DSS—An irrigation supporting system for water use efficiency improvement in precision agriculture: A maize case study. *Agric. Syst.* **2019**, *176*, 102646. [CrossRef]
24. Mehanna, M.H.M.; Pibars, S.K.; El-Noemani, A.-S. SIRMOD Model as a Management Tool for Basin Irrigation Method in Calcareous Soil. Impact of Water Regime and Phosphorus Fertilization and Their Interaction on the Characteristics of Rapeseed (*Brassica napus*) and Fatty Acid Profile of Extracted oil View project. A Simulation Model for Predicting Water Distribution under Self Compensating Gated Pipe Irrigation Technique for Small Holdings View Project. 2015. Available online: https://www.researchgate.net/publication/287251937 (accessed on 24 July 2023).
25. Rajput, T.B.S.; Patel, N.; Dinkar, D. Development of a Decision Support System, DOMIS, for Designing Micro Irrigation Systems. Available online: http://domis.iari.res.in (accessed on 24 July 2023).
26. Mannini, P.; Genovesi, R.; Letterio, T. IRRINET: Large Scale DSS Application for On-farm Irrigation Scheduling. *Procedia Environ. Sci.* **2013**, *19*, 823–829. [CrossRef]
27. Domínguez, A.; Martínez-López, J.A.; Amami, H.; Nsiri, R.; Karam, F.; Oueslati, M. Adaptation of a Scientific Decision Support System to the Productive Sector—A Case Study: MOPECO Irrigation Scheduling Model for Annual Crops. *Water* **2023**, *15*, 1691. [CrossRef]
28. Vuolo, F.; D'urso, G.; De Michele, C.; Bianchi, B.; Cutting, M. Satellite-based irrigation advisory services: A common tool for different experiences from Europe to Australia. *Agric. Water Manag.* **2015**, *147*, 82–95. [CrossRef]
29. Smith, M. CROPWAT: A computer program for irrigation planning and management. *Food Agric. Org.* **1992**, 55–62.
30. Richards, Q.; Bange, M.; Johnston, S. HydroLOGIC: An irrigation management system for Australian cotton. *Agric. Syst.* **2008**, *98*, 40–49. [CrossRef]
31. Belmonte, A.C.; Jochum, A.M.; García, A.C.; Rodríguez, A.M.; Fuster, P.L. Irrigation management from space: Towards user-friendly products. *Irrig Drain. Syst.* **2005**, *19*, 337–353. [CrossRef]

32. Montoro, A.; López-Fuster, P.; Fereres, E. Improving on-farm water management through an irrigation scheduling service. *Irrig. Sci.* **2011**, *29*, 311–319. [CrossRef]
33. Bergez, J.-E.; Nolleau, S. Maize grain yield variability between irrigation stands: A theoretical study. *Agric. Water Manag.* **2003**, *60*, 43–57. [CrossRef]
34. Saaty, T.L. Decision making with the analytic hierarchy process. *Int. J. Serv. Sci.* **2008**, *1*, 83–98. [CrossRef]
35. Duarte, A.C.; Mateos, L. How changes in cropping intensity affect water usage in an irrigated Mediterranean catchment. *Agric. Water Manag.* **2022**, *260*, 107274. [CrossRef]
36. Glen, J.J. Feature Article—Mathematical Models in Farm Planning: A Survey. *Oper. Res.* **1987**, *35*, 641–666. [CrossRef]
37. Ishizaka, A.; Nemery, P. *Multi-Criteria Decision Analysis: Methods and Software*; Wiley: Hoboken, NJ, USA, 2013.
38. Ishizaka, A.; Labib, A. Review of the main developments in the analytic hierarchy process. *Expert Syst. Appl.* **2011**, *38*, 14336–14345. [CrossRef]
39. Ishizaka, A.; Balkenborg, D.; Kaplan, T. Does AHP help us make a choice? An experimental evaluation. *J. Oper. Res. Soc.* **2011**, *62*, 1801–1812. [CrossRef]
40. Ishizaka, A.; Labib, A. Analytical hierarchy process and expert choice: Benefits and limitations. *Oper. Res. Insight* **2009**, *22*, 201–220. [CrossRef]
41. Ishizaka, A.; Lusti, M. How to Derive Priorities in AHP: A Comparative Study. *Cent. Eur. J. Oper. Res.* **2006**, *14*, 387–400. [CrossRef]
42. Ishizaka, A.; Pearman, C.; Nemery, P. AHPSort: An AHP-based method for sorting problems. *Int. J. Prod. Res.* **2012**, *50*, 4767–4784. [CrossRef]
43. Saaty, T.L.; Vargas, L.G. Models, Methods, Concepts & Applications of the Analytic Hierarchy Process. in International Series in Operations Research & Management Science. Springer US. 2012. Available online: https://books.google.it/books?id=FJLWhG5mWncC (accessed on 24 July 2023).
44. Triantaphyllou, E.; Mann, S. Using the analytic hierarchy process for decision making in engineering applications: Some challenges. *Int. J. Ind. Eng. Theory Appl. Pract.* **1995**, *2*, 35–44.
45. Saaty, R.W. The Analytic Hierarchy Process-What It Is and How It Is Used. *Math. Model.* **1987**, *9*, 161–176. [CrossRef]
46. Franek, J.; Kresta, A. Judgment Scales and Consistency Measure in AHP. *Procedia Econ. Financ.* **2014**, *12*, 164–173. [CrossRef]
47. Forman, E.H.; Gass, S.I. The Analytic Hierarchy Process: An Exposition. *Oper. Res.* **2001**, *49*, 469–486. [CrossRef]
48. Walker, W.R. Guidelines for Designing and Evaluating Surface Irrigation Systems, FAO. 1989. Available online: https://www.fao.org/3/t0231e/t0231e00.htm#Contents (accessed on 24 July 2023).
49. Forman, E.H. Facts and fictions about the analytic hierarchy process. *Math. Comput. Model.* **1993**, *17*, 19–26. [CrossRef]

agriculture

MDPI

Article

Factors Influencing the Perceptions of Smallholder Farmers towards Adoption of Digital Technologies in Eastern Cape Province, South Africa

Nasiphi Vusokazi Bontsa [1,*], Abbyssinia Mushunje [1] and Saul Ngarava [2]

[1] Department of Agricultural Economics and Extension, University of Fort Hare, P. Bag X1314, 1 King William's Town Road, Alice 5700, South Africa; amushunje@ufh.ac.za

[2] Copernicus Institute of Sustainable Development, Utrecht University, Princetonlaan 8a, 3584CB Utrecht, The Netherlands; s.ngarava@uu.nl

* Correspondence: 201213625@ufh.ac.za

Abstract: The objective of the study was to determine the factors that influence the perceptions of smallholder farmers towards the adoption of digital technologies. A purposively selected sample of 250 smallholder farmers who were cross-sectionally surveyed from Port St Johns and Ingquza Hill Local Municipalities in South Africa was used in the study. The Technology Acceptance Model (TAM) and the Attention, Desire, and Action (AIDA) model were used to analyse the data. The results showed that digital technologies were perceived to be expensive, cause a digital divide, and discouraged the use of Indigenous Knowledge even though they increased production. Positive perception towards digital technologies was associated with cattle rearing, with extreme negative perception for sheep and goat rearing. Educational level, employment status, monthly income, household size, being part of a cooperative (1% level), age, and source of income (10% level) were significant factors affecting smallholder farmers' perceptions of digital technologies. In conclusion, there are economic, social justice, and traditional perceptions towards digital technologies by smallholder farmers, with socio-economic factors affecting the perceptions. The study recommends providing low-cost digital technologies that promote Indigenous Knowledge, which should target the youth and young farmers with less education in small households who are full-time farmers with moderate-to-high incomes and are part of farmer groups/organisations.

Keywords: attention, interest, desire, and action (AIDA) model; digital technologies; perceptive index; technology acceptance model (TAM)

Citation: Bontsa, N.V.; Mushunje, A.; Ngarava, S. Factors Influencing the Perceptions of Smallholder Farmers towards Adoption of Digital Technologies in Eastern Cape Province, South Africa. *Agriculture* **2023**, *13*, 1471. https://doi.org/10.3390/agriculture13081471

Academic Editors: Roberto Henke and Filiberto Altobelli

Received: 15 June 2023
Revised: 21 July 2023
Accepted: 24 July 2023
Published: 25 July 2023

1. Introduction

Sustainable agricultural development can be achieved through the use of digital technologies, which can lead to achievement of Sustainable Development Goals (SDGs) 1, 2, 6, 8, 12, 14, and 15 [1]. Agriculture is an essential livelihood sector in Sub-Saharan Africa (SSA), accounting for 54% of the working population [2] and contributing 15% to Gross Domestic Product, even though highs of 50% in Chad and lows of 2% in South Africa have been recorded [3]. The OECD/FAO [3] highlighted that the agricultural sector in SSA has been currently increasing at 2.6% per year, fuelled by an increasing local and international interest in the farm land, the rise of a middle class, rural to urban migration, urbanisation, and population growth. Despite a steady increase, agricultural growth has not been ideal, with low productivity [4] fuelled by lack of use in appropriate technologies [3].

Digital technologies have been identified as a solution to transform the agricultural sector in SSA and livelihoods of the 250 million smallholder farmers in the region [2]. In South Africa, agriculture contributes USD 21 billion, employing 5.3% of the population [2]. Relative to the rest of SSA, South Arica has advanced utilisation of digital technologies in the agricultural sector, mainly focusing on precision agriculture, Agri-e-commerce, digital procurement, Agri-digital financial services, and digital advisory services [2]. This has been backed by advanced telecommunication infrastructure and digital penetration. Due to various reasons, such as costs, economies of scale, and lack of integration with Indigenous systems, there had been little usage of digital technologies by the 2.4 million farmers in the smallholder sector compared to the commercial sector [5,6]. Some of the constraints to the utilisation of digital technologies by smallholder farmers in South Africa has been due to the perceptions towards digital technologies. However, openness to digital technologies is determined by the combination of perception and socio-economic characteristics [7]. Caffaro et al. [8] identified that perceptions towards digital technologies relate to usefulness, increased productivity, cost reduction, efficiency, and workload reduction. A negative perception towards digital technology offers a barrier to its adoption. Despite the aforementioned issues, there is still limited understanding of the drivers of digital technologies which is itself driven by perceptions. The objective of this study was therefore to identify the drivers of digital technology perception in a country such as South Africa, hypothesising various socio-economic and institutional factors affecting this perception. The underlying question that this study seeks to answer is: What are the factors affecting digital technology perceptions by smallholder farmers in South Africa?

2. Literature Review

There are various determinants on the level of perception towards digital technologies which include demographics and socio-economic factors. A study by Pishnyak and Khalina [7] and Caffaro et al. [8] highlighted that perceived ease of use, usefulness, safety, reliability, and socio-economic characteristics had a positive impact on the propensity to use innovations. In particular, gender, age, educational level, and household income had an effect on perceptions towards digital technologies [7]. Pishnyak and Khalina [7] and Thomas et al. [9] highlighted that perceptions towards digital technologies are influenced by informal social networks through social capital. Some of the sources of information include other farmers and farmer organisations. A qualitative study by Reissig et al. [10] highlighted that there are farm and farmer characteristics affecting the perceived usefulness of digital technologies. Farm structure, including size and enterprise type, had an indirect impact on the usefulness of digital technology. Furthermore, there was a relationship between perceptions towards digital technologies and the user competences [10]. Perceptive usefulness is affected by digital technology characteristics relating to improved performance, effectiveness, productivity, control, and quality [11].

Kaur and Singh [12] found that factors such as farmers' age, education, and farming experience determined their perceptions towards digital technologies. However, gender, household size, land size, and income did not determine perceptions towards digital technology. These characteristics have an effect on the information received through the use of digital technologies [13]. Factors such as age, gender, marital status, education, and income were some of the socio-economic factors affecting perceptions towards digital technologies identified by Meijer et al. [14] and Pfeiffer [15]. According to Dissanayake [16], subjective norms affect the perceptions of farmers to technologies. In this instance, the relevance of how other farmers or neighbours feel about the technology and their attitude can help shape individual perceptions towards a technology. Technology usefulness and skills to be formed also help in shaping the perception towards a technology [16].

Even though various studies on the use of digital technologies have been conducted in South Africa, most have been adoption studies [5,17,18] with very few highlighting the factors affecting smallholder farmers' perceptions towards digital technologies. Some of these studies have also taken qualitative approaches, lacking in statistical rigour [19–22]. It is essential to acknowledge the perceptions of digital technologies by farmers to enable development of appropriate technologies [23]. This is informative to digital technology developers, extension officers, and research centres. This allows efficient budget and time allocations. It becomes imperative to also acknowledge the socio-economic circumstances of the farmers and how they shape their perception complex so that they can be taken on board in designing and disseminating digital technologies. According to Gerli et al. [24], transforming the perceptions towards digital technologies is more essential than the technology itself if there is going to be sustained adoption. Perception of digital technology is a barrier to its adoption [9]. The perceived characteristic of a digital technology determines its adoption. It is essential to understand farmer perceptions for effective development and dissemination. Farmer perceptions also help shape behaviour [16]. It is therefore against this background that this paper sought to determine the factors that influence the perceptions of smallholder farmers towards adoption of digital technologies in Port St Johns and Ingquza Hill Local Municipalities. In this study, empirical analysis will be used to analyse the factors that influence the perceptions of smallholder farmers towards adoption of digital technologies in the study area.

3. Materials and Methods

3.1. Description of Study Area

Port St Johns (PSJ) and Ingquza Hill (IH) Local Municipalities located in OR Tambo District Municipality of South Africa were the study sites that were used. Both Local Municipalities are Category B rural, having high poverty levels with reliance on social grants [25,26]. Agriculture accounts for 2.0% of Gross Value Added (GVA) in IH Local Municipality compared to 1.4% in PSJ. There are 53.9% of households engaged in agriculture in IH compared to 47.0% in PSJ, with PSJ also having a higher employment rate within agriculture at 5.4% compared to 4.1% in IH [25,26]. Port St Johns (PSJ) Local Municipality has a population of 166,779, with 18.9% having completed high school in 33,951 households on 1292.2 km^2 [27]. The median age of the population in PSJ is 17, with 54% of the population under the age of 18. Around 54.0% of the population in PSJ are female, with 59.0% of the households being female-headed [28]. The Local Municipality has 20 ward areas and 130 rural areas/villages [29]. Port St Johns (PSJ) has high smallholder maize production mainly through smallholder subsistence farmers with some using irrigation [30–32]. Ingquza Hill (IH) Local Municipality sits on 2479 km^2, housing 60,697 households with a population of 303,379 [33]. About 19.6% of the population in IH have completed secondary school and 14.0% are unemployed, with 31.0% of households having less than ZAR 10,000 annual income. The Local Municipality also has 54.0% of the population under the age of 18, with a mean age of 17, with 58.0% of the households being female-headed and 78.0% living in poverty [25,33]. The study areas were ideal because they are agro-based and characterised by the rural poor exhibiting high poverty rates.

3.2. Conceptual Framework

The study combined the Technology Acceptance Model (TAM) and the Attention, Interest, Desire, and Action (AIDA) model (Figure 1) [34]. The TAM investigates adoption decisions which are driven by behaviour, itself a construct of attitude towards the use of that technology. Attitude is a direct reflection of beliefs and perceptions in usefulness and ease of use. The AIDA model is one of the information-based rational choice models which show that users of digital technologies go through a series of cognitive and emotional steps in making a purchase and adoption decisions, or in a behaviour change process [34,35]. The steps involve attracting attention through creating interest (cognitive level), with the second step turning this interest into strong desire (affective level). The final step is taking action to move to that behaviour (behavioural level) [36,37]. The AIDA model prescribes agricultural digital technology provider behaviour in promoting their use by smallholder farmers. Greater competition amongst service providers, utilisation of multi-lingual, customised value-added services, and integration of Indigenous Knowledge were some of the service provider activities advocated by Maumbe [6] to enhance adoption of digital technologies by smallholder farmers in South Africa. In the current study, the cognitive and affective levels of the AIDA conceptual framework will shape the perceptions that smallholder farmers have towards digital technologies in the TAM. In conjunction with the socio-economic circumstances of the smallholder farmers, the perceptions will shape the behavioural level of the AIDA conceptual framework, enabling adoption of digital technologies. The current study was focused on the perceptions towards digital technologies, thereby focusing on the initial parts of the conceptual framework relating to perceived usefulness and perceived ease of use. This was inquired from the users of digital technologies from which an overall index was formulated, as will be indicated in Section 3.3.

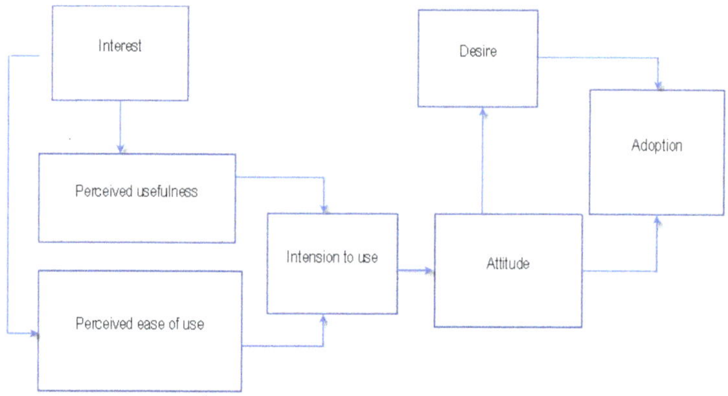

Figure 1. Conceptual framework combining the Technology Acceptance Model (TAM) and the Attention, Interest, Desire, and Action (AIDA) model. Source: Geng [34].

3.3. Study Design

A cross-sectional survey of a purposively selected sample of 250 smallholder farmers was used in the study. A Likert scale, Perceptive Index, and Tobit regression were used to analyse the data. The study used a Likert scale to measure the level of awareness and perception towards digital technologies by smallholder farmers. The 5-point scaled items that were used in the study are shown in Table 1. The respondents were asked to what extent they agreed with the perceptive questions. Internal validity was improved by pre-testing the questionnaire so that the questions measured what was intended, which was the perception, and any ambiguity in the questions was rectified. This reduced confusion and skip patterns.

Table 1. Perceptions of respondents towards digital technologies.

Perceptions of Adoption of Digital Technologies
Adoption of digital technologies can make farming easier
I have adequate knowledge of digital technologies
Use of digital technologies will be labour-saving
Use of digital technologies improves agricultural production
Through digital technologies smallholder farmers access information on time
Through digital technologies farmers access extension services easily
The use of digital technologies helps smallholder farmers to access the market
It is easy to access farm loans through digital technologies
Digital technologies are expensive compared to other agricultural innovations
The use of agricultural digital technologies improved household income
Digital technologies are user-friendly
Digital technologies are complicated
Digital technologies are the cause of the digital divide between smallholder and commercial farmers
Unequal access to digital technologies exists among smallholders
Digital technologies will discourage the use of Indigenous Knowledge and skills
All the digital technologies are suitable for smallholder farms
Use of digital technology will increase smallholder farmers' farming output
Use of digital technologies requires specific skills

Source: Authors own conceptualisation informed by the literature.

The Min-Max Normalisation as used by Ngarava [38] was used to standardise the perceptions towards digital technologies by smallholder farmers. The Min-Max Normalisation was used to produce an indicator which fell in the range of 0–1, using the following formula:

$$PI_{qi} = \frac{P_{qi(obs)} - P_{qi(min)}}{P_{qi(max)} - P_{qi(min)}}.$$

where PI_{qi} is the Perceptive Index of question i, $P_{qi(obs)}$ is the observed value of perceptive question i, $P_{qi(min)}$ is the global minimum value of question i (=1), and $P_{qi(max)}$ is the global maximum value of question i (=5). The overall PI_{qi} for each respondent was:

$$PI_{overall(j)} = \frac{\sum_{i=1}^{n} PI_{qi}}{n}$$

where n is the number of perception questions, which is 18.

The Tobit model used was modelled as follows [39–42]:

$$PI_i^* = x_i'\beta + \varepsilon_i,\ \varepsilon_i \sim N\left(0, \sigma^2\right)$$

$$PI_i^* = \begin{cases} PI_i^*,\ if\ PI_i^* > 0 \\ 0,\ if\ PI_i^* \leq 0 \end{cases}$$

where PI_i^* is the latent variable, which can only be observable when the values are greater than 0. Explanatory variables are depicted by x_i with a vector β and ε_i is the error term which is normally distributed. The following log-likelihood function L was maximised by estimating β and σ:

$$\max_{\beta,\sigma} \ln L = \sum_{PI_i>0} \frac{1}{\sigma}\rho\left(\frac{PI_i - x_i'\beta}{\sigma}\right) + \sum_{PI_i=0} \ln\left[1 - \tau\left(\frac{x_i'\beta}{\sigma}\right)\right]$$

where τ is the standard normal cumulative distribution function, ρ is the matching density function. The variables in Table 2 were used in the Tobit model.

Table 2. Variables used in the Tobit regression model for factors affecting the perceptions towards digital technologies by smallholder farmers in PSJ and IH Local Municipalities.

Variable	Explanation	Measurement	Expected Sign
Dependent			
PI	Perceptive Index	Truncated: 0 (negative)–1 (positive)	
Independent			
GEN	Gender	Nominal: 0—male, 1—female	−
AGE	Age (Years)	Nominal: 0—less than 40 years, 1—otherwise	−
MARST	Marital status	Nominal: 0—married, 1—not married	−
EDU	Education level	Nominal: 0—none, 1—otherwise	+
EMPL	Employment status	Nominal: 0—full-time farmer, 1—part-time farmer	−
SOUINC	Source of income	Nominal 0—social grant, 1—otherwise	+/−
MI	Monthly income (ZAR)	Nominal: 0—less than 1000, 1—otherwise	+
HHS	Household size	Nominal: 0—up to 5, 1—otherwise	+/−
FEN	Farming enterprise	Nominal: 0—crop production, 1—otherwise	+/−
TEN	Tenure	Nominal: 0—communal, 1—leased	+
FEX	Farming experience (Years)	Nominal: 0—less than 5 years, 1—otherwise	+
TR	Training	Nominal: 0—yes, 1—no	−
COOP	Part of cooperative member	Nominal: 0—yes, 1—no	−

The Tobit model was defined as follows:

$$PI = \beta_0 + \beta_1 GEN + \beta_2 AGE + \beta_3 MARST + \beta_4 EDU + \beta_5 SOUINC + \beta_6 MI + \beta_7 HHS + \beta_8 FEN + \beta_9 TEN + \beta_{10} FEX + \beta_{11} TR + \beta_{12} COOP + \varepsilon$$

4. Results

Empirical Results

Figure 2 shows the perceptions towards digital technologies by smallholder farmers in Port St Johns (PSJ) and Ingquza Hill (IH) Local Municipalities. The Cronbach's alpha value of 0.70 shows that the Likert measures of perceptions were reliable and valid. There was strong agreement with digital technologies being expensive compared to other agricultural technologies (48.4%), improves agricultural production, (43.2%), they are a cause of a digital divide (37.6%), and discourage use of Indigenous Knowledge (33.2%). However, there was strong disagreement to smallholder farmers having knowledge about digital technologies (42.0%).

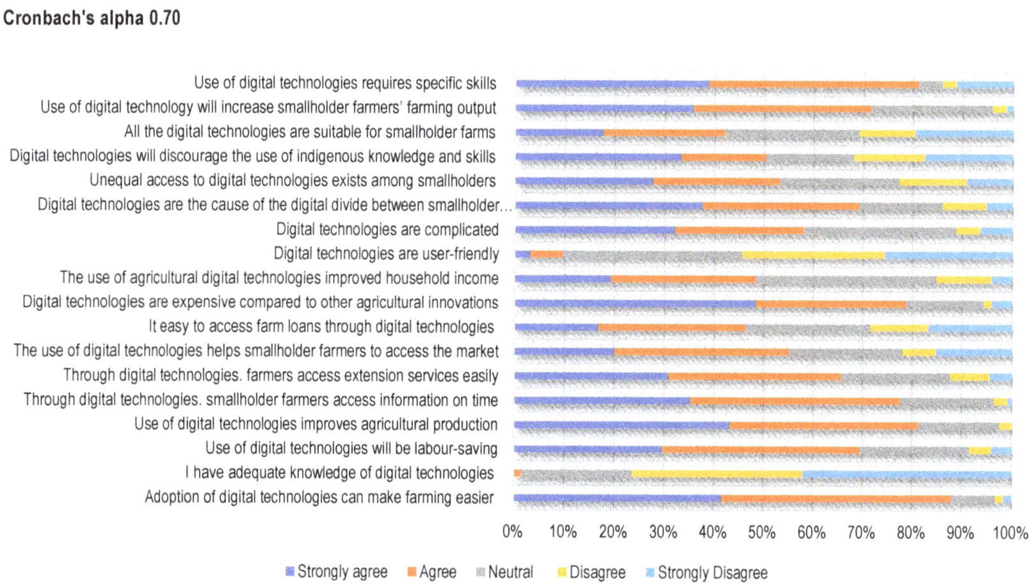

Figure 2. Perceptions towards digital technologies by smallholder farmers in Port St Johns and Ingquza Hill Local Municipalities.

The highest positive perception towards digital technologies for smallholder farmers in PSJ and IH Local Municipalities was associated with cattle rearing (23.53%), a combination of cattle and goat rearing (21.43%), and maize production (12.32%) (Figure A1). However, a certain level of negative perception was associated with all pig and chicken producers, as well as smallholder farmers involved with a combination of cattle and sheep rearing. In addition, extremely negative perceptions towards digital technologies by smallholder farmers had association with all farmers involved with rearing a combination of sheep and goats, a combination of cattle, sheep, and goats (44.0%), rearing goats (29.27%), and maize production (20.20%), respectively.

Extremely negative perceptions towards digital technologies by smallholder farmers in PSJ and IH Local Municipalities had association with 2–4 hectares of crop production (29.35%) and less than 1 hectare of animal rearing (25.66%) (Figure A2). Some level of positive perception towards digital technologies was associated with utilisation of less than 1 hectare of crop production (15.57%), 5–10 hectares (10.0%), and 2–4 hectares (7.61%). Positive perception was also associated with 2–4 hectares (13.33%) and less than 1 hectare (7.96%) of land use for animal rearing.

Figure A3 shows that extremely negative perceptions towards digital technologies by smallholder farmers in PSJ and IH Local Municipalities were associated with enterprises that sell less than 50 kg (42.86%), 150–200 kg (36.36%), and 250–300 kg (31.82%) of maize, respectively. Close to 23.08% of smallholder farmers who sell 150–200 kg and 15.15% of over 200 kg of cabbages have extremely negative perceptions towards digital technologies. Positive perception was exhibited by 27.27% of smallholder farmers that sell 150–200 kg of maize and 27.78% who sell less than 50 kg of cabbages.

There was association between positive perceptions of digital technologies and consumption of more than 200 kg (62.86%), 50–100 kg (9.80%), and 150–200 kg (9.38%) of maize, respectively (Figure A4). There was association with consumption of 50–100 kg (16.67%) and over 200 kg (12.50%) of cabbages. Most negative perceptions were associated with consumption of 50–100 kg of maize for 27.45% of smallholder farmers and consumption of less than 50 kg for 24.05%.

Table 3 shows the significance of the association between the perceptions towards digital technologies and the farming enterprise characteristics. There was a significant association between level of perception towards digital technologies and the type of livestock that are reared (1% level), land size that is devoted to crop and vegetable production (5% level), and land tenure (10% level). The low Spearman correlations (between 0.03 and 0.19) indicate very low levels of association.

Table 3. Cross tabulation of Port St Johns and Ingquza Hill Local Municipality farmers' perceptions towards digital technologies.

		Pearson χ^2	Spearman Correlation	Cramer's V	p-Value
Maize	Production	2.61	0.09	0.11	0.27
	Land size	5.40	−0.10	0.11	0.49
	Sale	17.89	−0.25	0.26	0.12
	Consumption	6.86	−0.08	0.13	0.33
Cabbage	Production	4.02	0.15	0.17	0.13
	Land size	1.92	−0.04	0.08	0.75
	Sale	10.13	0.16	0.27	0.12
	Consumption	8.50	−0.16	0.17	0.20
Livestock kept		35.93 ***	0.19	0.36	0.00
Total land size for crop and vegetables		14.88 **	0.14	0.18	0.02
Total land size for livestock		2.48	−0.13	0.13	0.29
Tenure of agricultural land		5.51	−0.03	0.15	0.06

, * represents 0.05 and 0.01 levels.

Table 4 shows the factors affecting the perception towards digital technologies by smallholder farmers from PSJ and IH Local Municipalities. The model was significant at the 1% level; however, with a low R^2 value indicating that there were other variables that were excluded from the model that affect the level of perceptions towards digital technologies. The results show that educational level (1% level), age, employment status, being part of a cooperative (5% level), source of income, monthly income, and household size (10%) were significant factors affecting the perception towards digital technologies by smallholder farmers.

Table 4 shows that an increase in age is associated with an increase in negative perceptions towards digital technologies by smallholder farmers in PSJ and IH Local Municipalities. Surprisingly, an increase in the educational levels were associated with an increase in the negative perceptions. The employment status results indicate that part-time farmers from PSJ and IH Local Municipalities are associated with negative perceptions towards digital technologies. An increase in monthly income increases the positive perception towards digital technologies by smallholder farmers, while an increase in the household size is associated with negative perceptions towards digital technologies. If smallholder farmers from PSJ and IH are not part of a farming cooperative, they are likely to have a negative perception towards digital technologies as exhibited.

Table 4. Factors affecting perceptions towards digital technologies by smallholder farmers in Port St Johns and Ingquza Hill Local Municipalities.

Variable	β	*Std Err*	*t*	$p > \|t\|$
Gender	0.55	1.47	0.35	0.66
Age	−1.24 **	0.79	−1.65	0.03
Marital status	−0.59	0.90	−0.72	0.50
Education level	−3.22 ***	0.95	−3.13	0.01
Employment status	−1.74 **	0.54	−2.97	0.04
Source of income	0.93 *	0.50	1.96	0.09
Monthly income	2.99 *	1.13	2.83	0.07
Household size	−2.96 *	0.98	−2.79	0.08
Farming enterprise	0.74	0.71	1.00	0.26
Tenure	−5.21	3.26	−1.59	0.14
Farming experience	−1.03	1.28	−0.84	0.44
Training	−0.44	0.50	−0.70	0.45
Part of cooperative member	−4.14 **	1.42	−2.85	0.04
Constant	78.91	4.80	18.33	0.03
		Summary statistics		
Sigma	15.43	0.62		
χ^2	45.45			
$p > \chi^2$	0.00			
Pseudo R^2	0.13			

*, **, *** represents 0.1, 0.05, and 0.01 levels, respectively.

5. Discussion

The results showed that digital technologies were regarded as expensive, cause a digital divide, and discourage use of Indigenous Knowledge, with farmers having little knowledge about them even though they improve agricultural production. Migiro and Kwake [43], Dlamini and Ocholla [5], as well as Mabaya and Porciello [20] concurred that the use of digital technologies was expensive. Strydom [44] attests that in developing countries, Africa included, the requirement of some digital technologies to access internet connectivity has increased their costs, making them unreachable to many. Thus, only 20% of the population in SSA have access to the internet [44]. This has been compounded by the profit-targeting private sector who have supplied and offered services to these digital technologies. This can work against smallholder farmers as big corporate takeovers consolidate the smaller farms [45,46]. However, this can be offset by the cost reductions that will be envisaged through using the digital technologies such as transaction costs, especially for profit-oriented small-scale farmers relative to subsistence-targeted farmers [47,48]. In addition, the expense of digital technologies has also been attributed to ancillary inputs, which are also expensive, such as electricity [44], servicing [49], and data [50,51]. However, authors such as McCampbell et al. [52] indicate that it is not the cost but rather the neglect of a bottom-up approach and social, political, and economic injustices in the design of the technologies. This will further exacerbate the digital divide.

Engås, Raja, and Neufang [53] indicated that the digital divide led to low adoption of digital technologies. In order to maximise the opportunities presented by digital technologies, there is a need to consider the digital divide in all three levels of capacities, competences, and inequality in access [54,55]. McCampbell, Schumann, and Klerkx [52] referred to the digital divide as a result of structural ethical issues, for example, limits in skills and access. According to Hackfort [56], the digital divide is deepened by digitalisation between small and large farms, as well as between farmers who are willing and unwilling to purchase digital technologies. Digital technologies need to be inclusive to reduce the risk of widening the digital divide through the exclusion of smallholder farmers based on their characteristics and circumstances [57]. However, according to Oosterlaken [58], the capability approach as prescribed by Sen [59,60] can inform the development and technologies that provide context in the form of human dignity and rights. In the current study, what appears as a digital divide might be informed by opportunities available for the smallholder farmers to value their way of doing things. This is based on their own capabilities.

Maumbe [6] highlighted that lack of integration with Indigenous Knowledge (IK) was a drawback in using digital technologies, with Dlamini and Ocholla [5] highlighting that instead of replacing Indigenous Knowledge, digital technologies can actually be used in conjunction, for instance, to record, store, and disseminate Indigenous Knowledge. Indigenous Knowledge management using digital technology is in its infancy in Africa, with language being costly and acting as a deterrent [61,62]. There should be prioritisation of improving food security through Indigenous Knowledge, which should be preserved, disseminated, and used. Kantiza et al. [63] highlighted the use of IK systems in Africa and that there is need to disseminate the good IK practices into areas where they are not currently available. This can be achieved through digital technologies. Franco et al. [64] go on to highlight that farmers themselves are able to adapt existing technologies to suit their contextual needs. This improves the appropriateness of technologies.

The results showed that livestock production on 2–4 hectares as well as maize crop production on less than 1 hectare were associated with positive perceptions towards the use of digital technologies. This was contrary to Pope and Sonka [65], who found economies of scale being associated with digital technology adoption. According to Bronson and Knezevic [66], various land sizes require different types of technologies. Farm sizes were highlighted by Annosi et al. [67] to have an effect on the use of digital technologies, with Tagarakis et al. [68] highlighting that utilisation of digital technology was limited by small farm sizes. Groher, Heitkämper, and Umstätter [69] indicated that new solutions that can be brought about by digital technologies are required for animal and environmentally friendly production systems, with adoption differing according to animal species. Furthermore, Kernecker et al. [70] found that crop farming systems had more adoption of digital technologies. Compared to the literature, this study reveals that there is a digital technology and farmer characteristic relationship, with small farms associated with low-tech technologies relative to large farms.

Smallholder farmers who sold 150–200 kg and consumed more than 200 kg of maize had positive perceptions towards the use of digital technologies. This was also true for smallholder farmers who sold less than 50 kg and consumed less than 50 kg of cabbages. According to Giua, Materia, and Camanz [71] as well as Omulo and Kumeh [72], digital technologies are useful when selling agricultural products, especially to obtain market and price information to increase their bargaining power. Expansion of the smallholder market will allow the use of more sophisticated digital technologies not just to market their produce but also to increase production to meet the market demand.

Increases in age, educational levels, household size, being part-time farmers, decreases in monthly income, and not being part of a farming cooperative were associated with negative perceptions towards digital technologies by smallholder farmers. da Silveira et al. [73] indicated that older farmers exhibit negative perceptions towards technologies due to resistance to change. They are sceptical due to lived experiences, including Indigenous Knowledge systems. Despite the fact that increases in age and educational levels improve digital technology literacy [74], surprisingly, an increase in education was associated with negative perceptions towards digital technologies. Firstly, it is worth noting that only 9.9% of the smallholder farmers in the study area had educational levels beyond secondary school. Secondly, the nature of the digital technologies used precludes the advantages of higher educational levels. Schulze Schwering, Bergmann, and Isabel Sonntag [75] attest that openness to technologies is related to educational levels. However, in the current study, the digital technologies being used are low tech and for everyday use, rendering higher education unnecessary. In fact, since the digital technologies are low tech, smallholder farmers with higher educational levels develop a negative perception. An increase in household size translates into an increase in digital technology-related costs such as data, especially given that most technologies being used in the study area include mobile phones. Thus, the smallholder farmers would develop a negative perception towards digital technologies as the household sizes increased with their associated costs. Gabriel and Gandorfer [76] indicated that practicing farming on a part-time basis is an exhibition and often results in less motivation, and reduced access to capital and time invested in the farming enterprise, which increases economic risk. Combined with the financial risk involved in digital technology investments, part-time smallholder farmers will be sceptical and have a negative perception towards digital technologies. The role of income in digital technology is a reinforcing one. An increase in disposable income allows access to digital technologies which results in positive perceptions towards digital technologies, and due to higher income allows access to improved digital technologies. Hence, there is a perpetual positive perception due to an increase in income. Farming cooperatives allow farmers to share ideas and information about digital technologies, thereby creating perceptions towards digital technologies. According to Kvam, Hårstad, and Stræte [77], farming cooperatives should play a central role in the adoption of digital technologies.

6. Conclusions

This study sought to determine the factors that influence the perceptions of smallholder farmers towards adoption of digital technologies. The conceptual framework used comprised a combination of the Technology Acceptance Model (TAM) and the Attention, Interest, Desire, and Action (AIDA) model. The Likert scale, Perceptive Index (PI), and a Tobit regression were used to analyse the data. The study showed that there were strong perceptions towards cost, inequality, and lack of adaptation to local knowledge exhibited by digital technologies, even though they increased production. Furthermore, a positive perception towards digital technologies was exhibited by smallholder farmers using small pieces of land for their agricultural enterprises, as well as those engaged in small-scale sales and marketing. This was informed and reinforced the use of low-tech digital technologies. Socio-economic factors such as age, educational levels, household size, being part-time farmers, monthly income, and being part of a farming cooperative affected the level of perception towards digital technologies by smallholder farmers in PSJ and IH Local Municipalities. This study concludes that there are socio-economic factors affecting the perception of digital technologies by smallholder farmers. All combined, these have influenced the use of low-tech digital technologies in PSJ and IH Local Municipalities.

7. Recommendations

Some of the recommendations that can be envisaged from this study include availing low-cost high-tech digital technologies for smallholder farmers in PSJ and IH Local Municipalities. The central government can subsidise the use of high-tech digital technologies for use by smallholder farmers. Lobbying for low-cost digital technologies from suppliers can also be used to avail low-cost technologies. This can also be a strategy used for ancillary and complementary services to access data, internet, and after-sales services. Another strategy that can be used to encourage the use of high-tech digital technologies is integrating the current low tech with high tech. Digital technology providers can be lobbied to take this route as the costs will be a compromise between the low and high ends. Availing cheaper options of digital technologies can avert the inequality that is associated with the use of digital technologies. Promotion of digital technology, especially to poor smallholder farmers, can also reduce the digital divide. Extension officers can play a significant role, as they are already embedded within the communities and have a database of not well-off farmers. In conjunction with traditional leaders and opinion leaders, as well as extension officers, there should be lobbying for technology developers to provide digital technologies that cater for Indigenous Knowledge. Central government funding can play a pivotal role in the development and dissemination of such technologies. Digital technologies that are promoted should also be cognisant of the small nature of the enterprises. In promoting the use of digital technologies, be it through print, electronic and social media, and even though extension, the target should be the youth and young farmers with less education in small households who are full-time farmers with moderate-to-high incomes and are part of farmer groups/organisations. This demographic already shows positive perceptions and thus will quickly embrace the use of digital technologies in PSJ and IH Local Municipalities. This however does not preclude the other demographics in the promotion of digital technologies. Areas of further study include focusing on the determinants of the component parts of interest, perceived usefulness, perceived ease of use, and intention to use as spelt out in the TAM and AIDA models.

8. Limitations of Study

The study had conceptual and methodological limitations. The perceptions could have been grouped according to the TAM and AIDA models to provide categorised factors affecting interest, perceived usefulness, perceived ease of use, and intention to use. Furthermore, external validity was limited because the study used a purposive design which was spatially and temporally limited. Any findings are difficult to extrapolate to other contexts.

Author Contributions: Conceptualisation, N.V.B., A.M. and S.N.; methodology, N.V.B.; validation, S.N. and A.M.; formal analysis, N.V.B. and S.N.; investigation, N.V.B.; writing—original draft preparation, N.V.B.; writing—review and editing, N.V.B.; visualisation, N.V.B.; supervision, A.M. and S.N.; project administration, N.V.B. All authors have read and agreed to the published version of the manuscript.

Funding: This research was funded through National Research Foundation (NRF), Grant number 75910.

Institutional Review Board Statement: Ethical clearance was obtained from the University of Fort Hare Research Ethics Committee, with reference number: MUS041SBON01.

Data Availability Statement: The data that support the findings of this study are available from the corresponding author, N.V.B., upon reasonable request.

Conflicts of Interest: The authors declare no conflict of interest.

Appendix A

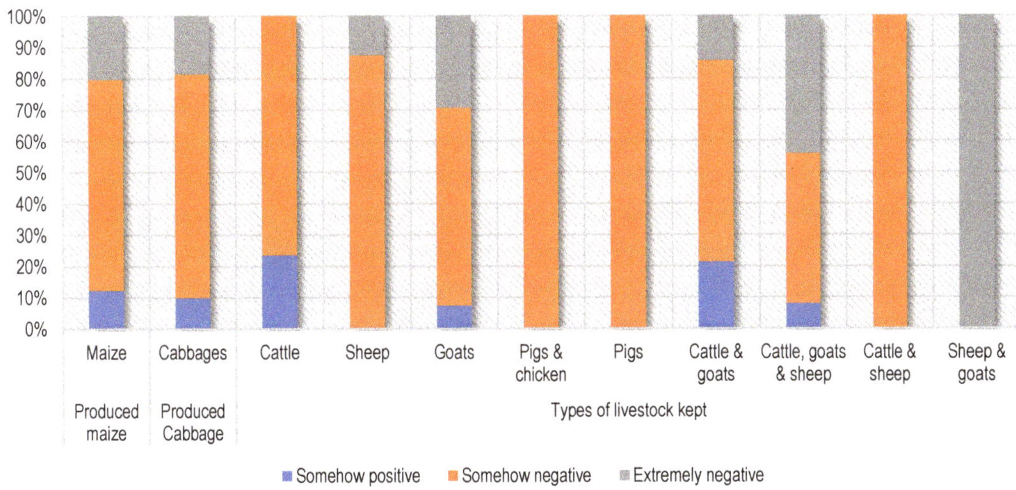

Figure A1. Perceptions towards digital technologies and the farming enterprises for smallholder farmers in Port St Johns and Ingquza Hill Local Municipalities.

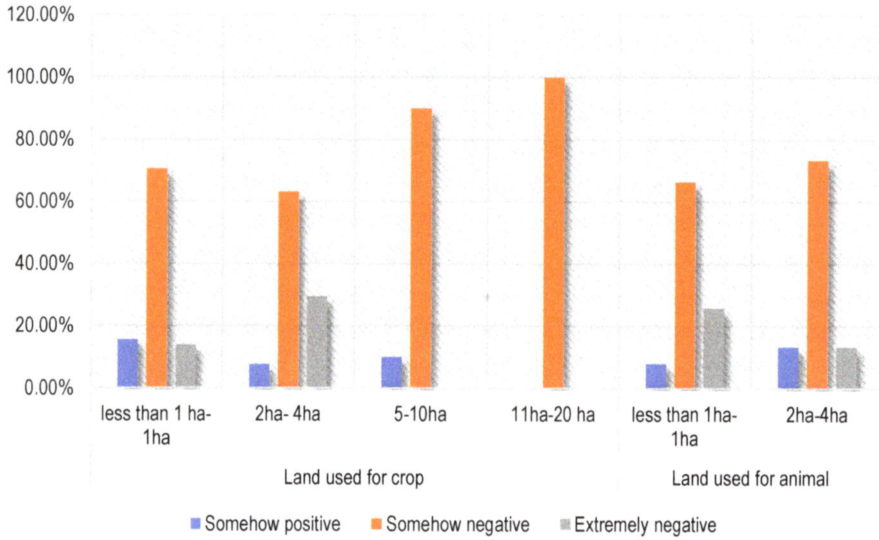

Figure A2. Perceptions towards digital technologies and farming land size for smallholder farmers in Port St Johns and Ingquza Hill Local Municipalities.

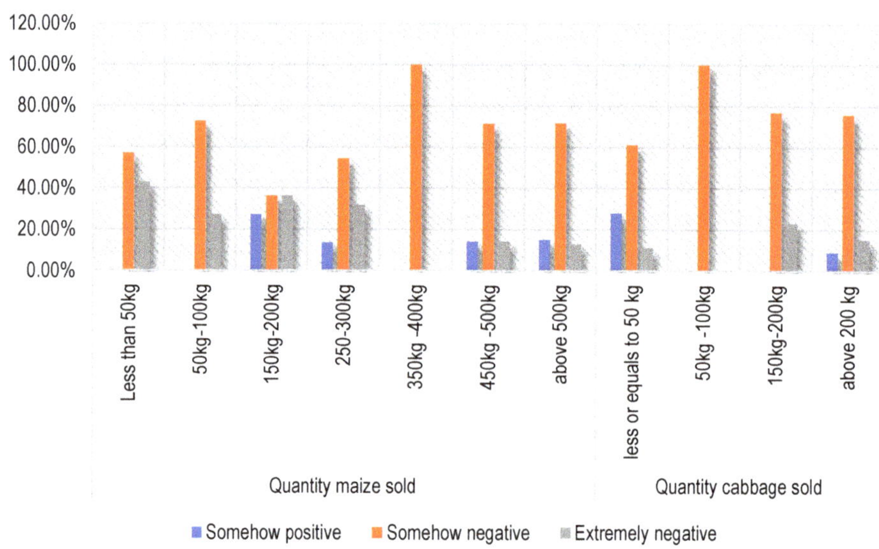

Figure A3. Perceptions towards digital technologies vis-à-vis propensity to sell maize and cabbage by smallholder farmers in Port St Johns and Ingquza Hill Local Municipalities.

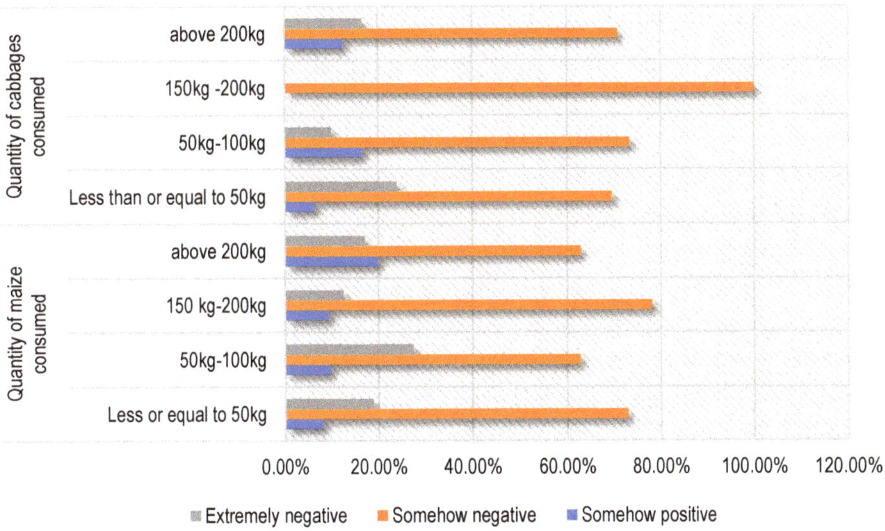

Figure A4. Perceptions towards digital technologies vis-à-vis propensity to consume maize and cabbage by smallholder farmers in Port St Johns and Ingquza Hill Local Municipalities.

References

1. Qin, T.; Wang, L.; Zhou, Y.; Guo, L.; Jiang, G.; Zhang, L. Digital technology-and-services-driven sustainable transformation of agriculture: Cases of China and the EU. *Agriculture* **2022**, *12*, 297. [CrossRef]
2. FAO; ITU. *Status of Digital Agriculture in 47 Sub-Saharan African Countries*; FAO: Rome, Italy, 2022.
3. OECD; FAO. Agriculture in Sub-Saharan Africa: Prospects and challenges for the next decade. In *OECD-FAO Agricultural Outlook 2016–2025*; OECD; FAO: Rome, Italy, 2016; Volume 181, pp. 1–95. ISBN 9789264253223.
4. Jayne, T.S.; Sanchez, P.A. Crop Yields on Existing Farmland. *Science* **2021**, *372*, 1045–1048. [CrossRef] [PubMed]
5. Dlamini, P.; Ocholla, D.N. Information and Communication Technology Tools for Managing Indigenous Knowledge in KwaZulu-Natal Province, South Africa. *Afr. J. Libr. Archeoloyg Inf. Sci.* **2018**, *28*, 137–153.
6. Maumbe, B. Mobile agriculture in South Africa: Implementation framework, value-added services and policy implication. *Int. J. ICT Res. Dev. Afr.* **2010**, *1*, 35–59. [CrossRef]
7. Pishnyak, A.; Khalina, N. Perception of new technologies: Constructing an innovation openness index. *Foresight STI Gov.* **2021**, *15*, 39–55. [CrossRef]
8. Caffaro, F.; Micheletti Cremasco, M.; Roccato, M.; Cavallo, E. Drivers of farmers' intention to adopt technological innovations in Italy: The role of information sources, perceived usefulness, and perceived ease of use. *J. Rural Stud.* **2020**, *76*, 264–271. [CrossRef]
9. Thomas, R.J.; O'Hare, G.; Coyle, D. Understanding technology acceptance in smart agriculture: A systematic review of empirical research in crop production. *Technol. Forecast. Soc. Chang.* **2023**, *189*, 122374. [CrossRef]
10. Reissig, L.; Stoinescu, A.; Mack, G. Why farmers perceive the use of e-government services as an administrative burden: A conceptual framework on influencing factors. *J. Rural Stud.* **2022**, *89*, 387–396. [CrossRef]
11. Adrian, A.M.; Norwood, S.H.; Mask, P.L. Producers' perceptions and attitudes toward precision agriculture technologies. *Comput. Electron. Agric.* **2005**, *48*, 256–271. [CrossRef]
12. Kaur, A.; Singh, R. Perception and Attitude of Agripreneurs toward Social Media Tools for Attaining Agribusiness Benefits. *Indian J. Posit. Psychol.* **2021**, *12*, 379–384.
13. Aldosari, F.; Al Shunaifi, M.S.; Ullah, M.A.; Muddassir, M.; Noor, M.A. Farmers' perceptions regarding the use of Information and Communication Technology (ICT) in Khyber Pakhtunkhwa, Northern Pakistan. *J. Saudi Soc. Agric. Sci.* **2019**, *18*, 211–217. [CrossRef]
14. Meijer, S.S.; Catacutan, D.; Ajayi, O.C.; Sileshi, G.W.; Nieuwenhuis, M. The role of knowledge, attitudes and perceptions in the uptake of agricultural and agroforestry innovations among smallholder farmers in sub-Saharan Africa. *Int. J. Agric. Sustain.* **2015**, *13*, 40–54. [CrossRef]
15. Pfeiffer, J.; Gabriel, A.; Gandorfer, M. Understanding the public attitudinal acceptance of digital farming technologies: A nationwide survey in Germany. *Agric. Hum. Values* **2021**, *38*, 107–128. [CrossRef]
16. Dissanayake, C.A.K.; Jayathilake, W.; Wickramasuriya, H.V.A.; Dissanayake, U.; Kopiyawattage, K.P.P.; Wasala, W.M.C.B. Theories and models of technology adoption in agricultural sector. *Hum. Behav. Emerg. Technol.* **2022**, *2022*, 9258317. [CrossRef]

17. Mdoda, L.; Mdiya, L. Factors affecting the using information and communication technologies (ICTs) by livestock farmers in the Eastern Cape Province. *Cogent Soc. Sci.* **2022**, *8*, 2026017. [CrossRef]
18. Makaula, Z. Information and communication technologies (ICT) towards agricultural development in rural areas: Case of smallholder farmers in Umzimvubu local municipality of the Eastern Cape Province in South Africa. *S. Afr. J. Agric. Ext.* **2021**, *49*, 81–90. [CrossRef]
19. Akinsola, O.S. ICT adoption for bridging South African black farmers ' knowledge gap. *Int. J. Agric. Sci. Technol.* **2014**, *2*, 39–47. [CrossRef]
20. Mabaya, E.; Porciello, J. Can digital solutions transform agri-food systems in Africa? Can digital solutions transform agri-food systems in Africa? *Agrekon* **2022**, *61*, 67–79. [CrossRef]
21. Mavilia, R.; Pisani, R. Blockchain for agricultural sector: The case of South Africa. *African J. Sci. Technol. Innov. Dev.* **2022**, *14*, 845–851. [CrossRef]
22. Zantsi, S.; Nkunjana, T. A review of possibilities for using animal tracking devices to mitigate stock theft in smallholder livestock farming systems in rural South Africa. *S. Afr. J. Agric. Ext.* **2021**, *49*, 162–182. [CrossRef]
23. Cavallo, E.; Ferrari, E.; Bollani, L.; Coccia, M. Attitudes and behaviour of adopters of technological innovations in agricultural tractors: A case study in Italian agricultural system. *Agric. Syst.* **2014**, *130*, 44–54. [CrossRef]
24. Gerli, P.; Clement, J.; Esposito, G.; Mora, L.; Crutzen, N. The hidden power of emotions: How psychological factors influence skill development in smart technology adoption. *Technol. Forecast. Soc. Chang.* **2022**, *180*, 121721. [CrossRef]
25. ECSECC. *Ingquza Hill Local Municipality Socio Economic Review and Outlook, 2017*; ECSECC: East London, South Africa, 2017. Available online: https://www.ecsecc.org/documentrepository/informationcentre/ngquza-hill-local-municipality_35291.pdf (accessed on 2 December 2022).
26. ECSECC. *Port St Johns Municipality Socio-Economic Review and Outlook*; ECSECC: East London, South Africa, 2017. Available online: https://www.ecsecc.org/documentrepository/informationcentre/port-st-jonhs-local-municipality_82667.pdf (accessed on 14 January 2023).
27. Wazimap Port St Johns. 2022. Available online: https://wazimap.co.za/profiles/municipality-EC154-port-st-johns/ (accessed on 1 December 2022).
28. GoSA Vaccine FAQs. 2021. Available online: https://www.gov.za/covid-19-coronavirus-vaccine-frequently-asked-questions (accessed on 13 July 2021).
29. OR Tambo DM. *Coastal Management Programme 2016–2021*; OR Tambo DM: Mthatha, South Africa, 2016.
30. Kambanje, A.; Ngarava, S.; Mushunje, A.; Taruvinga, A. Labour dynamics in climate and techno reliand small scale maize production. *J. Econ. Behav. Stud.* **2018**, *10*, 262–276. [CrossRef] [PubMed]
31. Kambanje, A.; Taruvinga, A.; Mushunje, A.; Mutengwa, C.; Ngarava, S. Determinants of Food Security Status amongst Smallholder Farmers Utilizing Different Maize Varieties in OR Tambo District, South Africa. *J. Soc. Sci. Res.* **2020**, *6*, 133–139. [CrossRef]
32. Ngcinela, S.; Mushunje, A.; Taruvinga, A.; Ngarava, S.; Mutengwa, C.S. Determinants of genetically modified (GM) maize adoption and the intensity of adoption in OR Tambo District Municipality, Eastern Cape Province, South Africa. *GM Crop. Food* **2019**, *10*, 1–11. [CrossRef] [PubMed]
33. Wazimap Ingquza Hill Local Municiality. 2022. Available online: https://wazimap.co.za/profiles/municipality-EC153-ngquza-hill/ (accessed on 1 December 2022).
34. Geng, L.; Li, Y.; Xue, Y. Will the interest triggered by virtual reality (VR) turn into intention to travel (VR vs. Corporeal)? The moderating effects of customer segmentation. *Sustainability* **2022**, *14*, 7010. [CrossRef]
35. Erdogdu, A.I. Social Marketing of Post-Pandemic Teleworking to Improve the Well-Being of White-Collar Workers in Instanbul. Master's Thesis, Middle East Technical University, Çankaya, Ankara, 2021.
36. Rawal, P. AIDA Marketing Communication Model: Stimulating a Purchase. *IRC's Int. J. Multidiscip. Res. Insocial Manag. Sci.* **2013**, *1*, 37–44.
37. Le, T.M.; Liaw, S.; Bui, T. The role of perceived risk and trust propensity in the relationship between negative perceptions of applying big data analytics and consumers' responses. *WEAS Trans. Bus. Econ.* **2020**, *17*, 426–435. [CrossRef]
38. Ngarava, S.; Mushunje, A.; Chaminuka, P. Qualitative Benefits of Livestock Development Programmes. Evidence from the Kaonafatso ya Dikgomo (KyD) Scheme in South Africa. *Eval. Program Plann.* **2020**, *78*, 101722. [CrossRef] [PubMed]
39. Michels, M.; Musshoff, O. A tobit regression model for the timing of smartphone adoption in agriculture. *Heliyon* **2022**, *8*, e11272. [CrossRef] [PubMed]
40. Tobin, B.Y.J. Estimation of relationships for limited dependent variables. *Econometrica* **1985**, *26*, 24–36. [CrossRef]
41. Greene, W.H. Econometric Analysis. In *Intellectual Property Management*; Physica-Verlag HD: Heidelberg, Germany, 2018. [CrossRef]
42. Wooldridge, J.M. *Introductory Econometrics*, 6th ed.; Cengage Learning: Boston, MA, USA, 2016; Volume 42; ISBN 9781305270107.
43. Migiro, S.O.; Kwake, A. Information needs and communication technology adoption in Africa: A comparative study of rural women in Kenya and South Africa. *J. Soc. Dev. Afr.* **2007**, *22*, 109–141.
44. Strydom, P.D.F. Digital technologies, employment and labour-market polarisation in Africa. In *Africa's Digital Future: From Theory to Action (The Future of International Trade and Development)*; Viviers, W., Parry, A., Jansen van Rensburg, S.J., Eds.; AOSIS: Cape Town, South Africa, 2021; pp. 101–127.

45. Patil, M.D.; Anatha, K.H.; Wani, S.P. Digital technologies for agricultural extension. In *Harnessing Dividends from Drylands: Innovative Scaling up with Soil Nutrients*; Raju, K.V., Wani, S.P., Eds.; CABI Digital Library: Oxfordshire, UK, 2016; pp. 331–334.
46. Jeanneaux, P. Towards the End of Farmers' Decision-Making Autonomy? 2018. Available online: https://agris.fao.org/agris-search/search.do?recordID=FR2021007992 (accessed on 4 October 2022).
47. Fourati-jamoussi, F.; Dantan, J.; Dubois, M.J.F.; Rizzo, D. Impact of Digital Technologies on Farms' Business Models—CahierCOSTECH. 2023. Available online: https://hal.science/hal-04010330v1/file/CosTECH-2023-02-Fourati-Jamoussi_et-al-impact-digital-technologies-farm-business-model.pdf (accessed on 7 February 2023).
48. Leeuwis, C.; Aarts, N. *Rethinking Adoption and Diffusion as a Collective Social Process: Towards an Interactional Perspective*; Campos, H., Ed.; Springer: Berlin/Heidelberg, Germany, 2020; ISBN 9783030509910.
49. Ayamga, M.; Lawani, A.; Akaba, S.; Birindwa, A. Developing institutions and inter-organizational synergies through digitalization and youth engagement in african agriculture: The case of "Africa Goes Digital". *Land* **2023**, *12*, 199. [CrossRef]
50. Sidibé, A.; Olabisi, L.S.; Doumbia, H.; Touré, K.; Niamba, C.A. Barriers and enablers of the use of digital technologies for sustainable agricultural development and food security: Learning from cases in Mali. *Elementa* **2021**, *9*, 00106. [CrossRef]
51. Born, L.; Chirinda, N.; Mabaya, E.; Afun-Ogidan, O.; Girvetz, E.H.; Jarvis, A.; Kropff, W. *Digital Agriculture Profile: South Africa*; FAO: Rome, Italy, 2021. Available online: https://www.fao.org/3/cb2506en/cb2506en.pdf (accessed on 14 January 2023).
52. McCampbell, M.; Schumann, C.; Klerkx, L. Good intentions in complex realities: Challenges for designing responsibly in digital agriculture in low-income countries. *Sociol. Rural.* **2022**, *62*, 279–304. [CrossRef]
53. Engås, K.G.; Raja, J.Z.; Neufang, I.F. Decoding technological frames: An exploratory study of access to and meaningful engagement with digital technologies in agriculture. *Technol. Forecast. Soc. Chang.* **2023**, *190*, 122405. [CrossRef]
54. Florez, M.; Melo, J.; Bourdon, I.; Piot-LePetit, I.; Gauche, K. Digital divide between Colombian and French agriculture. In Proceedings of the 27th Annual Americas Conference on Information Systems, AMCIS 2021, Virtual, 9–13 August 2021.
55. FAO. *Realizing the Potential of Digitalization to Improve the Agri-Food System: Proposing a New International Digital Council for Food and Agriculture. A Concept Note*; FAO: Rome, Italy, 2020.
56. Hackfort, S. Patterns of inequalities in digital agriculture: A systematic literature review. *Sustainability* **2021**, *13*, 12345. [CrossRef]
57. Renda, A. The age of foodtech: Optimizing the agri-food chain with digital technologies. In *Achieving the Sustainable Development Goals through Sustainable Food Systems*; Valentini, R., Sievenpiper, J.L., Antonelli, M., Dembska, K., Eds.; Springer: Cham, Switzerland, 2019.
58. Oosterlaken, I. Design for development: A capability approach. *Des. Issues* **2009**, *25*, 91–102. [CrossRef]
59. Sen, A. Capability and Well-Being. In *The Quality of Life*; Nussbaum, M., Sen, A., Eds.; Clarendon Press: Oxford, UK, 1993; pp. 30–53.
60. Sen, A. Development as Capability Expansion. *J. Dev. Plan.* **1989**, *19*, 41–58.
61. Achieng, M.S. Digitalisation of indigenous knowledge of agricultural practices: A strategy for food security in Africa. *J. Public Adm.* **2022**, *57*, 802–818.
62. Ghimire, P. Digitalization of indigenous knowledge in Nepal—Review article. *Acta Inform. Malays.* **2021**, *5*, 42–47. [CrossRef]
63. Kantiza, A.; Loytty, T.; Cherenet, E.; Sabimana, R.; Horak, P.; Osmancelebioglu, D.; Hajek, P.; Zadrazil, F.; Cerba, O.; Charvat, K.; et al. Digitalization of indigenous knowledge in African agriculture for fostering food security. In Proceedings of the 2021 IST-Africa Conference, IST-Africa 2021, Virtual Conference, 10–14 May 2021; pp. 1–13.
64. Franco, W.; Barbera, F.; Bartolucci, L.; Felizia, T.; Focanti, F. Developing intermediate machines for high-land agriculture. *Dev. Eng.* **2020**, *5*, 100050. [CrossRef]
65. Pope, M.; Sonka, S.; Quantifying the Eocnomic Benefits of on-Farm Digital Technologies. Farmdoc Daily. 4 March 2020. Available online: https://farmdocdaily.illinois.edu/2020/03/quantifying-the-economic-benefits-of-on-farm-digital-technologies.html (accessed on 3 August 2022).
66. Bronson, K.; Knezevic, I. The digital divide and how it matters for Canadian food system equity. *Can. J. Commun.* **2019**, *44*, 63–68. [CrossRef]
67. Annosi, M.C.; Brunetta, F.; Monti, A.; Nat, F. Is the trend your friend? An analysis of technology 4.0 investment decisions in agricultural SMEs. *Comput. Ind.* **2019**, *109*, 59–71. [CrossRef]
68. Tagarakis, A.C.; van Evert, F.K.; Kempenaar, C.; Ljubicic, N.; Milic, D.; Bengin, V.; Crnojevic, V. Opportunities for precision agriculture in Serbia. In Proceedings of the 14th International Conference on Precision Agriculture, Montreal, QC, Canada, 24 June–27 June 2018; pp. 1–12.
69. Groher, T.; Heitkämper, K.; Umstätter, C. Digital technology adoption in livestock production with a special focus on ruminant farming. *Animal* **2020**, *14*, 2404–2413. [CrossRef] [PubMed]
70. Kernecker, M.; Knierim, A.; Wurbs, A.; Kraus, T.; Borges, F. Experience versus expectation: Farmers' perceptions of smart farming technologies for cropping systems across Europe. *Precis. Agric.* **2020**, *21*, 34–50. [CrossRef]
71. Giua, C.; Materia, V.C.; Camanzi, L. Smart farming technologies adoption: Which factors play a role in the digital transition? *Technol. Soc.* **2022**, *68*, 101869. [CrossRef]
72. Omulo, G.; Kumeh, E.M. Technological Forecasting & Social Change Farmer-to-farmer digital network as a strategy to strengthen agricultural performance in Kenya: A research note on ' Wefarm ' platform. *Technol. Forecast. Soc. Chang.* **2020**, *158*, 120120. [CrossRef]

73. Da Silveira, F.; da Silva, S.L.C.; Machado, F.M.; Barbedo, J.G.A.; Amaral, F.G. Farmers' perception of barriers that difficult the implementation of agriculture 4.0. *Agric. Syst.* **2023**, *208*, 103656. [CrossRef]
74. Alant, B.P.; Bakare, O.O. A case study of the relationship between smallholder farmers' ICT literacy levels and demographic data w.r.t. their use and adoption of ICT for weather forecasting. *Heliyon* **2021**, *7*, e06403. [CrossRef]
75. Schulze Schwering, D.; Bergmann, L.; Isabel Sonntag, W. How to encourage farmers to digitize? A study on user typologies and motivations of farm management information systems. *Comput. Electron. Agric.* **2022**, *199*, 107133. [CrossRef]
76. Gabriel, A.; Gandorfer, M. Adoption of digital technologies in agriculture—An inventory in a european small-scale farming region. *Precis. Agric.* **2023**, *24*, 68–91. [CrossRef]
77. Kvam, G.T.; Hårstad, R.M.B.; Stræte, E.P. The role of farmers' microAKIS at different stages of uptake of digital technology. *J. Agric. Educ. Ext.* **2022**, *28*, 671–688. [CrossRef]

agriculture

MDPI

Article

Spatiotemporal Evolution and Spatial Convergence Analysis of Total Factor Productivity of Citrus in China

Yumeng Gu [1,2,3], Chunjie Qi [1,2,*], Fuxing Liu [1,3], Quanyong Lei [4] and Yuchao Ding [5]

1 College of Economics and Management, Huazhong Agriculture University, Wuhan 430070, China; guyumeng@webmail.hzau.edu.cn (Y.G.); liufuxing@webmail.hzau.edu.cn (F.L.)
2 Institute of Horticultural Economics, Huazhong Agriculture University, Wuhan 430070, China
3 Hubei Rural Development Research Center, Wuhan 430070, China
4 College of Business, Yancheng Teachers University, Yancheng 224002, China; leiqy928@163.com
5 Institute of Economics, Shanghai Academy of Social Sciences, Shanghai 200235, China; yiyouyc@163.com
* Correspondence: qichunjie@mail.hzau.edu.cn; Tel.: +86-15871782598

Abstract: In this study, the DEA–Malmquist index method was used to measure the total factor productivity of citrus in seven major mandarin-producing provinces and seven major tangerine-producing provinces in China from 2006 to 2020. Moran's I index was used to test the spatial correlation of total factor productivity of mandarin and tangerine, and its σ convergence and β convergence characteristics were explored using coefficient of variation and spatial panel models. The results show that from the perspective of time series evolution, the growth rate of total factor productivity of mandarin and tangerine in China slowed down year by year after reaching the maximum value in 2008. Technological progress was the main factor affecting the total factor productivity of citrus. The total factor productivity growth of tangerine was more stable than that of mandarin, and the pure technical efficiency index and scale efficiency change index of mandarin and tangerine were not stable. From the perspective of regional differences, the total factor productivity of China's main citrus-producing provinces all indicated positive growth, showing an increasing trend from east to west. The drivers of growth were mainly technological progress and scale efficiency. The regional differences in total factor productivity growth for mandarin were more obvious than for tangerine. The total factor productivity of mandarin and tangerine showed obvious spatial correlation characteristics; the positive spatial spillover effect was significant; and there were σ convergence, absolute β convergence, and conditional β convergence. Regional disparities in citrus industry development can be more objectively reflected by convergence analysis that takes spatial factors, economic and social factors, and other factors into account.

Keywords: citrus; total factor productivity; spatiotemporal evolution; Moran's I index; spatial convergence

Citation: Gu, Y.; Qi, C.; Liu, F.; Lei, Q.; Ding, Y. Spatiotemporal Evolution and Spatial Convergence Analysis of Total Factor Productivity of Citrus in China. *Agriculture* **2023**, *13*, 1258. https://doi.org/10.3390/agriculture13061258

Academic Editors: Roberto Henke and Filiberto Altobelli

Received: 12 May 2023
Revised: 3 June 2023
Accepted: 15 June 2023
Published: 16 June 2023

1. Introduction

China is one of the countries of origin of citrus and is the world's largest producer and seller of citrus [1]. Citrus agriculture dates back to over 4000 years ago in China. The size of China's citrus industry has grown steadily since the formation of the People's Republic of China. According to China's National Bureau of Statistics database, citrus planting area and production in China reached 2.832 million ha and 51.287 million tons, respectively, in 2020, overtaking apples to become China's largest fruit in terms of planting area and production. The citrus industry has become a pillar industry in agriculture in the hilly areas, reservoir areas, and underdeveloped areas of southern China [2]. It is critical in raising the revenue of fruit producers and contributing to the reduction in national poverty [3]. However, in comparison to other developed countries, China's citrus production per unit area has always been relatively low [4]. According to the FAO data, the unit area yield of Chinese citrus in 2020 was 14.88 t/ha, which was lower than the global average of 15.74 t/ha. With the annual increase in citrus planting area and yield in China, improving citrus total factor productivity is a priority due to the sector's importance to the Chinese economy. Analyzing the total factor productivity of Chinese citrus objectively is an important part of determining the entire production capacity of Chinese citrus. What are the trends in total factor productivity of citrus in China's primary citrus-producing areas? What has caused the increase in productivity in the citrus industry? Is it due to technological advancements or changes in efficiency? What are the differences in citrus TFP growth in different regions? Is there a spatial relationship between citrus TFP and region? Is the regional gap in citrus TFP narrowing? The answers to these questions will contribute to a thorough knowledge of the spatiotemporal evolution process of China's citrus total factor productivity, and such knowledge will have major theoretical and practical implications for encouraging the citrus industry's healthy and sustainable development in China.

Total factor productivity (TFP) refers to the portion of output growth minus the contribution of factor growth, and it is a comprehensive statistic employed in the neoclassical school's economic development theory to evaluate the contribution of pure technological improvement in production [5]. According to existing research, TFP measurement methods can be classified as parametric or non-parametric. Parametric methods include the Solow residual method, the random frontier production function method, and the trans-log production function method, among others, while non-parametric methods include the Malmquist index method, the Tornqvist index method, and the undesirable slacks-based measurement (SBM) method. Po et al. analyzed the productivity growth in China's agricultural sector over the period 1990–2003 [6]. Huang et al. measured the agricultural green TFP in China from 1998 to 2019 [7]. Namdari et al. calculated the energy use efficiency for citrus in Iran [8]. Xu et al. calculated the TFP of citrus in China [9]. He et al. found that the implementation of sustainable development policies has increased citrus TFP in China [10]. There are some Chinese researchers who used the DEA–Malmquist index method to measure the TFP of China overall and of some major citrus-producing areas and analyzed the trend of change [11–13]. These studies' conclusions are not always consistent due to diverse study locations and time lengths. However, one common feature is that they have not systematically analyzed the spatiotemporal evolution and convergence of citrus TFP. Based on the pioneering study of Baumol, Barro, and Sala-i-Martin on the convergence of economic growth [14,15], more studies on the convergence of agricultural TFP are being conducted. Many scholars have analyzed the convergence of agricultural production efficiency in different countries [16–20].

Although there have been many studies on citrus TFP, agricultural TFP measurement, and convergence analysis, there is still room for research to be expanded. First, the present literature focuses on citrus TFP measurement and analysis, ignoring the spatial link between regions in terms of citrus TFP. Second, research studies have focused mostly on the convergence analysis of TFP in agriculture and selected agricultural products, but there is a dearth of research on the convergence analysis of TFP in citrus.

In light of this, this study selects Chinese citrus production cost and revenue data from 2006 to 2020, measures the TFP of citrus in China using the DEA–Malmquist index method, and analyzes its spatial and temporal variation characteristics and patterns. The Moran's I index is used to evaluate the spatial correlation of TFP of citrus, and the neoclassical economic growth convergence theory is combined with spatial economics. Furthermore, the coefficient of variation is employed, and a spatial panel model is developed to investigate the σ convergence and β convergence characteristics of the TFP of citrus.

2. Materials and Methods

2.1. Materials

The output variables were chosen from the output value of citrus main products; the input variables were chosen from labor costs, land costs, fertilizer costs, pesticide costs, and other citrus costs. This could assure the consistency of material costs while also lowering the indicators based on the inclusion of diverse types of expenditures. In order to avoid omitting important control variables, we chose the following as the control variables:

level of economic development, measured using GDP per capita; mechanization level, measured using total mechanical power; fruit cultivation structure, measured based on the proportion of citrus planting area to total fruit planting area in each province; agricultural geographic agglomeration, measured based on the share of citrus cultivation area in each province to the national citrus cultivation area; agricultural financial support to agriculture, defined as the proportion of total financial expenditure on agriculture to total financial expenditure in each province; and urbanization rate, defined as the number of urban residents divided by the total population at the end of the year in each province. Based on data availability and the fact that the National Compilation of Cost and Benefit Information on Agricultural Products divides citrus and tangerine, this study divided citrus into two categories, mandarin and tangerine, and selected cost and benefit-related data from 2006 to 2020 for seven citrus-producing provinces (districts and cities), respectively. The data came from the National Agricultural Cost–Benefit Information Compilation in China, which ran from 2006 to 2020. In prior years, the data came from the National Compilation of Cost and Benefit Information of Agricultural Products, the National Bureau of Statistics, the International Monetary Fund database, and the China Population and Employment Statistical Yearbook. Missing data were interpolated using the moving average method and the trend forecasting method, and all variables involving prices were deflated using the corresponding fixed-base price indices.

2.2. Methods

2.2.1. Measurement of TFP

The Malmquist index method was proposed by Malmquist in 1953 [21], and Cave et al. first used it as a production efficiency function; this method was then combined with the DEA theory to create what is referred to as the DEA–Malmquist index method [22]. Fare et al. updated the DEA–Malmquist index method by incorporating technical efficiency, which is divided into technical change and efficiency change [23]. Many scholars favor this method because it employs distance functions to construct the optimal frontier, employs linear programming, does not require a specific functional form, and can decompose TFP changes. As a result, using the DEAP2.1 software and the following equations, this study measured and decomposed the DEA–Malmquist index for Chinese citrus production based on constant payoffs of scale and output orientation:

$$
\begin{aligned}
M_i(x^{t+1}, y^{t+1}; x^t, y^t) &= \left[\frac{D_i^t(x^{t+1}, y^{t+1})}{D_i^t(x^t, y^t)} \times \frac{D_i^{t+1}(x^{t+1}, y^{t+1})}{D_i^{t+1}(x^t, y^t)}\right]^{\frac{1}{2}} \\
&= \frac{D_i^{t+1}(x^{t+1}, y^{t+1})}{D_i^t(x^t, y^t)}\left[\frac{D_i^t(x^{t+1}, y^{t+1})}{D_i^{t+1}(x^{t+1}, y^{t+1})} \times \frac{D_i^t(x^t, y^t)}{D_i^{t+1}(x^t, y^t)}\right]^{\frac{1}{2}} \\
&= EFFCH(x^{t+1}, y^{t+1}; x^t, y^t) \times TECH(x^{t+1}, y^{t+1}; x^t, y^t) \\
&= PECH \times SECH \times TECH
\end{aligned}
\quad (1)
$$

In Equation (1), x^t and y^t represent the input and output vectors of the citrus industry in the period t, respectively; D_i^t is the distance function; $M_i(TFP)$ is the total factor productivity index; *EFFCH* is the technical efficiency change index; and *TECH* is the technological progress change index. The technical efficiency change index (*EFFCH*) can be further decomposed into pure technical efficiency index (*PECH*) and scale efficiency change index (*SECH*). A value of *EFFCH* greater than 1 represents an increase in technical efficiency, and a value of *TECH* greater than 1 represents a technological advancement or innovation; A value of *PECH* greater than 1 represents an increase in the level of technology, and vice versa; and a value of *SECH* greater than 1 represents a scale of production operation close to the optimal scale of production, and a scale deterioration if it is lower than 1.

2.2.2. Spatial Correlation Index

The spatial correlation index, which is frequently expressed as Moran's I index, can be used to examine if there is spatial autocorrelation in the TFP of citrus throughout the entire space, and its calculation formula is as follows:

$$Moran's\,\mathrm{I} = \frac{\sum\limits_{n=1}^{N}\sum\limits_{m=1}^{N}\omega_{nm}(x_n-\overline{x})(x_m-\overline{x})}{S^2\sum\limits_{n=1}^{N}\sum\limits_{m=1}^{N}\omega_{nm}} \quad (2)$$

where x_n and x_m are the index values of variable x on the geographical unit of region n and region m, respectively; $\overline{x}$ is the average of the index values in each region; ω_{nm} is the spatial weight matrix; ω_{nm} = 1 when n and m provinces are contiguous, and 0 otherwise; S^2 is the sample variance; and N is the total number of measured areas. In general, the range of the Moran's I index is -1 to 1. An index greater than 0 indicates positive spatial autocorrelation, and the closer the index value is to 1, the stronger the spatial correlation and clustering of similar attributes. An index less than 0 indicates negative spatial autocorrelation, and the closer the index value is to -1, the stronger the spatial correlation and agglomeration of different attributes. An index close to 0 indicates that the spatial distribution is random and there is no spatial autocorrelation [24].

2.2.3. Convergence Model

There are three common convergence models: σ convergence, absolute β convergence, and conditional β convergence. The convergence of σ reflects a decreasing trend in the deviation of the sample values in each region over time. This study aimed to investigate whether the *TFP* of citrus tends to be in a horizontal state with the passage of time. If the convergence coefficient of σ decreases gradually over time, the growth of citrus *TFP* has σ convergence. In this study, the coefficient of variation was used to measure the convergence of σ, and the formula is as follows:

$$\sigma_{\mathrm{t}} = \frac{\sqrt{\sum_{n=1}^{N}\left(TFP_{n,t}-\overline{TFP_t}\right)^2/N}}{\overline{TFP_t}} \quad (3)$$

$TFP_{n,t}$ is the citrus *TFP* of province n in year t; $\overline{TFP_t}$ is the average of the *TFP* for all provinces in year t; and N is the number of major citrus-producing provinces.

Furthermore, β convergence means that the growth rate disparity in citrus *TFP* between regions gradually narrows over time, eventually settling at a stable growth rate. Meanwhile, convergence can be classified as absolute β convergence or conditional β convergence. Absolute β convergence means that citrus *TFP* tends to converge across regions without taking into account factors that can have a significant impact on citrus *TFP*. The formula for absolute β convergence is as follows:

$$\ln\left(\frac{TFP_{n,t+1}}{TFP_{n,t}}\right) = \alpha + \beta\ln(TFP_{n,t}) + \mu_n + \eta_t + \varepsilon_{n,t} \quad (4)$$

$\ln\left(\frac{TFP_{n,t+1}}{TFP_{n,t}}\right)$ is the growth rate of citrus *TFP* of n province in t + 1 period. β is the coefficient of convergence, with a significant negative β indicating that the citrus *TFP* is showing absolute β convergence, and the convergence speed $\mathrm{V} = -\ln(|\beta|-1)/\mathrm{T}$. μ_n, η_n, and ε_t are the area effect, time effect, and random disturbance terms, respectively.

Considering the spatial correlation of citrus *TFP* and using the absolute β convergence model, the following three spatial measurement models were introduced: spatial lag model (*SAR*), spatial error model (*SEM*), and spatial Durbin model (*SDM*). The *SDM* model can be regarded as the general form of the other two models, and the spatial absolute β convergence formula is as follows:

$$SAR: \ln(\frac{TFP_{n,t+1}}{TFP_{n,t}}) = \alpha + \beta\ln(TFP_{n,t}) + \rho\sum_{m=1}^{N} \omega_{nm} \ln(\frac{TFP_{n,t+1}}{TFP_{n,t}}) + \mu_n + \eta_t + \varepsilon_{n,t} \quad (5)$$

$$SEM: \ln(\frac{TFP_{n,t+1}}{TFP_{n,t}}) = \alpha + \beta\ln(TFP_{n,t}) + \mu_n + \eta_t + u_{n,t}; u_{n,t} = \lambda\sum_{m=1}^{N} \omega_{nm}u_{n,t} + \varepsilon_{n,t} \quad (6)$$

$$SDM: \ln(\frac{TFP_{n,t+1}}{TFP_{n,t}}) = \alpha + \beta\ln(TFP_{n,t}) + \rho\sum_{m=1}^{N} \omega_{nm} \ln(\frac{TFP_{n,t+1}}{TFP_{n,t}}) + \gamma\sum_{m=1}^{N} \omega_{nm} \ln(TFP_{n,t}) + \mu_n + \eta_t + \varepsilon_{n,t} \quad (7)$$

ρ is the spatial lag coefficient, representing the effect of the growth rate of citrus *TFP* in neighboring provinces on a province. λ is the space error coefficient and represents the space effect in the random perturbation term $\varepsilon_{n,t}$. γ is the spatial lag coefficient of the independent variable, representing the influence of the citrus *TFP* of neighboring provinces.

The conditional β convergence model adds a series of control variables on the basis of the absolute β convergence model to examine whether citrus *TFP* has a convergence trend after controlling for the effects of factors that may have an important impact on citrus *TFP*. The formulae for conditional β convergence and spatial conditional β convergence are as follows:

$$\ln(\frac{TFP_{n,t+1}}{TFP_{n,t}}) = \alpha + \beta\ln(TFP_{n,t}) + \delta X_{n,t+1} + \mu_n + \eta_t + \varepsilon_{n,t} \quad (8)$$

$$SAR: \ln(\frac{TFP_{n,t+1}}{TFP_{n,t}}) = \alpha + \beta\ln(TFP_{n,t}) + \rho\sum_{m=1}^{N} \omega_{nm} \ln(\frac{TFP_{n,t+1}}{TFP_{n,t}}) + \delta X_{n,t+1} + \mu_n + \eta_t + \varepsilon_{n,t} \quad (9)$$

$$SEM: \ln(\frac{TFP_{n,t+1}}{TFP_{n,t}}) = \alpha + \beta\ln(TFP_{n,t}) + \delta X_{n,t+1} + \mu_n + \eta_t + u_{n,t} \quad u_{n,t} = \lambda\sum_{m=1}^{N} \omega_{nm}u_{n,t} + \varepsilon_{n,t} \quad (10)$$

$$SDM: \ln(\frac{TFP_{n,t+1}}{TFP_{n,t}}) = \alpha + \beta\ln(TFP_{n,t}) + \rho\sum_{m=1}^{N} \omega_{nm} \ln(\frac{TFP_{n,t+1}}{TFP_{n,t}}) + \gamma\sum_{m=1}^{N} \omega_{nm} \ln(TFP_{n,t}) + \delta X_{n,t+1} + \mu_n + \eta_t + \varepsilon_{n,t} \quad (11)$$

3. Spatial and Temporal Evolution of Citrus TFP

3.1. Time Series Evolution of Citrus TFP

Figures 1 and 2 depict the changes in TFP and the deconstruction of mandarin and tangerine production in China from 2007 to 2020. As shown in Figure 1, mandarin's TFP shows a fluctuating downward trend. It was 1 in 2007, increased to a maximum of 1.939 in 2008, and then gradually declined in the years that followed; specifically, the TFP of mandarin dropped to 1.107 in 2009; rebounded to 1.382 in 2010; fluctuated between 1.092 and 1.231 in 2011 to 2015; experienced a negative growth in 2016, with the TFP falling to 0.970 in 2017; rebounded to 1.001 in 2018 and further improved to 1.115 in 2019; and then fell again to 0.958 in 2020. Compared to the technological progress change (TECH), the TFP decreased while the TECH increased in 2014, and the TFP increased while the TECH decreased in 2019. The trends of the changes in these two indexes in other years are the same, with some differences in the magnitude of the increase or decrease of the changes. Thus, it can be concluded that there is a driving effect of mandarin's technological progress change (TECH) on the growth in TFP. The patterns of fluctuation in the technical efficiency change index (EFFCH), the pure technical efficiency change index (PECH), and the scale efficiency change index (SECH) are highly similar, exhibiting both increases and decreases. This suggests that the pure technical efficiency change index (PECH) and the scale efficiency change index (SECH) are the primary determinants of fluctuations in the overall technical efficiency change (EFFCH).

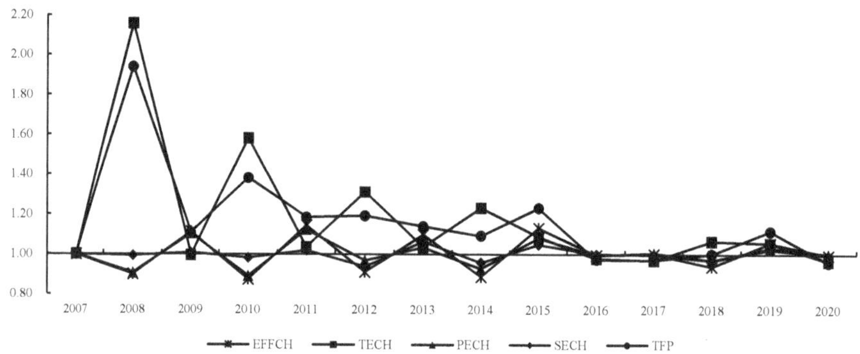

Figure 1. Changes in TFP and deconstruction of its components for China's mandarin production from 2007 to 2020.

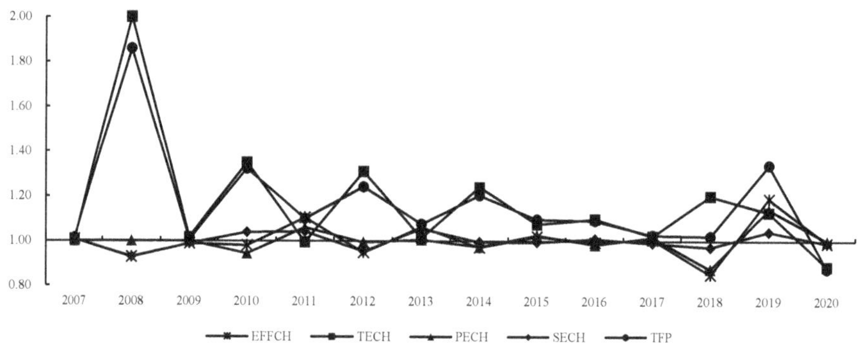

Figure 2. Changes in TFP and deconstruction of its components for China's tangerine production from 2007 to 2020.

Figures 1 and 2 reveal that the variations in total TFP and the deconstruction of its components for mandarin and tangerine are highly similar. Specifically, the fluctuations in the TFP and technological progress change (TECH) of tangerine are the most conspicuous, and their trends follow a similar pattern. In comparison, the TFP of tangerine is marginally higher than that of mandarin, recording a value of 1.014 in 2007, peaking at 1.859 in 2008, declining to 1.005 in 2009, recovering to 1.322 in 2010, and, subsequently, exhibiting a gradual downward fluctuation between 1.021 and 1.241 from 2011 to 2018, before increasing to 1.338 in 2019. This trend deviates from what is observed for mandarin. When comparing the technological progress change (TECH), it can be observed that, apart from the year 2016 when the technological progress change (TECH) increased and the TFP decreased, the trends in both indexes are highly similar across the years. Moreover, the impact of technological progress change (TECH) on the TFP growth of tangerine is more pronounced than the impact seen in mandarin. The technical efficiency change (EECH), the pure technical efficiency change (PECH), and the scale efficiency change (SECH) exhibit similar patterns of increase and decrease, suggesting that changes in pure technical efficiency change (PECH) and scale efficiency change (SECH) are also influential drivers of technical efficiency change (EECH) for tangerine.

The growth in TFP and the deconstruction of its components for mandarin and tangerine production in China exhibit four key characteristics. Firstly, the TFP and the technological progress change (TECH) show similar patterns for both mandarin and tangerine, indicating that technological progress plays a vital role in driving the growth in TFP. Secondly, the TFP for both mandarin and tangerine peaked in 2008 before fluctuating downwards. This might be attributed to the initiation of the national citrus industry's technological system in 2007 and the subsequent innovation chain that emerged around the citrus industry's production chain. This promoted the level of citrus technological progress in China. However, over time, as technology was internalized and breakthroughs became more challenging, the driving effect of technological progress weakened, leading to the fluctuation in TFP for both mandarin and tangerine.

Thirdly, the growth in TFP for tangerine is more stable compared to the growth in TFP for mandarin. Mandarin's TFP exhibited negative growth in 2016, 2017, and 2020, whereas tangerine's TFP only exhibited negative growth in 2020 due to the COVID-19 pandemic. This is because tangerine is more prevalent in Chinese citrus cultivation and shows stronger research capabilities than mandarin. Fourthly, the pure technical efficiency change (PECH) and the scale efficiency change (SECH) for both mandarin and tangerine fluctuate, indicating that the actual technical level and planting scale of Chinese citrus planting process are unstable. This may be attributed to the fact that citrus fruits are mainly grown in mountainous areas in China, and the proportion of continuous centralized planting is low. Additionally, citrus yellow dragon disease is a constant threat to production, making it difficult for technology to be implemented effectively and for resource allocation to reach a reasonable state.

3.2. *Regional Differences in TFP for Citrus*

Regional differences in natural environment and resource endowment have resulted in variations in the TFP and its specific components for mandarin and tangerine among different provinces in China. Comparing the average TFP and its components in different provinces is of great practical significance for identifying the key factors that determine regional differences in citrus output and for promoting regional synergistic development of the citrus industry. Table 1 presents the average TFP and its composition for the major mandarin- and tangerine-producing provinces in China from 2007 to 2020.

Table 1. Average TFP and its composition for the major mandarin- and tangerine-producing provinces in China from 2007 to 2020.

Classification	Provinces	EFFCH	TECH	PECH	SECH	TFP
Mandarin	Chongqing	0.998	1.176	1.000	0.998	1.173
	Guangxi	0.999	1.169	1.000	0.999	1.168
	Hunan	0.999	1.147	1.000	0.999	1.146
	Hubei	0.998	1.145	1.000	0.998	1.142
	Guangdong	0.999	1.135	1.000	0.999	1.134
	Jiangxi	1.000	1.123	1.000	1.000	1.123
	Fujian	1.000	1.116	1.000	1.000	1.116
Tangerine	Chongqing	0.999	1.161	1.000	0.999	1.159
	Hunan	1.000	1.157	1.000	1.000	1.157
	Hubei	0.998	1.148	1.000	0.998	1.147
	Jiangxi	1.000	1.141	1.000	1.000	1.141
	Zhejiang	1.001	1.131	1.000	1.001	1.133
	Guangdong	0.999	1.134	1.000	0.999	1.133
	Fujian	1.003	1.130	1.000	1.003	1.132

Table 1 highlights several key characteristics that contribute to the regional divergence in TFP for China's citrus industry. First, the average TFP of the major mandarin- and tangerine-producing provinces is positive, indicating a growth trend in the citrus industry. The province with the highest TFP growth rates for both mandarin and tangerine is Chongqing, while Fujian has the lowest TFP growth rates for both, although its rates are still positive.

Second, the regional divergence in mandarin's average TFP growth is more pronounced than that of tangerine. Chongqing has the highest mandarin TFP growth rate, while Fujian has the lowest. This suggests that the industrial development of tangerine in China is more balanced compared to mandarin.

Third, from a geographical perspective, the average TFP of mandarin and tangerine tends to increase from east to west, reflecting an "east citrus to west" trend. The central and western regions have relatively more arable land resources, fewer non-farming employment opportunities, and higher production economic efficiency, thus resulting in a higher TFP. Additionally, the main citrus-producing areas in the southeast coast are more susceptible to Huanglong disease, resulting in a lower TFP.

Fourth, technological progress change (TECH) and scale efficiency change (SECH) are the primary factors affecting the TFP growth of each citrus-producing province. The growth rate of technological progress change (TECH) is highest in Chongqing for mandarin and tangerine, while the province's scale efficiency change (SECH) growth rates are negative, indicating the importance of technological progress in TFP growth but also highlighting the need for optimizing production scale. In contrast, the growth rate of technological progress change (TECH) in Zhejiang for tangerine is lower than that of Guangdong, but the growth rate of scale efficiency change (SECH) is higher, leading to a higher TFP due to the optimization of production scale.

To gain further insight into regional differences in TFP in China's citrus industry, spatial distribution trends of TFP were analyzed using the Matlab 2021b software. The results are presented in Figure 3. The analysis revealed significant regional differences in the spatial distribution of TFP for both mandarin and tangerine, indicating obvious non-equilibrium characteristics. In terms of fitting surfaces, mandarin and tangerine are different. The TFP of Chinese mandarin shows a spatial distribution pattern of high in the northwest, depressed in the center, and low in the southeast. The TFP of Chinese tangerine shows a spatial distribution pattern of high in the northwest, convex in the middle, and low in the southeast. This is closely related to the acreage and research strength of both. In terms of fitting curves, in the east–west direction, both mandarin and tangerine show a trend of high in the east and low in the west, which is the same as the trend of "east citrus to west". In the north–south direction, mandarin shows a "U" shape, suggesting that the TFP of Chinese mandarin in the northern and southern regions is slightly higher than that in the central region during the same period. On the contrary, tangerine shows an inverted "U" shape, suggesting that the TFP of Chinese tangerine in the northern and southern regions is slightly lower than that in the central region during the same period.

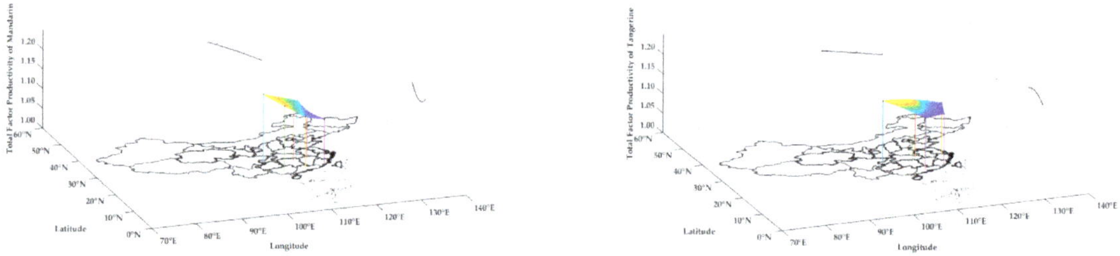

Figure 3. Spatial distribution of TFP of mandarin and tangerine in China, 2007–2020.

3.3. Spatial Correlation Analysis of Citrus TFP

We employed the Stata software to compute global Moran's I indices for both mandarin and tangerine based on their TFP panel data spanning from 2007 to 2020. Furthermore, we tested the significance level of the Moran's I indices. The findings are presented in Table 2.

Table 2. Global Moran's I of mandarin and tangerine TFP from 2007 to 2020.

Year	Mandarin			Tangerine		
	Moran's I	z	*p*	Moran's I	z	*p*
2007	−0.069	0.565	0.286	−0.299	−0.836	0.202
2008	−0.026	1.078	0.140	−0.192	−0.188	0.425
2009	−0.041	1.051	0.147	−0.053	1.049	0.147
2010	0.169 **	1.762	0.039	0.199 **	1.850	0.032
2011	0.109 *	1.462	0.072	0.136 *	1.595	0.055
2012	0.215 **	1.906	0.028	0.280 **	2.171	0.015
2013	0.208 **	1.880	0.030	0.075	1.187	0.118
2014	0.074	1.211	0.113	0.384 ***	2.615	0.004
2015	0.046	1.049	0.147	0.237 **	2.238	0.013
2016	−0.041	0.696	0.243	0.184 **	1.766	0.039
2017	−0.184	−0.088	0.465	−0.084	0.420	0.337
2018	−0.274	−0.733	0.232	0.118 **	2.303	0.011
2019	−0.356	−0.991	0.161	−0.185	−0.103	0.459
2020	−0.139	0.160	0.436	0.053 *	1.310	0.095

Note: "*", "**", and "***" indicate statistical significance at the 10%, 5%, and 1% level, respectively.

Table 2 shows that there is a positive spatial correlation in the TFP of mandarin in China from 2010 to 2013, and the global Moran's I index tends to increase over time. This indicates that the spatial correlation of mandarin's TFP in China has increased during the study period. The global Moran's I index of tangerine TFP is also positive and significant except for some years, suggesting that China's tangerine TFP also has a positive spatial correlation and the growth of tangerine TFP in neighboring provinces has gradually converged. Although the spatial correlation of China's citrus TFP has some volatility, there is an overall positive spatial correlation, indicating that a positive spatial spillover effect has gradually formed. This effect has become a new development trend, which highlights the importance of cooperation and driving effect at the spatial level.

It is essential to consider spatial factors when conducting the β convergence test since it cannot be assumed that each citrus-producing province is independent of each other. Neglecting the spillover effects of neighboring producing provinces on the convergence of citrus TFP could lead to biased convergence test results. Therefore, taking into account spatial factors is crucial to avoid such biases.

4. Convergence Analysis of TFP of Citrus

4.1. σ Convergence Test Result Analysis

As depicted in Figure 4, the σ value of the TFP of mandarin shows an increasing trend from 2007 to 2009, followed by a decreasing trend from 2009 to 2015, and eventually reaching its lowest value of 0.0005 in 2020, indicating that there is σ convergence in the TFP of mandarin. Similarly, the σ value of TFP of tangerine shows an increasing trend from 2007 to 2009 and then decreases steadily from 2009 to 2012. Although the σ value of TFP of tangerine increases in some years during the study period, the overall trend shows σ convergence. In summary, both mandarin and tangerine TFP values show σ convergence in China, suggesting that the disparities in TFP among the main producing province have gradually decreased over time.

4.2. Absolute β Convergence Test Result Analysis

Four models were used to estimate the absolute β convergence of TFP of mandarin and tangerine in China. These models included the absolute β convergence model (OLS), the spatial lag model (SAR), the spatial error model (SEM), and the spatial Durbin model (SDM). Additionally, a two-way fixed effects model was used to control for time and region effects during estimation. As shown in Table 3, the results indicate that there is significant absolute β convergence for both mandarin and tangerine TFP in China. The absolute β convergence coefficients of all models for both mandarin and tangerine are negative and pass the significance test at the 1% level, demonstrating a "catching-up effect" between the lower-producing provinces and the higher-producing provinces.

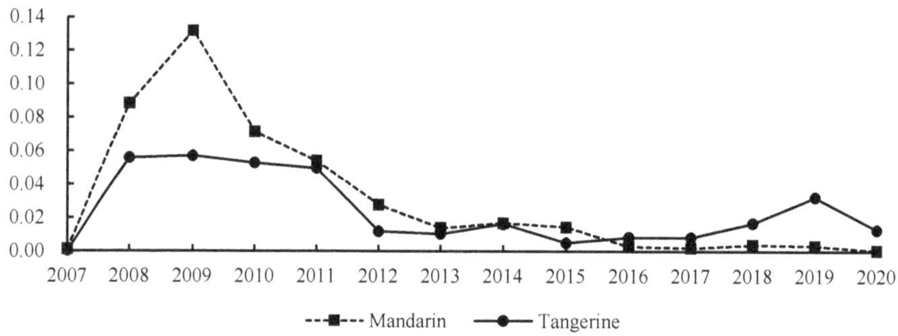

Figure 4. Convergence analysis of mandarin and tangerine TFP in China, 2007–2020.

Moreover, the ρ and γ of all models are positive and pass the significance test at the 1% level, indicating that the TFP of citrus in the main producing provinces is simultaneously affected by the positive spatial spillover effect of citrus TFP and the TFP growth rates in other regions. Therefore, the spatial spillover effect between main producing provinces cannot be ignored, and the spatial interaction between neighboring main producing provinces needs to be considered.

Additionally, when spatial factors are considered, the convergence speed of both mandarin and tangerine TFP is significantly faster. The absolute β convergence rate of mandarin TFP increases to 16.9%, 4.4%, and 4% and that of tangerine TFP increases to 8.6%, 3.9%, and 4%. This suggests that cross-regional flows of labor, land, and capital among the main citrus-producing provinces have enhanced the mutual influence of neighboring citrus-producing provinces and increased the spatial convergence effect.

However, there is a difference in the convergence speed of mandarin and tangerine TFP with the addition of spatial factors. The spatial lag model (SAR) and the spatial error model (SEM) estimate the convergence rate of mandarin TFP to be 16.9% and 4.4%, respectively, which values are both higher than that of tangerine at 8.6% and 3.9%. This may be due to differences in economic development, mechanization level, fruit cultivation structure, agricultural geographic agglomeration, financial support for agriculture, and urbanization rate in each region. Therefore, further conditional β convergence analysis is needed to account for these differences.

Table 3. Absolute β convergence of mandarin and tangerine TFP.

Coefficient	Mandarin				Tangerine			
	OLS	SAR	SEM	SDM	OLS	SAR	SEM	SDM
β	−1.699 *** (0.0848)	−1.094 *** (0.105)	−1.540 *** (0.0885)	−1.571 *** (0.0899)	−1.697 *** (0.0900)	−1.301 *** (0.102)	−1.576 *** (0.0904)	−1.568 *** (0.0939)
ρ or λ	—	0.343 *** (0.0727)	0.705 *** (0.0544)	0.687 *** (0.0565)	—	0.296 *** (0.0594)	0.614 *** (0.0654)	0.613 *** (0.0655)
γ	—	—	—	1.191 *** (0.126)	—	—	—	0.939 *** (0.149)
V	0.026	0.169	0.044	0.040	0.026	0.086	0.039	0.040
Time effect	YES	YES	YES	YES	YES	YES	YES	YES
Individual effect	YES	YES	YES	YES	YES	YES	YES	YES
R^2	0.919	0.605	0.675	0.697	0.928	0.760	0.781	0.781

Note: "***" indicate statistical significance at the 1% level. "YES" indicates that time and individual fixed effects have been controlled.

4.3. Conditional β Convergence Test Result Analysis

Table 4 presents the results of the conditional β convergence analysis of TFP for mandarin and tangerine, using the same models as in the absolute β convergence analysis. The results indicate that there is significant conditional β convergence for both mandarin and tangerine TFP in China. Even after controlling for various economic and social factors, such as economic development level, mechanization level, fruit cultivation structure, agricultural geographic agglomeration, financial support for agriculture, and urbanization rate, the conditional β convergence coefficients of all models for both mandarin and tangerine remain statistically significantly negative at the 1% level, suggesting that the "catch-up effect" of TFP still exists. Additionally, the TFP values of citrus in the whole country and in the main producing provinces continue to converge to a uniform steady-state equilibrium value in the long run.

Table 4. Conditional β convergence of mandarin and tangerine TFP.

Coefficient	Mandarin				Tangerine			
	OLS	SAR	SEM	SDM	OLS	SAR	SEM	SDM
β	−1.716 *** (0.0861)	−1.574 *** (0.0872)	−1.646 *** (0.0752)	−1.638 *** (0.0807)	−1.701 *** (0.0935)	−1.557 *** (0.0883)	−1.674 *** (0.0797)	−1.598 *** (0.0877)
ρ or λ	—	0.0997 * (0.0604)	0.309 *** (0.117)	0.198 * (0.122)	—	0.151 *** (0.0534)	0.349 *** (0.106)	0.260 ** (0.107)
γ	—	—	—	0.342 * (0.212)	—	—	—	0.308 (0.207)
V	0.024	0.040	0.031	0.032	0.025	0.042	0.028	0.037
Control variables	YES	YES	YES	YES	YES	YES	YES	YES
Time effect	YES	YES	YES	YES	YES	YES	YES	YES
Individual effect	YES	YES	YES	YES	YES	YES	YES	YES
R^2	0.925	0.725	0.663	0.535	0.930	0.869	0.841	0.541

Note: "*", "**", and "***" indicate statistical significance at the 10%, 5%, and 1% level, respectively. "YES" indicates that time and individual fixed effects have been controlled.

Consistent with the absolute β convergence analysis, the ρ and γ values of all models are positive and significant at the 10% level, indicating that the growth in TFP for mandarin and tangerine is not only influenced by their initial levels in a given region, but also by the positive spatial spillover effects of TFP and growth rates in neighboring regions. Furthermore, compared to the absolute β convergence analysis, the β convergence rates estimated by all models are reduced in the conditional β convergence analysis. Specifically, the convergence rates of mandarin and tangerine β coefficients estimated by the spatial lag model (SAR) decrease by 12.9% and 4.4%, respectively, while the decreases in the other models are minimal. This suggests that the control variables used in the analysis are effective and scientifically reasonable.

Finally, after considering the control variables, the convergence rate difference between mandarin and tangerine β coefficients estimated by each model is reduced, indicating that the conditional β convergence analysis takes a more comprehensive set of factors into consideration and produces more reasonable results than the absolute β convergence analysis. Overall, the results suggest that various economic and social factors play an important role in explaining the differences in TFP between regions, and that spatial spillover effects should be taken into account when designing policies to promote regional development and reduce regional disparities in citrus production.

5. Conclusions and Policy Implications

5.1. Conclusions

In this study, we used data on the production costs and revenues of Chinese citrus from 2006 to 2020 and the DEA–Malmquist index method to calculate the TFP and its specific components for citrus production in China overall and in its major producing provinces. We analyzed the spatiotemporal evolution characteristics of Chinese citrus

TFP and used Moran's I index and spatial convergence model to investigate the spatial correlation and convergence of citrus TFP. The main research findings are listed below.

First, from the perspective of temporal evolution, the change in the trend of Chinese citrus TFP from 2007 to 2020 is basically consistent with technological progress change (TECH), and technological progress is the main factor affecting citrus TFP. The growth rate of Chinese citrus TFP reached its maximum in 2008 and has been slowing down year by year. There are some differences in the TFP between mandarin and tangerine, with the former showing more stability in growth. The unstable pure technical efficiency change (PECH) and scale efficiency change (PECH) due to the high proportion of small-scale planting and the spread of diseases and pests, such as the Huanglong disease, have limited the growth in citrus TFP.

Second, from the perspective of regional differences, the TFP in major producing provinces have all increased, with Chongqing having the highest TFP of citrus. The regional differences in TFP growth for mandarin are more significant than for tangerine, which is related to the development of Chinese citrus with mandarin as the main product. Influenced by the flow of production factors and the Huanglong disease, the TFP of mandarin and tangerine shows an increasing trend from east to west, which is consistent with the trend of "citrus moving westward" in the production layout of Chinese citrus. The main sources of TFP growth for each major producing province of citrus are technological progress and scale efficiency.

Third, from the perspective of convergence characteristics, Chinese citrus TFP exhibits both σ convergence, absolute β convergence, and conditional β convergence. In terms of σ convergence, the σ value of citrus TFP shows a significant downward trend overall. In terms of absolute β convergence, the TFP of citrus in each major producing province is simultaneously affected by the positive spatial spillover effect of TFP and the growth rates of TFP of citrus in other regions, and the introduction of spatial effects into the convergence model significantly accelerates the convergence speed of mandarin and tangerine TFP, with mandarin having a higher convergence speed than tangerine. In terms of conditional β convergence, after adding the control variables, the convergence and positive spatial spillover effects of TFP of mandarin and tangerine are still significant, but the convergence speed decreases, and the difference in β convergence speed between mandarin and tangerine narrows.

5.2. Policy Implications

First of all, there is a need to further promote the national citrus industry's technological system construction; increase scientific research investment in citrus breeding, planting, and processing; develop advanced and applicable technology according to the natural environment and resource endowment of each citrus-producing province; and improve the TFP of citrus by promoting technological progress. There is also a need to cultivate and develop new agricultural business entities; promote moderate-scale operation; and gradually form a new agricultural business system based on family contracting, with large professional households, family farms, farmers' cooperatives, and leading agricultural industrialized enterprises as the backbone and other organizational forms as the supplement, while strengthening the prevention and control of citrus pests and diseases, such as the Huanglong disease, to improve the technical efficiency and scale efficiency of Chinese citrus production.

Additionally, using the positive spatial spillover effect of TFP of citrus, there is a need to increase the learning opportunities of underdeveloped areas of citrus industry development from developed areas; promote new technologies, such as labor-saving cultivation and water–fertilizer integration in advanced areas, through technology and management experience exchange; improve orchard mechanization and give full play to the role of radiation demonstration in advanced citrus-planting areas, while driving the latter development with the former development; and promote overall regional coordination for healthy and sustainable development.

Finally, the convergence effect of TFP of citrus should be valued; the allocation of scientific and technological inputs among citrus production regions should be optimized; regional differences should highlighted while the existence of cross-regional flows of factors, such as labor, land, and capital, should be strengthened; and institutional guarantees should be provided for effective cross-regional cooperation to provide conditions for narrowing the regional gap in the development of the citrus industry.

Author Contributions: Conceptualization, Y.G.; methodology, Y.G. and Y.D.; software, F.L.; validation, Q.L.; formal analysis, F.L.; resources, Y.G.; data curation, F.L. and Q.L.; writing—original draft preparation, Y.G.; writing—review and editing, Y.G. and Q.L.; visualization, Y.G. and Y.D.; supervision, C.Q.; project administration, C.Q.; and funding acquisition, C.Q. All authors have read and agreed to the published version of the manuscript.

Funding: This research was funded by the China Agriculture Research System (CARS-26-06BY); the Major Consulting Project of Jiangxi Research Institute for the Development Strategy of China Engineering Science and Technology (2022-DFZD-37); Major Science and Technology Special Program of Yunnan Provincial Science and Technology Department (202102AE090020).

Institutional Review Board Statement: Not applicable.

Data Availability Statement: The data presented in this study are available from the first author upon request.

Conflicts of Interest: The authors declare no conflict of interest.

References

1. Deng, X.X. The development trend of citrus industry and the adjustment of citrus variety structure in Guilin. *South. Hortic.* **2020**, *31*, 1–4. (In Chinese)
2. Qi, C.J.; Gu, Y.M.; Zeng, Y. Progress of citrus industry economy in China. *J. Huazhong Agric. Univ.* **2021**, *40*, 58–69. (In Chinese)
3. He, Y.; Li, Z.P.; Fang, G.Z.; Qi, C.J. Study on spatial correlation effect of citrus price in main producing areas—Based on VAR model and SNA model. *Chin. J. Agric. Resour. Reg. Plan.* **2023**, *44*, 173–184. (In Chinese)
4. Fang, G.Z.; Qi, C.J.; Lei, Q.Y. Calculation of total factor productivity of citrus in China and the difference of region—Based on the DEA-Malmquist index method. *Chin. J. Agric. Resour. Reg. Plan.* **2019**, *40*, 29–34. (In Chinese)
5. Shi, Z. How to understand and calculate the growth of "total factor productivity"—Econometric analysis of evaluating a specific technical and economic problem. *J. Quant. Technol. Econ.* **1988**, *4*, 68–71. (In Chinese)
6. Chen, P.C.; Yu, M.M.; Chang, C.C.; Hsu, S.H. Total factor productivity growth in China's agricultural sector. *China Econ. Rev.* **2008**, *19*, 580–593. [CrossRef]
7. Huang, X.; Feng, C.; Qin, J.; Wang, X.; Zhang, T. Measuring China's agricultural green total factor productivity and its drivers during 1998–2019. *Sci. Total Environ.* **2022**, *829*, 154477. [CrossRef]
8. Namdari, M.; Kangarshahi, A.A.; Amiri, N.A. Input-output energy analysis of citrus production in Mazandaran province of Iran. *Afr. J. Agric. Res.* **2011**, *6*, 2558–2564.
9. Xu, X.; Yang, J.X. Change analysis of citrus total factor productivity in China. *Acta Agric. Zhejiangensis* **2018**, *30*, 470–478.
10. He, Y.; Chen, W. Evaluation of Sustainable Development Policy of Sichuan Citrus Industry in China Based on DEA–Malmquist Index and DID Model. *Sustainability* **2023**, *15*, 4260. [CrossRef]
11. Li, D.H.; Guo, J.Y.; Zhu, S.B. On all-factor productivity, technological advances and efficiency in China's orange industry. *J. Jiangxi Agric. Univ. Soc. Sci. Ed.* **2010**, *9*, 43–47. (In Chinese)
12. Xiong, W.; Qi, C.J. DEA analysis on citrus production efficiency in Hubei province. *J. Huazhong Agric. Univ. Soc. Sci. Ed.* **2012**, *5*, 36–40.
13. Tang, L.R. The Research on Input and Output Efficiency of Citrus in China. Master's Thesis, Northwest A&F University, Yangling, China, 2012.
14. Barro, R.J.; Sala-i-Martin, X. Convergence. *J. Political Econ.* **1992**, *100*, 223–251. [CrossRef]
15. Baumol, W.J. Productivity growth, convergence, and welfare: What the long-run data show. *Am. Econ. Rev.* **1986**, *76*, 1072–1085.
16. Mcerlean, S.; Wu, Z. Regional agricultural labour productivity convergence in China. *Food Policy* **2003**, *28*, 237–252. [CrossRef]
17. Gong, B. Agricultural productivity convergence in China. *China Econ. Rev.* **2020**, *60*, 101423. [CrossRef]
18. Poudel, B.N.; Paudel, K.P.; Zilberman, D. Agricultural productivity convergence: Myth or reality? *J. Agric. Appl. Econ.* **2011**, *43*, 143–156. [CrossRef]
19. Rezitis, A.N. Agricultural productivity and convergence: Europe and the United States. *Appl. Econ.* **2010**, *42*, 1029–1044. [CrossRef]
20. Suhariyanto, K.; Thirtle, C. Asian agricultural productivity and convergence. *J. Agric. Econ.* **2001**, *52*, 96–110. [CrossRef]
21. Malmquist, S. Index numbers and indifference surfaces. *Trab. De Estadística* **1953**, *4*, 209–242. [CrossRef]

22. Caves, D.W.; Christensen, L.R.; Diewert, W.E. The economic theory of index numbers and the measurement of input, output, and productivity. *Econ. J. Econ. Soc.* **1982**, *50*, 1393–1414. [CrossRef]
23. Färe, R.; Grosskopf, S.; Norris, M.; Zhang, Z. Productivity growth, technical progress, and efficiency change in industrialized countries. *Am. Econ. Rev.* **1994**, *84*, 66–83.
24. Xue, P.F.; Li, G.J.; Luo, Q.Y.; Liu, S.; Chen, Y. Study on the Regional Difference and Spatial Structure of Agricultural Science and Technology Resources in China. *J. Agrotech. Econ.* **2021**, *5*, 108–120.

agriculture

MDPI

Article

A Step toward Water Use Sustainability: Implementing a Business Model Canvas for Irrigation Advisory Services

Alessandra Santini [1], Antonella Di Fonzo [2], Elisa Giampietri [1,*], Andrea Martelli [2], Orlando Cimino [2], Anna Dalla Marta [3], Maria Carmela Annosi [4], Francisco José Blanco-Velázquez [5], Teresa Del Giudice [6] and Filiberto Altobelli [2,*]

1 Department of Land, Environment, Agriculture and Forestry (TESAF), University of Padova, 35020 Legnaro, Italy
2 CREA Research Centre for Agricultural Policies and Bioeconomy, 00184 Rome, Italy
3 Department of Agriculture, Food, Environment and Forestry, University of Florence, 50144 Florence, Italy
4 Department of Social Sciences, Business Management and Organization Division, Wageningen Universiteit, 6707 KN Wageningen, The Netherlands
5 Evenor-Tech, 41018 Seville, Spain
6 Department of Agricultural Sciences, University of Naples Federico II, 80055 Naples, Italy
* Correspondence: elisa.giampietri@unipd.it (E.G.); filiberto.altobelli@crea.gov.it (F.A.)

Abstract: Some major future global challenges are linked to more efficient use of water for irrigation to respond to the growing water scarcity coupled with the increasing food demand. Although irrigation advisory services (IASs) are considered effective instruments to increase water use efficiency in agriculture, their diffusion remains limited. This is due to several constraints mainly linked to their low accessibility and high costs. To overcome the bottlenecks associated with IASs' adoption, this paper proposes a business model (BM) as a tool for scaling up IASs within a business perspective, with the aim of encouraging the diffusion of this technology while enhancing the associated environmental and social benefits. Drawn from the experience of the OPERA project, we structured the business model taking advantage of the opinion of relevant stakeholders and IASs' potential users to identify specific limitations and understand their needs. It turned out that farmers are willing to adopt IASs but require that the service is easily accessible, with high-quality information that are delivered at an affordable cost. Indeed, here a BM with an innovative way to produce and deliver value is proposed. The value proposition is built upon key features namely, integration, customization, accessibility, and sustainability that reflect users' needs and preferences. Our BM also provides a detailed revenues strategy that guarantees the financial sustainability of IASs. To design and represent our BM, the "Business Model Canvas ©" has been adopted. We concluded that an innovative and well-structured BM has the potential to leave the IASs profitable and capable to ensure environmental and social sustainability.

Keywords: business model; irrigation advisory services; sustainable agriculture; irrigation; water use efficiency

Citation: Santini, A.; Di Fonzo, A.; Giampietri, E.; Martelli, A.; Cimino, O.; Dalla Marta, A.; Annosi, M.C.; Blanco-Velázquez, F.J.; Del Giudice, T.; Altobelli, F. A Step toward Water Use Sustainability: Implementing a Business Model Canvas for Irrigation Advisory Services. *Agriculture* **2023**, *13*, 1081. https://doi.org/10.3390/agriculture13051081

Academic Editor: Janusz Gołaszewski

Received: 10 March 2023
Revised: 14 May 2023
Accepted: 17 May 2023
Published: 18 May 2023

1. Introduction

Climate change impacts are worsening the scarcity of water resources by negatively affecting precipitation [1], consequently increasing the frequency and intensity of droughts in many areas worldwide [2]. In this context, agriculture is expected to experience the greatest impact since it accounts for 70% of global freshwater withdrawals [3]. Dealing with water scarcity to ensure that agricultural production keeps pace with a growing global demand for food [4] requires more efficient and sustainable irrigation management, from field to watershed, based on accurate knowledge and information [2]. Indeed, improving water use efficiency through minimizing water losses aims at increasing yields through water-management optimization [5], in line with the current priority of producing more with less [6].

Irrigation advisory services (IASs) (i.e., achieving efficiency in irrigation water use) represent a potential response to this emerging need [7,8]. IASs can optimize water management in agriculture thanks to tools and techniques ranging from traditional irrigation scheduling that advises farmers on when and how to irrigate, to Earth-observation (EO), satellite-based technologies, the internet of things (IoT) and unmanned aerial vehicles (UAVs). These latter are able to estimate different variables regarding different components of the water cycle, from the atmosphere to water bodies, land, soil, and vegetation [9]. IASs can be seen as a sustainable innovation, due to the fact that both environmental and social considerations are taken into account in their development and use [10].

Although IASs are considered effective to increase water use efficiency and productivity and improve the decision-making process with regard to irrigation practices [11], their diffusion remains limited [12]. Accordingly, several constraints tied to the functioning and organization of IASs contributed to slowing down their adoption.

Business models (BMs)—namely the firms' strategies to create, capture, and share value—are often employed to spread the diffusion of new technologies [13]. Innovations in the business model are required to generate a change in the ways a firm creates and delivers value, resulting in a significant improvement in the value proposition [14]. Adopting an innovative business potentially allows for the transformation of the characteristics of sustainable and innovative technologies into economic value [15]. Additionally, a different BM may overcome the market barrier linked with technologies diffusion [16]. Lastly, a BM clearly outlines a revenue strategy to guarantee the economic sustainability of every business.

In this context, BMs may represent a powerful tool for widespread adoption and commercialization of IASs. BM concepts have been previously applied to other irrigation technologies such as smart irrigation networks [17], pivot irrigation systems [18], or solar pumps [19,20]. As opposite, to the best of our knowledge, there is a lack of BM application to IASs, as the literature focused mainly on quantifying the benefits associated with the adoption of the service [9,21] and on the investigation of potential users' preferences and willingness to pay [8].

Thus, the aim of this paper is to propose an innovative BM to support the successful development of IASs, encouraging the diffusion of this technology, while enhancing environmental and social benefits associated with their adoption. Drawn from the experience and the results of the OPERA (operationalizing the increase of water use efficiency and resilience in irrigation) EU research project, a BM is structured taking advantage of the opinion of relevant stakeholders and IASs' potential users to identify specific limitations and understand their needs. As a matter of fact, our BM is a tool proposed to give a first operationalization to all the results and outcomes of the OPERA project. In particular, this paper designs a BM for IASs by applying a business model canvas (BMC) developed by Osterwalder and Pigneur [22] for BM representation. By doing so, this paper does not propose any innovation for the IAS as a product, rather—as a novel contribution—it focuses on developing a value proposition and explaining how it can generate revenues to spur its market potential and increase its diffusion among potential users.

2. Background

2.1. Determinants of the Adoption of Technological Innovations in Irrigation

Nowadays, major technological innovations in irrigation are dedicated to scheduling irrigation interventions by handling a large amount of information [23], as for IASs. The adoption of these technologies is complex and may be affected by several factors. However, the literature on identifying these factors is still limited with some rare exceptions that, however, relate to the broader precision agriculture (PI) technologies. For instance, a major limitation is represented by the high costs of these technologies [24] which provide information thanks to expensive sources such as satellites, remote sensing, or sensors. Despite the usefulness of the information delivered by these sources, data may suffer from low temporal and spatial accuracy, consequently influencing their reliability [23].

Moreover, information is usually delivered neglecting the users' skills to effectively manage it. Accordingly, Galioto et al. [23] (p. 4) state that a "farmer's skills and financial capacity, coupled with his/her networking capacity and opportunity to consult service providers are considered the main factors conditioning the adoption of precision agriculture". The lack of tariffs on water use, such as water pricing, is an additional factor that negatively influences the choice of adopting PI [12]. As a matter of fact, regulatory instruments also include rules of use (i.e., turns and quotas) that may favour the adoption of more efficient irrigation technologies [25].

Regarding IASs, specific limitations have been identified by Smith and Muñoz [26]. First, they found that the complexity of data and information provided by IASs require specific knowledge and may not easily be translated into operational advice for farmers that often lack specific skills to understand the information provided. Furthermore, this complexity may result in no user-friendly interfaces. Second, the purpose of IASs is often not in line with farmers' interests and priorities; they are usually more interested in increasing productivity rather than water efficiency, as suggested also by Levidow et al. [27]. Third, most IASs are developed within specific grants with a limited time duration, and this negatively affects their adoption. Fourth, public and private financial resources to sustain the service are limited and IASs sustainability can only be guaranteed by the willingness and ability to pay farmers. Lastly, a lack of communication and trained users may reduce the adoption among potential beneficiaries.

2.2. The Business Model

Scaling-up the diffusion of IASs and overcoming their major limitations require a fundamental reconsideration of the services offered, to which users are offered and in which way. This also implies a reconsideration of the costs of sustaining the services and the revenues that are generated by those services. To propose innovations in the way IASs may sell their services, earn money, and deliver value for their users, a business model perspective has been adopted, intended as a system of interdependent activities that enable the IASs to create value [28].

The concept of the business model came to the spotlight in the 1900s to communicate business ideas to potential investors within a limited time frame [29]. Nowadays, the main purpose of any business model is to describe "how a firm does business". As a conceptual tool, a BM provides great support in assessing the performance, management, communication, and innovation of a business [30]. Chesbrough and Rosenbloom [15] defined BM as a framework to convert technical potential into economic value. Richardson [31] proposed a widely accepted framework for BMs to describe how the firm captures, creates, and delivers value. Additionally, Zott and Amit [28] considered BM as a bundle of specific activities that depicts the way a company "does business" with its customers, partners and vendors. The majority of concepts of business found in the literature closely link BMs to value creation for firms and customers [13,22,32]. This paper adopts the definition proposed by Osterwalder [33] (p. 15), according to which BM can be defined as "a conceptual tool that contains a set of elements and their relationships and allows expressing a company's logic of earning money. It is a description of the value a company offers to one or several segments of the customers and the architecture of the firm and its network of partners for creating, marketing, and delivering this value and relationship capital, to generate profitable and sustainable revenue streams". This definition fits perfectly into our aim of designing an innovative BM capable of describing the essential elements needed to enable the value proposition to be transferred as benefits to the customers, guaranteeing the economic sustainability of IASs.

3. Material and Methods

3.1. Business Model Canvas

To design the BM, this study applies the tool "Business Model Canvas" (BMC) proposed by Osterwalder and Pigneur [22]. BMC is a visual map that represents the elements of a business model in a holistic manner, also displaying potential interconnections among these and showing how they can impact on value creation [34]. The tool is based on nine interconnected axes where each represents a specific and essential BM component as: the customer segments, the value proposition for the customers, the channels to communicate with the customers, the customers' relationships, key resources, key activities, the definition of the key partnerships, costs, and revenues. By providing a visual representation of each element of the business system, BMC is particularly helpful to explore potential innovations with an inside-out approach [35,36]. Notably, BMC has represented an effective tool to address sustainability issues [34,37]. Nowadays, the success of a business increasingly encompasses the ability to design business models that are able to deal with challenges and opportunities linked to the society and its transition towards sustainability. Hence, the integration of social and environmental issues while creating and delivering value is fundamental.

3.2. Data Collection

This paper builds on the results obtained within the OPERA project to understand both the limitations behind the adoption of an IAS and the needs of stakeholders involved in the agricultural sector (i.e., farmers, water managers, and policymakers). Data were collected in six countries (i.e., France, Spain, Italy, The Netherlands, South Africa, and Poland), namely the six case studies of the OPERA project (Table 1). Each of them presents different climatic conditions; however, water scarcity conditions and the urgency to increase water use efficiency are the common denominator at all sites.

Table 1. Location of the OPERA case studies and stakeholders involved.

Country	Pilot Area	Stakeholders Involved
France	Crau aquifer area	Farmers, irrigation association, water agency, aquifer syndicate, municipalities, and farmers associations.
Italy	Campania	Farmers, regional government, land and water reclamation authorities, farmer associations, local policy makers, and legislators.
Poland	sub-catchment Zglowiaczka	Farmers, regional agricultural advisory centre, and local policymakers.
South Africa	Breede River valley	Farmers, officials in the water sector, consultants, and academics.
Spain	Andalusia	Farmers associations, irrigation associations, local policymakers, and NGOs.
The Netherlands	Reusel	Meteo service and Farmers.

Stakeholders were surveyed to identify their needs, perceived barriers to adopting IASs, and their preferences towards the characteristics of IASs. These data allowed for the definition of the BMC elements to develop an innovative business model for IASs, designing a value proposition fully tailored to potential customers' needs and capable of overcoming all the possible barriers to IASs' adoption.

More specifically, two different data collections were conducted along with the project. The first questionnaire was administered between 2017 and 2018 to a total of 123 stakeholders involved in water management, including farmers, water users' associations (WUAs), and local authorities in the six case studies of the project (see Table 1). The main aim of the survey was to identify the needs and demands of the agricultural sector. To do so, respondents were asked to elicit the main important actions to increase their competitiveness, their propensity to adopt new technologies in irrigation, and how to increase their water use efficiency. Data were collected through a structured questionnaire with closed-ended questions. Results show that stakeholders are interested in improving the sustainability of the production process and are willing to adopt new technologies for increasing water use efficiency. However, respondents from each case study identified the cost of sensors or advice tools as a limitation to adopt new technologies. The full results of the questionnaire are available in the "D1.1 Assessment of user requirements of the sector" of OPERA at this link http://opendata.waterjpi.eu/dataset/2a2a87e0-5c84-42cd-a9da-ecac0bbb9257/resource/09d7444c-c5e2-4473-835b-9c28f27d20d3/download/d1.1_report_stakeholder_opera.pd (accessed on 11 October 2022) [38].

A second data collection took place between 2018 and 2019 involving only farmers as respondents. From the six case studies of OPERA, 108 farmers were surveyed using face-to-face interviews. The questionnaire was composed of two parts. The first part consisted in a strengths, weaknesses, opportunities, and threats (SWOT) analysis that was implemented to understand the most important internal and external factors influencing IASs' adoption. As suggested by many authors, e.g., [39,40], the SWOT analysis represents a valid methodology for examining problems related to water management in agriculture. This method is largely adopted because it has the potential to clarify the present conditions with respect to the strengths and weaknesses (i.e., internal factors), and the future implications from opportunities and threats (i.e., external factors). Through the SWOT analysis, it was possible to understand farmers' opinions about IASs implementation within the agricultural sector. The strengths section considers the gains and advantages of the adoption of the IAS. Results show that water saving is considered the main strength followed by costs reduction. Weaknesses represent the obstacles to adopting IAS: among these, the results show the low ability to use electronic devices for water management. Opportunities are related to external benefits generated for farmers when adopting IAS. Here, farmers identified different benefits such as the improvement of agricultural productivity and an increased competitive advantage in the market. Lastly, threats represent the elements that may friction the diffusion of the IAS at farm levels. Major threats for farmers are their low level of education and the perceived lack of public financial support. Table 2 resumes the main findings of the SWOT analysis conducted within the OPERA project.

The second part of the questionnaire was used to investigate farmers' preferences for IASs through a choice experiment (CE), by examining several IASs alternatives based on five different attributes plus a monetary option to investigate their willingness to pay for IASs. The results, which are widely described by Altobelli et al. [8], show that the adoption of IASs is positively influences by the time length of forecasts (more days available in the future are preferred), and the need to record water information (longer time intervals needed to record information are preferred). On the other hand, the adoption is negatively influenced by the price, the duration (time length) of the contract, and the frequency of satellite data availability for crop monitoring. The full results of the SWOT and CE analysis are described in the "D4.1 Report on socio-economic assessment" of OPERA [41].

To sum up, the results from these data collections represent the foundation for designing the BMC for IASs, which is at the core of this work that draws from stakeholders' and farmers' opinions of our value proposition for the customers, the cost structure, and the revenue strategy. Based on these three main elements we defined what are the key activities to be implemented, the resources needed, the possible partners, and the communication and marketing channels (Table 3).

Table 2. SWOT analysis results. Source: adapted from OPERA project—Report on socioeconomic assessment.

Internal Parameters (Present)		External Parameters (Future)	
Strength	**Weakness**	**Opportunity**	**Threat**
S.1 Water savings	**W.1** Low use of electronic devices for water management	**O.1** Improving agricultural productivity	**T.1** Social aspect, education
S.2 Cost reduction	**W.2** Negative perception of information provided by IASs	**O.2** New market and consumers	**T.2** Lack of funding
S.3 Capacity and competence	**W.3** Lack of funds for IASs implementation	**O.3** Increase water management through ITC	**T.3** Lack of institutional mechanisms to link rural communities
S.4 Good network of land reclamation and irrigation consortia		**O.4** Reducing environmental pollution	
S.5 Innovation development			
S.6 Provisioning of water measurements			

Table 3. Description of the data collection elements taken to implement the BMC.

Data Collection	Scope	Elements of the Questionnaires	Elements of the BM
Stakeholders questionnaire	Identifying the sector needs to increase water use efficiency	Q1. Actions to increase the competitiveness of farm into the market	Value proposition; Customer segments.
		Q2. Interest in adopting new technologies	Value proposition; Customer segments.
		Q.3 Main limitation on improving irrigation efficiency	Value proposition; Customer segments.; Revenues strategy
		Q4. Preference options that an irrigation support tool should include	Value proposition; Cost structure; Customer segments; Revenues strategy
SWOT analysis	Understand farmers' opinions about IASs implementation within the agricultural sector	Strengths, Weaknesses, Opportunities and Threats of IASs	Value proposition
CE and WTP	Investigate farmers' preferences for IASs	IASs attribute preferences	Value proposition; Cost structure
		Willingness to pay for IASs	Revenues strategy

4. Results and Discussion

4.1. Business Model Development

This section outlines the business model development through the description of key components of the BMC. The stakeholders involved in the OPERA project were essential for our research purposes and represented a relevant source of knowledge to understand IASs' limitations and, consequently, to design the BM. Indeed, the BM design is intended to propose a new way to create and deliver value within IASs that may be helpful to overcome the identified limitations linked to the adoption and diffusion of IASs.

First, potential IASs' users and their needs are identified. Then, the value proposition and the elements that create value in the service are described. A strategy to generate revenues that outweigh costs is proposed to ensure the IASs' financial sustainability. The other axes of the BMC are briefly described through a graphic visualization of the canvas (Figure 1).

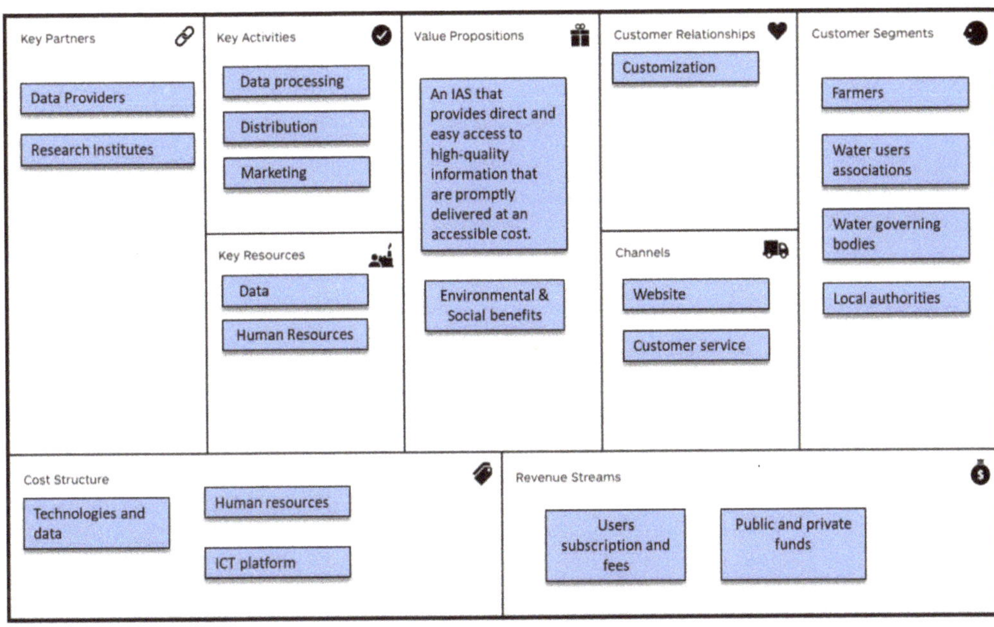

Figure 1. The nine elements of BMC for OPERA-related IASs.

4.1.1. Customer Segments and Their Needs

The first step in developing our business model was to define potential adopters of IASs. As a matter of fact, a financially sustainable IAS relies on the willingness to pay of its users. The OPERA potential users are all those who manage water resources for irrigation purposes and those who take decisions about water resources management. Generally, three main user segments, their interests and needs can be identified, as shown in Figure 2. Farmers have to decide on how much and when to irrigate to maximize crop productivity and profits. In addition, managers of water users' associations have to monitor the irrigation water consumption over seasons to comply with the exploitation plan, while regional authorities require a spatially distributed monitoring of the water-exploitation plan on the irrigation schemes, aquifers, or river basins. All these actors require access to a great amount of information coming from different sources to properly manage water resources.

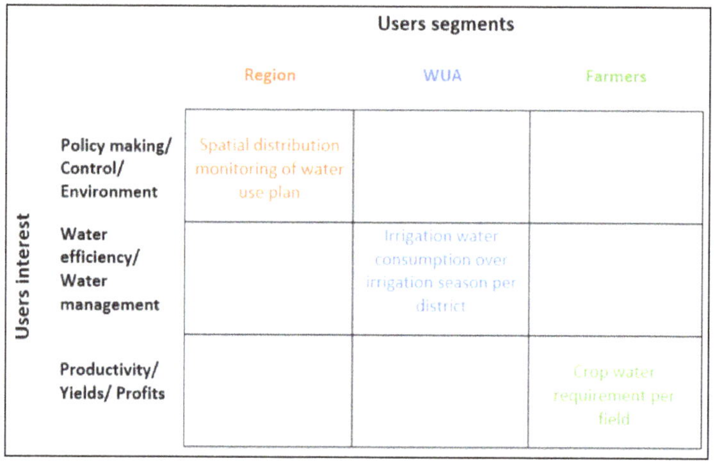

Figure 2. User needs: description of the user segments (region, water users' associations—WUA, farmers) and relationship with their interests.

However, the availability of data for water management alone is not sufficient to guarantee the increase of water use efficiency in irrigation. Much depends on how these data are delivered and used. According to the results from the stakeholders' need assessment, the most important characteristic of an IAS should be the "affordable cost", followed by "direct access to the information", "easy use of information", and "regularity in the delivery of information". Farmers and other water users are inclined to adopt new technologies to increase water efficiency in irrigation together with a proper irrigation strategy, as technologies can increase their market competitiveness, improve the sustainability of the production process, and lower the costs. However, stakeholders consider several constraints to the market expansion of IASs in the private sector: affordability (cost of the technology relative to the farmer's income level, awareness (knowledge about the technology), accessibility (options for obtaining the technology), and lack of customization (capacity to match farmer needs with technological solutions). Based on this, it is possible to conclude that users prefer IASs characterized by a direct and easy access to high-quality information that are promptly and constantly delivered at an accessible cost.

4.1.2. The Value Proposition

The value proposition represents the core of any business model. According to Osterwalder [33], value propositions are products and services that create value for a specific customer segment. The ambition was to develop a service that could effectively support water users in decision-making for irrigation-management purposes, particularly under the anticipation of climate variability and critical moments of water scarcity. However, it is fundamental that sustainable innovation meets user needs to be adopted successfully [42]. Hence, in our value proposition, value is created to address the specific needs of the users. To this purpose, the IAS will be built upon some key features: **integration**, **customization**, **accessibility**, and **sustainability**.

The main aspect to consider is integration. the IAS will make use of numerous available technologies (e.g., Earth-observation or sensor) consequently providing different data outcome and information. Data will be collected and integrated into an information and communication technology (ICT) solution to be used by final users. The wide availability of data will allow the users to choose the information that is most useful to them to meet their water management needs. Indeed, the needs of users are very different depending on the socioeconomic context and the climatic and environmental conditions. This translates into a need for customization of the service that starts with the possibility of choosing from a wide range of information and continues with a payment service based on the concept of pay for what you use. As high costs have been recognized as one of the main limitations in adopting IASs, a pay-for-what-you-use approach may be able to limit the cost for the users. This means that users will not be forced to pay a fee for the whole irrigation advisory service; instead, they will pay to access the information they need when they need it. Lowering the costs means increasing the accessibility of the service by reaching more and more users. The concept of accessibility also refers to the possibility of accessing the information provided by IASs. As a matter of fact, among the major limitations identified by stakeholders, there is the complexity of information that may result in not user-friendly interfaces and in difficulties to interpret and effectively use the data. Hence, the IAS will be built to offer an intuitive interface that allows the users to use the products in the easiest way and to visualize complex information in a simple graphic vest. Additionally, the data will be prior processed and returned to the users as accessible and ready-to-use practical irrigation advice. Practically, the service will be an ICT solution accessible through a smartphone application, with a user-friendly interface. The tool will work with a great amount of processed data providing (i) the ability to explore data from a catalogue by choosing them based on specific needs; (ii) the possibility of evaluating the scenarios resulting from the choices; and (iii) the possibility to receive daily weather bulletins and alerts when extreme events are coming. Finally, integration, customization and accessibility are essential to reach the sustainability of IAS (Figure 3). Thus, here the IAS is conceived to be an ICT tool for the optimization of irrigation management, by increasing water use efficiency and reducing production costs for farmers, but also to ensure sustainability throughout all the service processes. Sustainability is intended in its environmental, social, and financial terms. From the environmental perspective, the widespread diffusion of the IAS among farmers and other water users will result in a more sustainable agriculture able to produce more with less water resources, minimizing the negative environmental impact of irrigation. Improving water efficiency also brings social benefits by optimizing the access to water among different stakeholders and making water available for purposes different from agriculture (e.g., more water for citizens), thus ensuring increasing water security for all. Moreover, the increased water efficiency and productivity will result in increased agricultural outputs that can be translated into greater food security. As for the case of Gebrezgabhe et al., 2021 [19], our value proposition is intended not only to attract customers to adopt IASs but also to encourage sustainable water management practices in line with the perspective of a sustainable BMC [34].

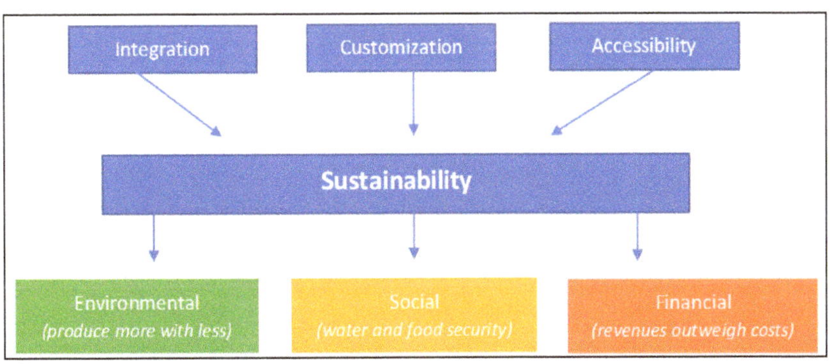

Figure 3. Workflow of sustainability in the value proposition.

4.1.3. Revenues and Financial Sustainability

Since financial sustainability is fundamental to the deployment of IASs, a separate paragraph describes our revenues-generation strategy. Among the main limitations in adopting IASs, financial constraints, namely the lack of funding to sustain the IAS and the high costs of the service, are the most relevant, resulting in low affordability for the users. Most IAS are developed under research projects and consequently can rely on limited funding over time. In these cases, the service is offered for free—or almost for free—to the users. Instead, where private enterprises provide the service, the cost of the IAS is much higher with respect to the farmer's income level. A financially sustainable IAS should depend on the willingness to pay of its potential users rather than external funding. For this reason, IASs must operate a shift from service for free to service for fee. On the other hand, the high costs associated with the provision of this service represent an adoption barrier. However, the result of our WTP analysis shows that farmers are willing to pay to introduce an irrigation advisory system that results in an economic advantage over their current situation. The effects of better irrigation management can be successful in terms of increasing farmers' income and to diminish the energy costs incurred by the management of water bodies.

For all these reasons, a revenue strategy that is based on the concept of pay for what you use is proposed, namely pay as you go (PAYG). Business models based on PAYG give users the ability to pay for only what they use as they need it (e.g., the information they need at any time they need it) and can afford it. In addition, they can also choose to pay a fixed fee for a contract that can be monthly or yearly. The contract can be customized with features that reflect the users need (e.g., interval of information delivery, amount of information, time length of forecasts, and use of scenario for the decision making). All these features will shape the total amount of the fee so that all those who manage water for irrigation purposes can access the service even with basic functionalities. This payment model enables a win-win situation in which the risks are minimized both for users and providers. Indeed, users can access modern irrigation technologies at an affordable cost while IASs providers reduce the operational risk and recover its cost. PAYG models are widely adopted for irrigation technologies, especially when farmers' access to finance is limited [43,44].

Additionally, IAS may rely on external funding for extra revenues and benefits. As a matter of fact, access to financing represents a great challenge to scaling-up agricultural and irrigation technologies. Different finance mechanisms can include national direct and indirect support programs, such as credit guarantee funds, value chain financing, and price smoothing.

5. Conclusions

IASs improve water efficiency in irrigation, gaining increased productivity, and reducing costs for farmers, together with environmental and social benefits. However, their widespread adoption has been limited among users and managers of water for irrigation. Surveying relevant stakeholders and potential users within the OPERA project revealed that the constraints in adopting IASs may arise either from the user's side, including farmers' low skills and knowledge of new technologies, and the services side, such as the low accessibility of information or the high costs. Despite the existing limitations, farmers are willing to adopt IASs. Nevertheless, they require that the service is easily accessible, with high-quality information that are delivered at an affordable cost. BM seems to be a crucial tool to support IASs' strategic organization through the representation of the elements that the service may innovate to create, deliver, capture, and exchange value with its customers. The proposed BM is innovative in the sense that it is tailored to the needs of potential users and has the potential to overcome the IASs diffusions barriers. The implementation of this innovative business model will ensure that IAS becomes financially sustainable, without the need for continued public funding, but only relying on the willingness to pay its users. Further, a well-structured business that leaves the service profitable and that is capable to ensure environmental and social sustainability while providing water use efficiency, will attract potential investors, including governments and public agencies, to fund new schemes. This research contributed to expanding the literature on business models and their relationship with sustainable innovations for irrigation. Despite business models seeming to be promising tools to support the diffusion and commercialization of IASs, further research is needed to empirically analyze the response of the users and the markets to new ways to create value.

Limitations of the Study

This study represents novel research on IASs by taking a step forward in the application of BMC to potentially scale up IASs through innovation and business perspectives. However, it has several limitations. First, the lack of previous studies related to the topic limited the possibility of conducting a systematic literature review and created some difficulties in discussing our results. Indeed, the concept of BM has been rarely applied to IASs. A second limitation can be found in the study regarding the design of the BM. As a matter of fact, it was not possible to segment our value proposition accordingly to the different potential users that have been identified. This occurred because, during the data collection, results were not classified according to stakeholders' categories (i.e., farmers, authorities, WUAs, etc.). Hence, the value proposition has been designed to be as inclusive as possible to take into account the needs of all the customer segments. Lastly, our study represents a theoretical exercise with the scope of understanding enablers and barriers of IASs to propose an innovative business strategy capable to increase their diffusion. The application of the BM to a specific case study is missing. Further research should address this issue by implementing a BM in a real IAS company.

Author Contributions: Conceptualization, A.S., F.A., A.D.M., T.D.G., O.C. and M.C.A.; methodology, A.S., F.A., A.D.M., T.D.G. and O.C.; validation, E.G., F.A. and O.C.; investigation, F.A., T.D.G., O.C., A.D.M. and A.S.; data curation, A.S. and F.A.; writing—original draft preparation, A.S., F.A. and E.G.; writing—review and editing, A.D.F., A.M., F.J.B.-V. and A.S.; visualization, A.S.; supervision, F.A.; project administration, F.A. and A.D.M.; funding acquisition, F.A. and A.D.M. All authors have read and agreed to the published version of the manuscript.

Funding: This research received no external funding.

Institutional Review Board Statement: Not applicable.

Data Availability Statement: Data are available on request from the corresponding author.

Acknowledgments: This study was carried out as part of the OPERA project. The authors would like to thank the EU and The Ministry of Economic Affairs (The Netherlands), CDTI (Spain), MINECO (Spain), ANR (France), MIUR (Italy), NCBR (Poland) and WRC (South Africa) for funding, in the frame of the collaborative international consortium OPERA financed under the ERA-NET Cofund WaterWorks2015 Call. This ERA-NET is an integral part of the 2016 Joint Activities developed by the Water Challenges for a Changing World Joint Program Initiative (Water JPI).

Conflicts of Interest: The authors declare no conflict of interest. The partners of the OPERA project were informed and involved in the work, facilitating the access, and the use of data.

References

1. IPCC. *AR5 Synthesis Report: Climate Change 2014*; No. 2; IPCC: Geneva, Switzerland, 2014; Volume 218. Available online: https://www.ipcc.ch/report/ar5/syr/ (accessed on 26 April 2022).
2. HLPE. Water for food security and nutrition. A report by the High Level Panel of Experts on Food Security and Nutrition of the Committee on World Food Security. 2015. Available online: www.fao.org/cfs/cfs-hlpe (accessed on 26 April 2022).
3. FAO. *The State of Food and Agriculture 2020*; FAO: Rome, Italy, 2020. [CrossRef]
4. FAO. The State of Food and Agriculture: Climate Change, Agriculture and Food Security. 2016. Available online: www.fao.org/publications (accessed on 24 April 2022).
5. FAO. *Coping with Water Scarcity: An Action Framework for Agriculture and Food Security*; Food and Agriculture Organization of the United Nations: Rome, Italy, 2012; p. 79. Available online: http://www.fao.org (accessed on 24 October 2022).
6. Giordano, M.; Turral, H.; Scheierling, S.M.; Treguer, D.O.; McCornick, P.G. *Beyond 'More Crop per Drop': Evolving Thinking on Agricultural Water Productivity*; International Water Management Institute: Washington, DC, USA, 2017. [CrossRef]
7. Altobelli, F.; Lall, U.; Dalla Marta, A.; Caracciolo, F.; Cicia, G.; D'Urso, G.; Del Giudice, T. Willingness of farmers to pay for satellite-based irrigation advisory services: A southern Italy experience. *J. Agric. Sci.* **2018**, *156*, 723–730. [CrossRef]

8. Altobelli, F.; Marta, A.D.; Heinen, M.; Jacobs, C.; Giampietri, E.; Mancini, M.; Cimino, O.; Trestini, S.; Kranendonk, R.; Chanzy, A.; et al. Irrigation Advisory Services: Farmers preferences and willingness to pay for innovation. *Outlook Agric.* **2021**, *50*, 277–285. [CrossRef]
9. Vuolo, F.; D'Urso, G.; De Michele, C.; Bianchi, B.; Cutting, M. Satellite-based irrigation advisory services: A common tool for different experiences from Europe to Australia. *Agric. Water Manag.* **2015**, *147*, 82–95. [CrossRef]
10. Larson, B. *Sustainable Development Research Advances*; Nova Science Publishers: New York, NY, USA, 2007. Available online: https://books.google.it/books?hl=it&lr=&id=_9uzFtt4yqsC&oi=fnd&pg=PR7&ots=gb6jEMZiS6&sig=L2BvBWQEpKz3Beev_k5L0nrk1lU&redir_esc=y#v=onepage&q&f=false (accessed on 26 April 2022).
11. Martín De Santa Olalla Mañas, F.; Brasa Ramos, A.; Fabeiro Cortés, C.; Fernández González, D.; López Córcoles, H. Improvement of irrigation management towards the sustainable use of groundwater in Castilla-La Mancha, Spain. *Agric. Water Manag.* **1999**, *40*, 195–205. [CrossRef]
12. Giannakis, E.; Bruggeman, A.; Djuma, H.; Kozyra, J.; Hammer, J. Water pricing and irrigation across Europe: Opportunities and constraints for adopting irrigation scheduling decision support systems. *Water Supply* **2016**, *16*, 245–252. [CrossRef]
13. Teece, D.J. Business Models, Business Strategy and Innovation. *Long Range Plann.* **2010**, *43*, 172–194. [CrossRef]
14. Bocken, N.M.P.; Short, S.W.; Rana, P.; Evans, S. A literature and practice review to develop sustainable business model archetypes. *J. Clean. Prod.* **2014**, *65*, 42–56. [CrossRef]
15. Chesbrough, H.; Rosenbloom, R.S. The role of the business model in capturing value from innovation: Evidence from Xerox Corporation's technology spin-off companies. *Ind. Corp. Chang.* **2002**, *11*, 529–555. [CrossRef]
16. Bohnsack, R.; Pinkse, J.; Kolk, A. Business models for sustainable technologies: Exploring business model evolution in the case of electric vehicles. *Res. Policy* **2014**, *43*, 284–300. [CrossRef]
17. Amani, M.; Isaai, M.T.; Farhadi, A. A Novel Business Model for a Cyber-Physical System(CPS)-Based Smart Irrigation Network. *Mark. Sci. Technol. J.* **2022**, *1*, 165–179. Available online: https://www.mstj.net/article_252210.html (accessed on 24 October 2022).
18. De Wet, C. Development of Internet of Things Technology and Business Model for a Centre Pivot Irrigation System (Powasave) to Optimise Power and Water Consumption. Available online: https://scholar.sun.ac.za/bitstream/handle/10019.1/123798/dewet_internet_2021.pdf?sequence=2 (accessed on 11 October 2022).
19. Gebrezgabher, S.; Leh, M.; Merrey, D.J.; Kodua, T.T.; Schmitter, P. Solar photovoltaic technology for small-scale irrigation in ghana: Suitability mapping and business models. *IWMI Res. Rep. 178* **2021**. [CrossRef]
20. Otoo, M.; Lefore, N.; Schmitter, P.; Barron, J.; Gebregziabher, G. *Business Model Scenarios and Suitability: Smallholder Solar Pump-Based Irrigation in Ethiopia. Agricultural Water Management—Making a Business Case for Smallholders*; International Water Management Institute: Colombo, Sri Lanka, 2018. [CrossRef]
21. Bonfante, A.; Monaco, E.; Manna, P.; De Mascellis, R.; Basile, A.; Buonanno, M.; Cantilena, G.; Esposito, A.; Tedeschi, A.; De Michele, C.; et al. LCIS DSS—An irrigation supporting system for water use efficiency improvement in precision agriculture: A maize case study. *Agric. Syst.* **2019**, *176*, 102646. [CrossRef]
22. Osterwalder, A.; Pigneur, Y. *Business Model Generation: A Handbook for Visionaries, Game Changers, And Challengers*; John Wiley Sons: Hoboken, NJ, USA, 2010. Available online: https://books.google.it/books?hl=it&lr=&id=UzuTAwAAQBAJ&oi=fnd&pg=PP1&dq=canvas+osterwalder+pigneur+2010&ots=yYGNwjG41s&sig=NI27RFBeYBzCrywutWk1FWn9Lts&redir_esc=y#v=onepage&q=canvasosterwalderpigneur2010&f=false (accessed on 26 April 2022).
23. Galioto, F.; Raggi, M.; Viaggi, D. Assessing the potential economic viability of precision irrigation: A theoretical analysis and pilot empirical evaluation. *Water* **2017**, *9*, 990. [CrossRef]
24. Sadler, E.J.; Evans, R.G.; Stone, K.C.; Camp, C.R. Opportunities for conservation with precision irrigation. *J. Soil Water Conserv.* **2005**, *60*, 371–378.
25. Molle, F. Can Water Pricing Policies Regulate Irrigation Use? In Proceedings of the 13th World Water Congress, Montpellier, France, 1–4 September 2008.
26. Smith, M.; Muñoz, G. Irrigation Advisory Services for effective water use: A review of experiences. In Proceedings of the FAO-ICID Workshop "Irrigation Advisory Services and Participatory Extension in Irrigation Management", Montreal, QC, Canada, 24 July 2002.
27. Levidow, L.; Zaccaria, D.; Maia, R.; Vivas, E.; Todorovic, M.; Scardigno, A. Improving water-efficient irrigation: Prospects and difficulties of innovative practices. *Agric. Water Manag.* **2014**, *146*, 84–94. [CrossRef]
28. Zott, C.; Amit, R. Business Model Design: An Activity System Perspective. *Long Range Plann.* **2010**, *43*, 216–226. [CrossRef]
29. Zott, C.; Amit, R.; Massa, L. The Business Model: Recent Developments and Future Research. *J. Manag.* **2011**, *37*, 1019–1042. [CrossRef]
30. Osterwalder, A.; Pigneur, Y.; Tucci, C.L.; Osterwalder, A.; Pigneur, Y. Clarifying Business Models: Origins, Present, and Future of the Concept. *Commun. Assoc. Inf. Syst.* **2005**, *16*, 1. [CrossRef]
31. Richardson, J. The business model: An integrative framework for strategy execution. *Strateg. Chang.* **2008**, *17*, 133–144. [CrossRef]
32. Geissdoerfer, M.; Bocken, N.M.P.; Hultink, E.J. Design thinking to enhance the sustainable business modelling process—A workshop based on a value mapping process. *J. Clean. Prod.* **2016**, *135*, 1218–1232. [CrossRef]
33. Osterwalder, A. The Business Model Ontology a Proposition in a Design Science Approach. Ph.D. Thesis, Universite de Lausanne Ecole des Hautes Etudes Commerciales, Lausanne, Switzerland, 2004.

34. Joyce, A.; Paquin, R.L. The triple layered business model canvas: A tool to design more sustainable business models. *J. Clean. Prod.* **2016**, *135*, 1474–1486. [CrossRef]
35. Baden-Fuller, C. Strategic Innovation, Corporate Entrepreneurship and Matching Outside-in to Inside-out Approaches to Strategy Research. *Br. J. Manag.* **1995**, *6*, S3–S16. [CrossRef]
36. Chesbrough, H.W.; Garman, A.R. How open innovation can help you cope in lean times. *Harv. Bus. Rev.* **2009**, *87*, 68–76. [CrossRef] [PubMed]
37. França, C.L.; Broman, G.; Robèrt, K.-H.; Basile, G.; Trygg, L. An approach to business model innovation and design for strategic sustainable development. *J. Clean. Prod.* **2017**, *140*, 155–166. [CrossRef]
38. Blanco-Velázquez, F.J.; Anaya-Romero, M. OPERA WP1 D1.1 Assessment of User Requirements of the Sector-Identifying Sector Needs to Increase Resource Use Efficiency. 2018. Available online: http://opendata.waterjpi.eu/dataset/2a2a87e0-5c84-42cd-a9da-ecac0bbb9257/resource/09d7444c-c5e2-4473-835b-9c28f27d20d3/download/d1.1_report_stakeholder_opera.pd (accessed on 11 October 2022).
39. Kallioras, A.; Pliakas, F.; Diamantis, I.; Kallergis, G. SWOT analysis in groundwater resources management of coastal aquifers: A case study from Greece. *Water Int.* **2010**, *35*, 425–441. [CrossRef]
40. Mainali, B.; Ngo, H.H.; Guo, W.; Pham, T.T.N.; Johnston, A. Feasibility assessment of recycled water use for washing machines in Australia through SWOT analysis. *Resour. Conserv. Recycl.* **2011**, *56*, 87–91. [CrossRef]
41. Altobelli, F.; Cimino, O.; Marta, A.D. OPERA WP4 D4.1 Report on Socio-Economic Assessment. 2019. Available online: http://opendata.waterjpi.eu/dataset/2a2a87e0-5c84-42cd-a9da-ecac0bbb9257/resource/32392134-35a4-499b-8fb2-76085e41a93a/download/d4.1-report-on-socio-economic-assessment.pdf (accessed on 11 October 2022).
42. Long, T.B.; Blok, V.; Poldner, K. Business models for maximising the diffusion of technological innovations for climate-smart agriculture. *Int. Food Agribus. Manag. Rev.* **2017**, *20*, 5–23. [CrossRef]
43. Ockwell, D.; Atela, J.; Mbeva, K.; Chengo, V.; Byrne, R.; Durrant, R.; Kasprowicz, V.; Ely, A. Can Pay-As-You-Go, Digitally Enabled Business Models Support Sustainability Transformations in Developing Countries? Outstanding Questions and a Theoretical Basis for Future Research. *Sustainability* **2019**, *11*, 2105. [CrossRef]
44. Mashnik, D.; Jacobus, H.; Barghouth, A.; Wang, E.J.; Blanchard, J.; Shelby, R. Increasing productivity through irrigation: Problems and solutions implemented in Africa and Asia. *Sustain. Energy Technol. Assess.* **2017**, *22*, 220–227. [CrossRef]

agriculture

MDPI

Article

Who Is Going Green? Determinants of Green Investment Intention in the Saudi Food Industry

Meqbel M. Aliedan [1], Mansour A. Alyahya [1], Ibrahim A. Elshaer [1,2] and Abu Elnasr E. Sobaih [1,3,*]

1 Management Department, College of Business Administration, King Faisal University, Al-Ahsa 31982, Saudi Arabia; maliedan@kfu.edu.sa (M.M.A.); malyahya@kfu.edu.sa (M.A.A.); ielshaer@kfu.edu.sa (I.A.E.)
2 Faculty of Tourism and Hotel Management, Suez Canal University, Ismailia 41522, Egypt
3 Faculty of Tourism and Hotel Management, Helwan University, Cairo 12612, Egypt
* Correspondence: asobaih@kfu.edu.sa

Abstract: The Kingdom of Saudi Arabia (KSA) has witnessed major transformations in social, economic, and environmental aspects since the inauguration of Saudi Vision 2030 in April 2016. In alignment with this, the leadership of KSA has inaugurated green initiatives that pave the way for green investment opportunities in different industries within KSA. However, there was limited, if any, research about green investment intention and behaviour in KSA. This research tests an expanded model of the Theory of Planned Behaviour (TPB) to investigate the determinants of green investment intention in the Saudi food industry. A questionnaire survey was electronically directed to 550 fresh agricultural and food sciences graduates in public KSA universities. The results of PLS-SEM showed significant positive influences of the attitude, perceived behavioural control, green investment knowledge, and green consumption commitment on the green investment intention of potential investors. However, the results confirmed a negative influence of subjective norms on green investment intention. The results also confirmed a moderating role of religiosity on the relationship between attitude, perceived behavioural control, green consumption commitment, and green investment intention. The results send some important messages to scholars and policymakers in higher education regarding the foundation of green investment among their graduates, which are elaborated.

Keywords: green investment intention; theory of planned behaviour; religiosity; green consumption commitment; green investment knowledge

Citation: Aliedan, M.M.; Alyahya, M.A.; Elshaer, I.A.; Sobaih, A.E.E. Who Is Going Green? Determinants of Green Investment Intention in the Saudi Food Industry. *Agriculture* **2023**, *13*, 1047. https://doi.org/10.3390/agriculture13051047

Academic Editors: Roberto Henke and Filiberto Altobelli

Received: 10 March 2023
Revised: 27 March 2023
Accepted: 10 May 2023
Published: 12 May 2023

1. Introduction

The global food system is facing several challenges related to food security, environmental sustainability, social equity, and economic viability [1–3]. These challenges are driven by the increasing demand for food due to increasing population, water scarcity, climate change, changes in consumption patterns, and the depletion of natural resources [1]. To address these challenges, sustainable food production practices have been developed to reduce environmental impacts and improve the livelihoods of smallholder farmers. In the context of Saudi Arabia, the food industry is an important sector that plays a significant role in the country's economy [4]. However, the industry is faced with numerous environmental challenges that influence its sustainability, e.g., water scarcity, desertification, and land degradation [5]. Green investment intention has emerged as a viable strategy to promote sustainable food production practices in the Saudi food industry. The Saudi government has launched several initiatives to promote the eco-system. These initiatives provide financing and incentives for sustainable agriculture, which can attract green investments and promote sustainable food production practices. This research investigates the determinants of green investment intention to promote sustainable agricultural food production in the Saudi food industry. In the context of the Saudi food industry, green investments can promote sustainable food production by financing the adoption of sustainable practices such as conservation agriculture and organic farming. These practices can promote soil health, increase crop yield, reduce greenhouse gas emissions, and enhance biodiversity.

The initiation of the Kingdom of Saudi Arabia (KSA) Vision 2030 promoted considerable transformations in the social, economic, and governmental landscape [6]. The Saudi Vision 2030 is a comprehensive national improvement strategy with three main objectives: a vibrant economy, a thriving society, and an ambitious nation [7]. The Vision has led to enormous changes in the economy's operation by promoting non-oil sectors instead of comprehensively relying on oil [7]. The Vision focuses on generating a diverse, thriving, and sustainable economy to supplement the quality of lives of citizens [7]. Environmental sustainability is one of the key issues in the Vision. Hence, there are several green initiatives promoted by the leadership of KSA. This was started by the King Salman Renewable Green Energy initiative in 2016, the National Renewable Energy Program in 2017, the launch of the National Environment Strategy in 2108, and the "Let's Make it Green" Campaign in 2020 [8,9]. In 2021, The Middle East Green Initiative and Saudi Green Initiative were inaugurated by the Crown Prince and Prime Minister of KSA, Mohamed Bin Salman [8,9]. These initiatives aim to create a greener future for all and meet the challenges of climate change [8]. The Saudi Green Initiative has three main objectives, i.e., reducing emissions, greening Saudi, and protecting the land and sea [9]. The first objective is to reduce emissions by accelerating the green energy transition in KSA and mitigating the impacts of climate change [4]. The key performance indicator is net zero emissions by 2060 [10]. The second objective is to rehabilitate 40 million hectares of land by planting 10 billion trees across KSA [11]. The third objective is to protect 30% of KSA's land and sea [12]. The Saudi leadership has a clear vision to champion climate actions in KSA and abroad through SGI and MGI, respectively. However, these initiatives are in progress, and achievement reports have not been announced yet.

Despite the clear vision of KSA's leadership, green investment in KSA is still in its infancy stage. The Crown Prince argued that these green initiatives paved the way for new investment opportunities in green KSA, which need collaboration from all stakeholders. Green investment is vital to ensure the proper implementation and success of these national initiatives and advance the national agenda. Hence, there is a need to understand the determinants of green investment intention in KSA to support the Saudi Vision 2030. This is especially true for the Saudi food industry since KSA suffers from tough weather and barren land; hence, it relies heavily on imports to meet the need of its citizens. One of the objectives of the Saudi government is to ensure food security for the Kingdom, which can be achieved while ensuring environmental sustainability [6]. The Saudi food market was USD 14.16 billion in 2022 and is expected to grow to USD 30.47 billion by 2029 [13].

Green investment refers to investment activities that primarily conserve natural resources and adopt environmentally friendly business practices [14]. This type of investment has positive consequences on the environment, such as reducing greenhouse gas and air pollution while maintaining the quality of production and consumption [14]. Green investment adopts new initiatives and technologies that ensure environmental sustainability [15]. Green investment is a key factor in the growth of renewable energy in China [15]. Green investment is boosted by political support and economic growth [15], which is the case of KSA, with economic prosperity. Despite the importance of green investment, studies on the antecedents of green investment intention remain limited in general, and in developing and Islamic countries, in particular, remain very limited [16]. This study bridges this knowledge gap and investigates the determinants of green investment in the KSA food industry that are critical to the economy and Saudi Vision 2030. This also contributes to the achievement of sustainable development, which also becomes critical in today's business environment [17–19].

The current study tests an expanded model of the Theory of Planned Behaviour (TPB) [20] to investigate the causes of green investment intention in the KSA food industry. The study tests the influence of the three determinants of green TPB: attitude towards behaviour (ATB), subjective norms (SNs), perceived behavioural control (PBC), as well as the effect of green investment knowledge [16] and green consumption commitment [21] of the agricultural and food sciences graduates in KSA universities on their green investment intention in the food industry. These graduates are expected to be potential investors. The study tests the moderation effect of religiosity on such links. The role of religiosity was considered because KSA is categorised as an Islamic society [22]; hence, this study examined whether this would have an impact on the above-mentioned relationships. The level of green investment knowledge and sustainable agriculture practices can influence green investment intention [16]. Investors' knowledge and understanding of sustainable agriculture practices can enhance their willingness to invest in environmentally sustainable projects. Therefore, promoting awareness and education about sustainable agriculture practices can increase green investment intention in the Saudi food industry. Moreover, consumers' willingness to choose sustainably produced food can create a demand for sustainable farming practices [21], which can incentivise farmers to adopt sustainable practices.

To fulfil the purpose of this research, the next section reviews the literature in relation to the effect of TPB, i.e., ATB, SNs, and PBC, as well as green consumption commitment and green investment knowledge on green investment intention. It also considers the moderation role of religiosity in this relationship. The next section presents the research methods adopted for data collection and data analysis. The results of the collected data are then presented and discussed. The implications of the study are highlighted, and the study conclusion is presented.

2. Literature Review

2.1. *Theory of Planned Behaviour and Green Investment*

Ajzen [20] established the TPB framework to enhance the early developed Theory of Reasoned Action (TRA) by adding PBC as an originator of behavioural intention. TPB argues that ATB, SNs, and PBC are positively correlated with an individual's intention or motivation to undertake a certain behaviour. It also argues that behavioural intention is the main determinant of an individual's definite behaviour. ATB is the assessment of the behaviour of interest, positively or negatively. Positive ATB is associated with behavioural intention [21]. SNs include the influence perceived by an individual's network or people surrounding him/her, such as their peers, family, and teachers [20]. PBC refers to an individual's perceptions of his/her ability to perform a behaviour of interest [21]. Behavioural intention refers to the motivation or readiness of an individual to perform a given behaviour [20]. The current study adopts TPB for understanding investment intention among university graduates in a green context. TPB framework has been adopted for examining human behaviour in different contexts, such as entrepreneurial intention among higher education [23,24], food waste intention [25], excessive food buying [26], fast food buying intention [27], healthy food item choices [28], state-branded food product purchase [29], or consumer behaviour regarding organic menus [30].

The TPB framework was also extensively adopted to predict individual green purchase intentions and behaviour. For example, Paul et al. [31] used TPB to examine customers' green purchase intentions in India and found that both ATB and PBC have a positive effect on customers' intentions to purchase green items, whereas SNs have no effect on green purchase intention. Moreover, Chen and Tung [32] tested TPB to predict customers' intention to stay at green hotels in Taiwan. The results showed that ATB, SNs, and PBC have a positive effect on customers' intentions to choose green hotels. Similarly, Han et al. [33] reported the same findings among US customers. The study of Moon et al. [34] found that green ATB and green SNs explained a substantial amount of variance in green purchase intentions among university students in Pakistan [34]. Judge et al. [35] predicted consumers' intention to buy housing with sustainability certification using TPB. The results confirmed that TPB variables predict consumers' intention to buy housing with sustainability certification.

In the green investment context, there were some attempts by scholars to investigate the drivers of green investment. For example, the study of Chan et al. [36] used TPB to test the green investment intention among undergraduate students in Kuala Lumpur,

Malaysia. The findings showed that the three variables of TPB (ATB, SNs, PBC) are predictors of green investment intention. A study on investment intention in wind energy projects in Germany [37] found that SNs and PBC positively affect wind energy investment intention, whereas ATB has no effect on green investment intention. Yee et al. [38] examined investment intention toward renewable energy in Malaysia and found that TPB constructs have an indirect effect through the evaluation of the regulatory framework. A recent study on the determinants of green investment intention in Muslim nations, i.e., Malaysia [17], showed that ATB and PBC are among the key determinants of green investment intention. Based on these arguments and the TPB framework, we assume hypotheses (H):

H1: *Agricultural and food science graduates' green attitudes positively affect their green investment intention.*

H2: *Agricultural and food science graduates' green subjective norms positively affect their green investment intention.*

H3: *Agricultural and food science graduates' green perceived behavioural control positively affects their green investment intention.*

2.2. Green Consumption Commitment, Green Knowledge, Green Investment Intention

Research [16,21] has shown that there are other variables that could be determinants of green investment intention, such as green consumption commitment and green investment knowledge. Green consumption commitment refers to an individual's preferences for products or services with green characteristics [39]. Hence, those individuals with green consumption commitment are ready to devote more time and money to these green products and/or services [40]. Previous research found that individuals who pay more attention to environmental issues are more likely to be more concerned about green products and services [41]. However, a high food consumption culture among consumers encourages negative ecological practices such as food waste [25] or excessive buying behaviour [26]. Recent research [21] on green entrepreneurship intention found a moderating effect of green consumption commitment on the relationships between green entrepreneurship intention and actual behaviour. Other research found that consumption profile influences wind energy investment intention as a green source of energy [37].

A relationship was also established between knowledge of green investment and green investment intention and behaviour [16]. Recent research on risky investment intention [42] in KSA has shown that financial knowledge has a direct effect on risky investment intention among university graduates and an indirect influence through TPB constructs. Another study [43] found a significant effect of investment awareness on investment intention. This investment knowledge of graduates is shaped by university education support [23,24]. A recent study [17] found that knowledge of green investment is a predictor of green investment among university students in Malaysia. Thus, we assume that:

H4: *Agricultural and food science graduates' green consumption commitment positively affects their green investment intention.*

H5: *Agricultural and food science graduates' knowledge of green positively affects their green investment intention.*

2.3. The Role of Religiosity

Religiosity refers to religious values and ideals that many individuals or groups hold and practice [44]. It also can be defined as a commitment to a certain religion [45]. KSA is categorised as an Islamic society, where its citizens believe that the Holy Quran is "the Message of God" and his prophet Mohamed is "the Messenger of Islam". Islam is one of the heavenly religions that guide the attitude and behaviour of individuals and groups [46]. There is no doubt that religiosity drives human behaviour toward positive attitude and practice as the orders of God, whom they believe and trust [47]. It was confirmed that religiosity guides an individual's ethical practices [48] and lifestyle [49].

There is growing research on the role of religiosity in encouraging positive environmental practices. While the study of Liobikienė et al. [50] found no link between religiosity and naturally friendly behaviour, another study [51] found that the doctrinal ethical tenet of religiosity guides consumers to nature conservation. It was confirmed that religiosity positively influences the attitude of owner–managers toward environmental sustainability [52]. Religiosity was found to be associated with positive environmental behaviour, such as lower rates of smoking initiation [53]. Religiosity has a significant influence on pro-environmental behaviour [54]. Research confirmed that Islamic values predict consumers' green buying intention and behaviour [55]. Wang et al. [56] found that religiosity has an indirect influence on pro-environmental intention through frugality consciousness and connectedness to nature. Osman et al. [17] found that Islamic religious values are the most significant predictor of green investment intention among university students in Malaysia. This study makes a first attempt to test the moderation effect of religiosity on the link between determinants and green investment intention (Figure 1).

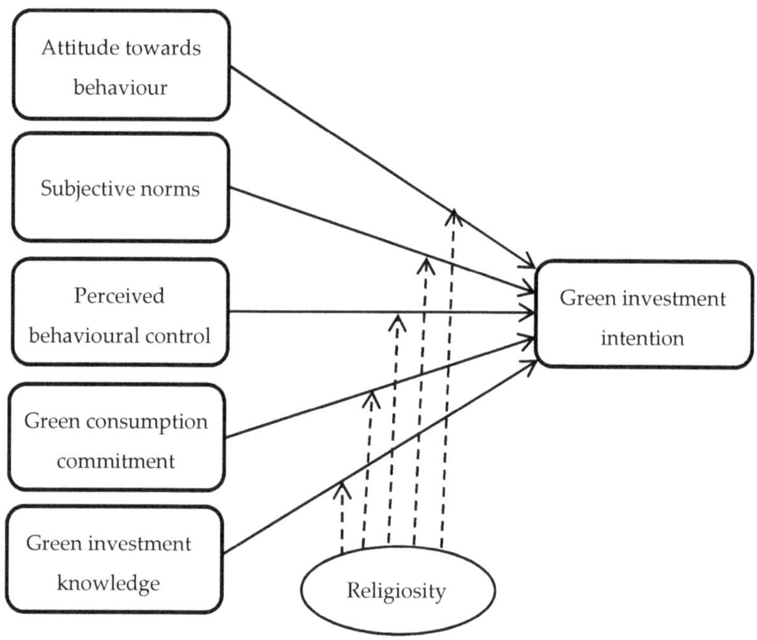

Figure 1. Determinants of green investment intentions model.

As highlighted earlier in the introduction, the current study draws on the Theory of Reasoned Action (TRA) and TPB [20] to test an expanded model of TPB. In this context, the study tests the moderating effect of religiosity on the relationship between TPB constructs, green investment knowledge, and green consumption intention on green investment intention. We hypothesise that:

H6: *Religiosity moderates the relationship between agricultural and food science graduates' attitude and their green investment intention.*

H7: *Religiosity moderates the relationship between agricultural and food science graduates' subjective norms and their green investment intention.*

H8: *Religiosity moderates the relationship between agricultural and food science graduates' perceived behavioural control and their green investment intention.*

H9: *Religiosity moderates the relationship between agricultural and food science graduates' green consumption commitment and their green investment intention.*

H10: *Religiosity moderates the relationship between agricultural and food science graduates' knowledge of green investment and their green investment intention.*

3. Methods

3.1. Study Measures

The survey form has three parts. The initial part explains the goals of the study and offers directions for filling out the form. The second part solicits personal data from the respondents, such as age and gender. Lastly, the third part includes the primary research inquiries, which employ a 7-point Likert scale, where 1 means "strongly disagree" and 7 means "strongly agree". We assessed the intention towards investing in green projects by utilising three modified statements from Chen's [57] research. The participants were informed to express their level of concurrence or disagreement with statements related to their willingness to regularly invest in eco-friendly (green) projects and encourage others to do the same. Additionally, they were asked about their plans to invest in green projects in the future. The items that measure GII showed good reliability with a value equal to 0.891. The measures of TPB typically include attitude and subjective norm [17,18]. These measures are widely employed in different fields and are usually assessed through self-reported scales where individuals evaluate the agreement or disagreement with the questions related to each construct. The measures were slightly modified to match the study context, where green attitude was measured using three items from Mohd Suki [58] and showed good internal consistency in our study (a = 0.972). Similarly, green subjective norms were measured using three items derived and modified from Gopi and Ramayah [59] (a = 0.969). Green PBC was measured using four items (a = 0.959) as employed by Amin, Rahman, and Razak [60].

From Jaffar and Musa [61], four items that measure green investment knowledge were employed in our study and showed good and adequate reliability with a score equal to 0.987. Similarly, green consumption commitment was measured using four items adopted from Zeithaml et al. [62]. Finally, religiosity as a moderating variable was measured using three items (a = 0.836) from Jaafar and Musa [61]. All measures with related items are presented in Table 1.

Table 1. Respondents descriptions.

Respondents Profile (n = 550)		**Frequencies**	**%**
University name	University of Mohammad ibn Saud Islamic	165	30%
	University of King Faisal	137	25%
	University of King Khaled	137	25%
	University of Umm Al Qura	111	20%
Gender type	Female	281	51%
	Male	269	49%
Age range	<1 Years	165	30%
	21–<25 Years	358	65%
	>25 Years	27	05%

The questionnaire underwent testing by university professors (15) and graduates (13) to confirm its consistency, clarity, and user-friendliness. We implemented measures to ensure the confidentiality of respondents' information. Since research surveys are susceptible to Common Method Variance (CMV), Harman's single-factor was undertaken with Exploratory Factor Analysis (EFA) to discover potential CMV. The EFA findings revealed that CMV was not a problem since one single variable clarified only 41% of the variance in the endogenous one, which is below the 50% threshold recommended by Nunnally [63].

3.2. Participants and Methods of Data Collection

The research team conducted a random survey of graduates in agriculture and food science from national universities located in various provinces of the Kingdom of Saudi Arabia. A digital survey was distributed to national universities. The team leveraged their connections with professors and lecturers to disseminate the survey via official university emails and other social networking sites (i.e., WhatsApp groups). Contribution was voluntary, and the questionnaire introduction clearly communicated its purpose and the privacy of all collected data. We sent out a survey to graduates who may be interested in investing, and they received the survey in November and December 2022. A total of 600 forms were distributed, and 537 had usable responses, resulting in a total of 550 (537 + 13 pilot study graduates) with a response rate of 91.6%. We did not have any issues with late answers. A *t*-test showed no significant differences in the means, which confirmed that there was no bias in the responses [64].

3.3. Data Analysis Procedures

This research utilised PLS-SEM with the SmartPLS vs. 4-software [65]. PLS-SEM is considered a non-parametric technique that calculates the variance in latent dimensions [66] and is commonly used in management science. Smart PLS-SEM is usually employed to investigate the connections between different variables. Following Leguina's [67] suggestion, we evaluated the suggested theoretical model in two main stages: (1) first for convergent and discriminant validity, then (2) for hypothesis confirmation.

4. Research Results

4.1. Descriptive Analysis Results

The surveyed students had almost equal representation of males and females, with 90% of them between the ages of 17 and 25. A total of 30% were from Mohammad ibn Saud Islamic University, 25% of the participants were enrolled at King Faisal University, 25% at King Khaled University, and 20% at Umm Al-Qura University. The answers to the survey questions varied, with mean scores ranging from 4.33 to 5.60 and standard deviation values between 1.083 and 1.818, which suggests that the responses were not grouped around the mean. Additionally, the variance inflation value was lower than 0.5 for the survey items, meaning that multicollinearity was not a concern.

4.2. Outer Model Evaluation

To ensure the validity and reliability of the research, a number of benchmarks (indices) were employed per the recommendations of Hair et al. [66] and Kline [68], including the composite reliability (CR) value, internal consistency reliability (a) value, convergent validity index, and discriminant validity index.

4.2.1. Convergent Validity Results

To evaluate the convergent validity of the employed scale, a number of criteria were used, including Cronbach's alpha (a), reliability, composite reliability (C.R.), loadings, and Average Variance Extracted (AVE). As shown in Table 2, the C.R. and (a) values for all the scales employed surpassed the threshold value of 0.7, indicating an appropriate level of internal reliability [66]. These values were as follows: DII (a = 0.81, C.R = 0.82); green attitude (a = 0.97, C.R = 0.97); green subjective norms (a = 0.96, C.R = 0.97); green perceived behaviour control (a = 0.95, C.R = 96); green investment knowledge (a = 95, C.R = 98); green consumption commitment; and religiosity (a = 0.97, C.R = 0.97).

Table 2. Psychometric properties.

Abbr.	SFL	α	C.R	AVE
Green Attitude		0.972	0.975	0.946
In my opinion, opting for green investments is a smart choice.	0.962			
I believe that the performance of green investments is usually dependable.	0.985			
I have faith that the assertions made about green investments are usually credible.	0.971			
Green investment intention		0.819	0.820	0.734
I plan to make frequent investments in green projects.	0.868			
I intend to promote green investments to my friends and family.	0.848			
I have plans to invest in green projects in the near future.	0.854			
Green consumption commitment		0.973	0.973	0.925
My future goal is to create eco-friendly products.	0.980			
I am keen to suggest green products to my friends and acquaintances.	0.971			
I speak positively about environmentally-friendly products to others.	0.973			
I would motivate others to develop green products.	0.923			
Green investment knowledge		0.987	0.988	0.963
I am aware of the availability of eco-friendly investments.	0.990			
I engage in green investments because they align with my environmental values.	0.984			
I invest in green projects because they offer greater environmental advantages compared to other options.	0.990			
Green investments have the potential to yield long-term benefits.	0.961			
Green perceived behavioural control		0.959	0.967	0.891
I am capable of taking part in green investments.	0.920			
It would be effortless for me to engage in green investments	0.956			
I trust my ability to select the type of eco-friendly investment that suits me.	0.947			
I am interested in investing in green initiatives.	0.952			
Religiosity		0.836	0.860	0.753
Shariah-compliant financial institutions and organisations that offer green investments are available.	0.881			
Green investments provided by Islamic financial institutions and agencies do not involve interest-based transactions.	0.913			
My religious beliefs inspire me to partake in eco-friendly investments.	0.806			
Green subjective norms		0.969	0.973	0.942
The majority of individuals whose perspectives I esteem would endorse my involvement in green investments.	0.975			
People who hold significance in my life believe that I should engage in eco-friendly investments.	0.982			
My loved ones, who hold great importance in my life, support my decision to invest in green initiatives.	0.954			

The research constructs were found to be reliable, as each factor had an SFL "standardised factor loading" value that was greater than 0.70. Convergent validity was also established by assessing AVE values to a cutoff point of 0.5 [69]. To test the scale discriminant validity, the Fornell–Larcker criterion, the cross-loading matrixes, and the heterotrait–monotrait method ratios (HTMT) were used, as recommended by Leguina [67].

4.2.2. Discriminant Validity Results

To test the discriminant validity of the research factors, cross-loadings, the Fornell–Larcker criterion, and the heterotrait–monotrait ratio were used. Each latent unobserved variable's outer loading was higher than its cross-loading, as shown in Table 3 [67], indicating that discriminant validity was established. Furthermore, Table 4 shows that the AVE scores on the diagonal were higher than the inter-variable correlations, giving more evidence of adequate discriminant validity [68]. Lastly, the HTMT scores should be lower than the 0.90 cutoff point, as recommended by Leguina [67], and the reference value in Table 4 was satisfied. All of these results confirmed that the research constructs have high discriminant validity, and the hypotheses were then evaluated with the outer structural model.

Table 3. Cross-loading matrix.

	1	2	3	4	5	6	7
1. Green attitude							
G.Attude_1	0.96	0.11	0.26	0.03	0.22	0.23	−0.01
G.Attude_2	0.99	0.06	0.24	−0.04	0.18	0.14	−0.06
G.Attude_3	0.97	0.05	0.23	−0.03	0.19	0.12	−0.07
2. Green investment intention							
G.Invst_1	0.19	0.38	0.87	0.26	−0.09	0.38	0.40
G.Invst_2	0.24	0.31	0.85	0.23	−0.19	0.36	0.28
G.Invst_3	0.21	0.43	0.85	0.40	−0.06	0.27	0.32
3. Green consumption commitment (G_Comtt)							
G_Comtt_1	0.09	0.97	0.42	0.33	−0.07	0.30	0.34
G_Comtt_2	0.06	0.92	0.43	0.33	−0.15	0.21	0.33
G_Comtt_3	−0.02	0.32	0.34	0.99	0.15	0.26	0.41
G_Comtt_4	−0.02	0.30	0.35	0.98	0.10	0.21	0.41
4. Green investment knowledge							
G_Invst_knw_1	0.08	0.98	0.42	0.29	−0.14	0.26	0.33
G_Invst_knw_2	0.06	0.97	0.41	0.32	−0.07	0.28	0.34
G_Invst_knw_3	0.09	0.97	0.42	0.33	−0.07	0.30	0.34
G_Invst_knw_4	0.06	0.92	0.43	0.33	−0.15	0.21	0.33
5. Green perceived behavioural control							
Per_Beh_1	0.14	0.25	0.32	0.28	0.05	0.92	0.19
Per_Beh_2	0.16	0.28	0.35	0.27	0.03	0.96	0.22
Per_Beh_3	0.18	0.25	0.41	0.16	−0.12	0.95	0.19
Per_Beh_4	0.17	0.25	0.40	0.19	−0.08	0.95	0.21
6. Religiosity							
Reliogisty_1	−0.05	0.26	0.28	0.41	−0.15	0.13	0.88
Reliogisty_2	−0.01	0.34	0.40	0.34	−0.12	0.20	0.91
Reliogisty_3	−0.07	0.29	0.33	0.33	−0.22	0.22	0.81
7. Green subjective norms							
Subj_Nrms_1	0.21	−0.13	−0.12	0.11	0.98	−0.03	−0.17
Subj_Nrms_2	0.18	−0.13	−0.14	0.10	0.98	−0.07	−0.20
Subj_Nrms_3	0.20	−0.06	−0.10	0.18	0.95	−0.01	−0.14

Table 4. Fornell–Larcker matrix and HTMT matrix.

		Fornell–Larcker Criterion							HTMT Results						
		a	b	c	d	e	f	g	a	b	c	d	e	f	g
a.	Green Attitude	0.97													
b.	Green Consumption Commitment	0.07	0.96						0.07						
c.	Green Investment Intention	0.25	0.43	0.85					0.28	0.48					
d.	Green Investment Knowledge	−0.01	0.32	0.34	0.98				0.03	0.33	0.38				
e.	Green Subjective Norms	0.20	−0.11	−0.12	0.12	0.97			0.20	0.11	0.14	0.13			
f.	Perceived Behavioral Control	0.17	0.27	0.39	0.23	−0.03	0.94		0.17	0.28	0.44	0.24	0.08		
g.	Religiosity	−0.04	0.34	0.39	0.41	−0.18	0.21	0.86	0.05	0.38	0.46	0.45	0.20	0.23	

4.3. Inner Model Evaluation

The study employed SmartPLS 4′s inner model to examine the hypotheses. The aim was to evaluate the capability of the study model to clarify and anticipate the variations in endogenous variables triggered by exogenous variables [69]. Additionally, to assess the model's goodness of fit (GoF), we utilised the equation introduced by Chin [70]. This equation computes GoF by obtaining the square root of the R^2 multiplied by the average of all AVE values. Our GoF analysis yielded a score of 0.59, which suggests a substantial level of model fit, as recommended by Wetzels et al. [71]. To confirm the goodness of fit (GoF) of the research model, the value of the endogenous variables should be at least 0.10.

The R^2 value of the endogenous latent variable GII in our study was 0.401, which exceeded the recommended scores and gave more evidence that the study model adequately fits the study's empirical data. Additionally, the Stone–Geisser Q2 statistic had a value of 0.376 for GII, which was more than zero, indicating an acceptable result [70]. Furthermore, the SRMR score should be lower than the value of 0.08, and the NFI value had to be greater than 0.90 to safeguard adequate model fit to the data [67].

The study yielded an SRMR value of 0.040; this result shows that the calculated residual value obtained by fitting the variance–covariance matrix of the proposed model to the observed sample data's variance–covariance matrix is less than the predetermined threshold of 0.08, as suggested by Hair et al. [66] and Kline [68]. Additionally, the NFI score surpassed the recommended threshold of 0.90, which indicates a good fit. Furthermore, the f^2 values, which quantify the change in R^2 after removing an exogenous variable, were also computed. The findings demonstrated that the exogenous variables had a minimal impact on the GII (green attitude, $f^2 = 0.093$; consumption commitment, $f^2 = 0.053$; green investment knowledge, $f^2 = 0.031$; green perceived behaviour control, $f^2 = 0.046$; and green subjective norms, $f^2 = 0.017$). This implies that removing any exogenous variables from the model would only result in a slight alteration in the main model, as suggested by Cohen [72].

Once a satisfactory model fit was established, a 5000 bootstrapping repetition was employed in SmartPLS4 to determine the path coefficient and *t*-value for the study's proposed interrelationships and moderation paths, which are presented in Table 5 and Figure 2. The study suggested and evaluated ten hypotheses, with five being direct relationships and the other five involving moderation. The PLS-SEM findings revealed that GII was positively and significantly influenced by green attitude ($\beta = 0.259$, $t = 6.327$, $p < 0.001$), green consumption commitment ($\beta = 0.221$, $t = 4.712$, $p < 0.001$), green investment knowledge ($\beta = 0.169$, $t = 4.559$, $p < 0.001$), and green perceived behaviour control ($\beta = 0.187$, $t = 5.389$, $p < 0.001$), hence corroborating hypotheses H1, H2, H3, and H4. However, contrary to expectations, subjective norms had a significant but negative effect on GII ($\beta = -0.111$, t-value = 2.740, $p < 0.001$), which resulted in the rejection of H5.

Table 5. Hypotheses evaluation.

	Hypotheses	Beta (β)	*t*-Value	*p*-Value	Results
H1	Green Attitude → Green Investment Intention	0.259	6.327	0.000	Accepted
H2	Green Consumption Commitment → Green Investment Intention	0.221	4.712	0.000	Accepted
H3	Green Investment Knowledge → Green Investment Intention	0.169	4.559	0.000	Accepted
H4	Green Perceived Behavioral Control → Green Investment Intention	0.187	5.389	0.000	Accepted
H5	Green Subjective Norms → Green Investment Intention	−0.111	2.740	0.006	Not Accepted
H6	Religiosity × Green Subjective Norms → Green Investment Intention	0.047	1.253	0.210	Not Accepted
H7	Religiosity × Green Perceived Behavioral Control → Green Investment Intention	0.124	2.892	0.005	Accepted
H8	Religiosity × Green Attitude → Green Investment Intention	0.120	2.687	0.004	Accepted
H9	Religiosity × Green Consumption Commitment → Green Investment Intention	0.082	2.360	0.018	Accepted
H10	Religiosity × Green Investment Knowledge → Green Investment Intention	0.057	1.676	0.094	Not Accepted

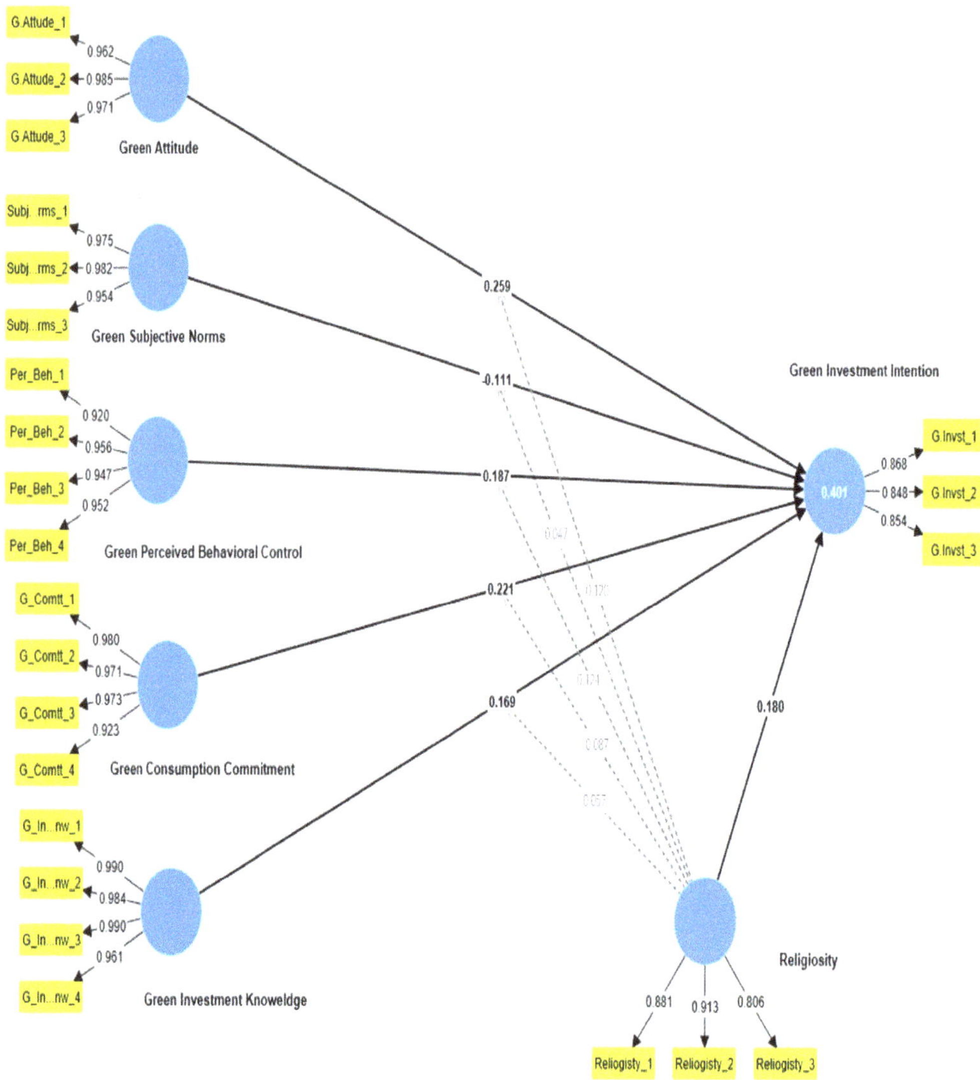

Figure 2. The research model.

The study's findings also provide information on how religious beliefs moderate the relationships being examined. The results displayed in Table 5 indicate that there was no significant impact of religiosity on the green subjective norms–GII path ($\beta = 0.47$, $t = 1.253$, $p = 0.210$), nor on the link between green investment knowledge and GII ($\beta = 0.057$, $t = 1.676$, $p = 0.094$). Consequently, hypotheses H6 and H10 were not supported. Conversely, religiosity did have a significant moderating effect on the green perceived behaviour control–GII path ($\beta = 0.124$, $t = 2.892$, $p < 0.01$), the green attitude–GII path ($\beta = 120$, $t = 2.687$, $p < 0.01$), and the green consumption commitment–GII path ($\beta = 0.082$, $t = 2.360$, $p < 0.05$). Therefore, Hypotheses H7, H8, and H9 were supported, as seen in Table 5, Figures 2 and 3.

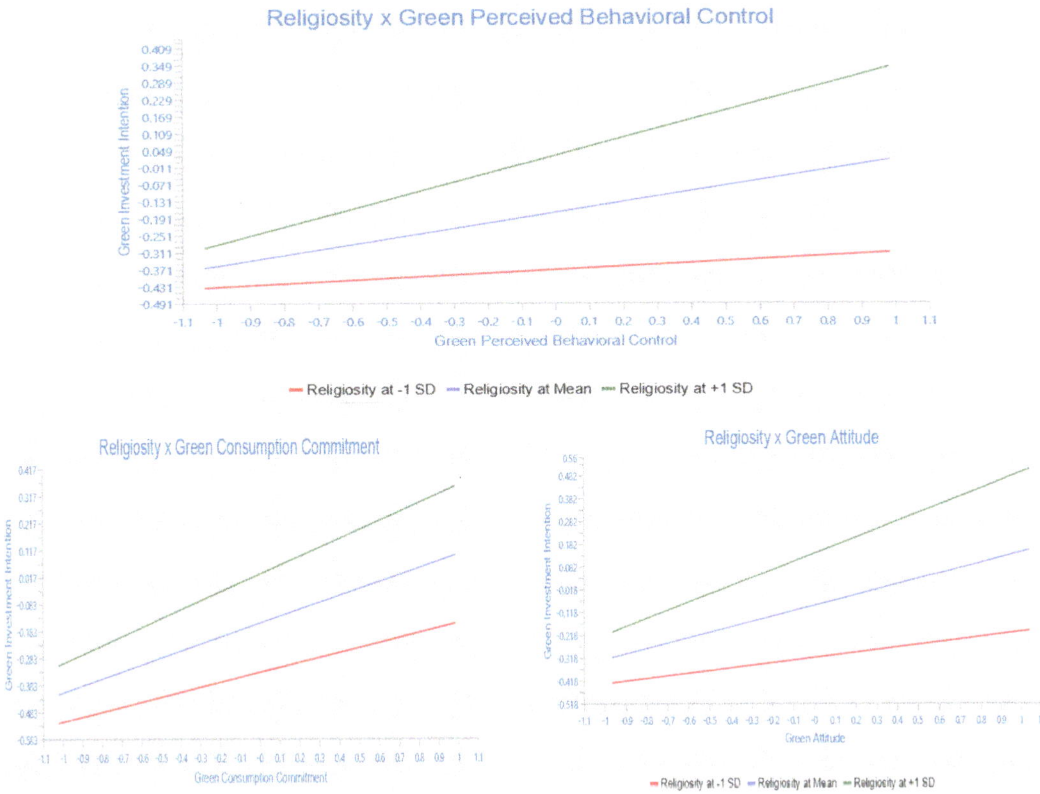

Figure 3. Slope Analysis.

5. Discussion

Environmental sustainability draws the consideration of policymakers and academics in the context of KSA, especially after instating Vision 2030. This research is among new attempts that investigate the determinants of green investment intention among graduates of agriculture and food science. The study examined an expanded model of TPB that incorporates green ATB, SNs, and PCB with green investment knowledge and green consumption commitment as key determinants of green investment intention. The research examined the moderation role of religiosity in these relationships.

The findings of PLS-SEM supported the TPB framework [20,21] that green ATB and PBC positively and significantly influence green investment intention. These results mean that positive green ATB and green PBC are predictors of green investment intention. These findings are in line with Chan et al. [36], who also found that ATB and PBC are predictors of green investment intention. They also support the work of Osman et al. [17] that ATB and PBC are key determinants of green investment intention. These results confirm that graduates hold positive green ATB. They feel that green investment is a wise idea, green investment performance is generally reliable, and green investment claims are generally trustworthy. The results also mean that graduates perceive themselves as able to participate in green investment. They find it easy to participate in green investment, and they have control in choosing a green investment and want to do this. On the other side, the findings confirmed the negative effect of green SNs on green investment intention. The results confirm that the social influence on green investment is significantly negative. Saudi is categorised as a collective society [73]; hence, its citizens are highly influenced by the opinion of their family members and friends. The results confirm that social networks and people whose opinions graduates value would not approve of their participation in green investment. These people do not think that graduates should participate in green investment.

The results showed that green consumption commitment among agriculture and food science graduates positively influenced their intention toward green investment. This means that graduates have preferences for green products and services. This positive commitment toward green consumption stimulates graduates to encourage others to establish green products. Such green commitment consumption encourages positive environmental intention and behaviour [21]. Additionally, supporting previous literature reviews [17,43], the current research found that green investment knowledge has a significant positive influence on green investment intention. The results confirm that agriculture and food science graduates have knowledge of green investment, which encourages them to participate in green investment. They believe that green investment is beneficial in the long term.

With regard to the moderation effect of religiosity, the results showed that religiosity has a significant moderating effect on the relationship between green PBC, green ATB, green consumption commitment, and green investment intentions. These results mean that religiosity has the ability to enhance these relationships; hence, it could stimulate green investment intention. On the other hand, religiosity was found to have no significant influence on the relationship between green SNs, green investment knowledge, and green investment intention. In other words, religiosity failed to change the influence of green SNs and green knowledge on green investment intention. Despite this, religiosity still had an impact on green investment intention by moderating the effect of green PBC, green ATB, and green consumption commitment on green investment intentions.

The results have important implications for scholars and policymakers. The results confirmed the direct effect of green ATB, green PBC, green knowledge, and green consumption commitment on green investment intention among agriculture and food science graduates, which has great implications for the growth and sustainability of the Saudi food industry. It is, therefore, important that policymakers pay more attention to these factors to stimulate green investment intention, hence, ensuring sustainable development [18,19]. The results also confirmed the moderating effect of religiosity on the relationship between green ATB, green PBC, green consumption commitment, and green investment intention. It is important that policymakers promote green social influence since this was found to have a significant negative influence on green investment intention. This social influence could be created by university education support given to students and graduates to encourage them to engage in entrepreneurship and investment [23], particularly green investment. University incubation support can also play an important role in stimulating green investment intention [24]. In addition, media activities, including social media, could also be undertaken to highlight the value of green investment for society. The current study highlighted that the role of government is important in stimulating green investment not just through regulation [17] but also through the education system, which has a significant effect on graduates' green ATB, green SNs, green PBC, green knowledge, and green consumption commitment. This green investment intention is the significant predictor of actual green investment, which has implications for sustainable KSA.

6. Conclusions

The research investigated a more comprehensive version of TPB, which included green ATB, SNs, and PCB, along with green investment knowledge and green consumption commitment as important factors affecting people's willingness to make green investments. The study also looked at how religiosity affects the relationship between these variables. Data were collected from 550 fresh graduates from agriculture and food science programs in four national universities in KSA (Mohammad ibn Saud Islamic University, King Faisal University, King Khaled University, and Umm Al-Qura University). PLS-SEM was employed as the main data analysis technique to analyse the collected data. The results of the PLS-SEM analysis indicated that green attitude, green consumption commitment, green investment knowledge, and green perceived behaviour control had a positive and significant impact on GII (Green Investment Intention). However, contrary to the hypothesis, subjective norms had a significant negative effect on GII. Moreover, the findings indicated that the impact of green perceived behaviour control (PBC), green attitude towards behaviour (ATB), and green consumption commitment on green investment intentions is significantly moderated by religiosity. Religiosity can strengthen these connections and consequently promote the intention to invest in green initiatives. However, religiosity was found to have no meaningful impact on the connection between green social norms (SNs), green investment knowledge, and green investment intention. In other words, religiosity did not alter the influence of green social norms and green knowledge on green investment intention.

The current research focused on fresh graduates of agriculture and food science programs using a self-reporting study. Further research could examine the intention of current investors in the Saudi food industry and their intention to turn their current business green. The research has not examined the effect of gender on these links, which could have different results [74]. Hence, further research could examine these results with a wider research sample and examine the role of gender. Other research could examine the influence of personality traits on green investment intention. Additionally, the influence of green legislation on green investment intention and behaviour could be examined.

Author Contributions: Conceptualisation, M.M.A., M.A.A., I.A.E. and A.E.E.S.; methodology, I.A.E. and A.E.E.S.; software, I.A.E.; validation, I.A.E. and A.E.E.S.; formal analysis, I.A.E. and A.E.E.S.; investigation, I.A.E. and A.E.E.S.; resources, I.A.E.; data curation, M.M.A., M.A.A., I.A.E. and A.E.E.S.; writing—original draft preparation M.M.A., M.A.A., A.E.E.S. and I.A.E.; writing—review and editing, M.M.A., A.E.E.S., M.A.A. and I.A.E.; visualisation, I.A.E.; supervision, A.E.E.S. and I.A.E.; project administration, M.M.A., M.A.A., A.E.E.S. and I.A.E.; funding acquisition, M.M.A., M.A.A., A.E.E.S. and I.A.E. All authors have read and agreed to the published version of the manuscript.

Funding: This research was funded by the Deputyship for the Research and Innovation, Ministry of Education in Saudi Arabia, through project number INST044.

Institutional Review Board Statement: The study was conducted according to the guidelines of the Declaration of Helsinki and approved by the Deanship of scientific research ethical committee, King Faisal University (project number: INST044, date of approval: 1 November 2022).

Informed Consent Statement: Informed consent was obtained from all subjects involved in the study.

Data Availability Statement: Data are available upon request from researchers who meet the eligibility criteria. Kindly contact the first author privately through email.

Conflicts of Interest: The authors declare no conflict of interest.

References

1. Godfray, H.C.J.; Crute, I.R.; Haddad, L.; Lawrence, D.; Muir, J.F.; Nisbett, N.; Pretty, J.; Robinson, S.; Toulmin, C.; Whiteley, R. The future of the global food system. *Philos. Trans. R. Soc. B Biol. Sci.* **2010**, *365*, 2769–2777. [CrossRef] [PubMed]
2. Myers, S.S.; Smith, M.R.; Guth, S.; Golden, C.D.; Vaitla, B.; Mueller, N.D.; Dangour, A.D.; Huybers, P. Climate change and global food systems: Potential impacts on food security and undernutrition. *Annu. Rev. Public Health* **2017**, *38*, 259–277. [CrossRef] [PubMed]
3. Clark, M.A.; Domingo, N.G.; Colgan, K.; Thakrar, S.K.; Tilman, D.; Lynch, J.; Azevedo, I.L.; Hill, J.D. Global food system emissions could preclude achieving the 1.5 and 2 C climate change targets. *Science* **2020**, *370*, 705–708. [CrossRef] [PubMed]
4. Sobaih, A.E.E.; Hasanein, A.; Gharbi, H.; Abu Elnasr, A.E. Going Green Together: Effects of Green Transformational Leadership on Employee Green Behaviour and Environmental Performance in the Saudi Food Industry. *Agriculture* **2022**, *12*, 1100. [CrossRef]
5. Baig, M.B.; Straquadine, G.S.; Aldosari, F. Revisiting extension systems in Saudi Arabia: Emerging reasons and realities. *J. Exp. Biol. Agric. Sci.* **2017**, *5*, S160–S164. [CrossRef]
6. The Saudi Vision 2030. An Overview. Available online: https://www.vision2030.gov.sa/v2030/overview/ (accessed on 3 January 2023).
7. The Saudi Vision 2030. Progress and Achievement. Available online: https://www.vision2030.gov.sa/v2030/achievements/ (accessed on 3 January 2023).
8. The Green Initiatives. Our Vision. Available online: https://www.greeninitiatives.gov.sa/ (accessed on 3 January 2023).

9. The Saudi Green Initiative. SGI: Steering Saudi Arabia towards a Green Future. Available online: https://www.greeninitiatives.gov.sa/about-sgi/ (accessed on 3 January 2023).
10. The Saudi Green Initiative. Paving the Way to Net Zero Emissions by 2060. Available online: https://www.greeninitiatives.gov.sa/about-sgi/sgi-targets/reducing-emissions/reduce-carbon-emissions/ (accessed on 5 January 2023).
11. The Saudi Green Initiative. SGI Target: Plant 10 Billion Trees across Saudi Arabia. Available online: https://www.greeninitiatives.gov.sa/about-sgi/sgi-targets/greening-saudi/plant-10-billion-trees/ (accessed on 5 January 2023).
12. The Saudi Green Initiative. SGI Target: Protecting 30% of Saudi Arabia's Land and Sea. Available online: https://www.greeninitiatives.gov.sa/about-sgi/sgi-targets/protecting-land-and-sea/protecting-saudi-arabia-s-land-and-sea/ (accessed on 5 January 2023).
13. Business Insights, The Saudi Arabia Food Market Size. Available online: https://www.fortunebusinessinsights.com/saudi-arabia-food-service-market-106896 (accessed on 5 January 2023).
14. Shipochka, R. Investor's Behaviour towards Green Investments and How Ireland Invest in Sustainable Green Energy Projects—Is Green the New Bubble. Ph.D. Thesis, Dublin Business School, Dublin, Ireland, 2013.
15. Eyraudn, L.; Clements, B.; Wane, A. Green investment: Trends and determinants. *Energy Policy* **2013**, *60*, 852–865. [CrossRef]
16. Bright, I. Green Investment: Which Industries Are Genuinely Green? 2008. Available online: https://www.tropical-rainforest-animals.com/green-investment.html#priority (accessed on 5 January 2023).
17. Osman, I.; Maâ, M.; Muda, R.; Husni, N.S.A.; Alwi, S.F.S.; Hassan, F. Determinants of behavioural intention towards green investments: The perspectives of Muslims. *Int. J. Islam. Bus.* **2019**, *4*, 16–38. [CrossRef]
18. Manioudis, M.; Meramveliotakis, G. Broad strokes towards a grand theory in the analysis of sustainable development: A return to the classical po litical economy. *New Political Econ.* **2022**, *27*, 866–878. [CrossRef]
19. Tomislav, K. The concept of sustainable development: From its beginning to the contemporary issues. *Zagreb Int. Rev. Econ. Bus.* **2018**, *21*, 67–94.
20. Ajzen, I. The theory of planned behavior. *Organ. Behav. Hum. Decis. Process.* **1991**, *50*, 179–211. [CrossRef]
21. Amankwah, J.; Sesen, H. On the Relation between Green Entrepreneurship Intention and Behavior. *Sustainability* **2021**, *13*, 7474. [CrossRef]
22. Nevo, J. Religion and national identity in Saudi Arabia. *Middle East. Stud.* **1998**, *34*, 34–53. [CrossRef]
23. Aliedan, M.M.; Elshaer, I.A.; Alyahya, M.A.; Sobaih, A.E.E. Influences of University Education Support on Entrepreneurship Orientation and Entrepreneurship Intention: Application of Theory of Planned Behavior. *Sustainability* **2022**, *14*, 13097. [CrossRef]
24. Ayad, T.; Sobaih, A.E.E.; Elshaer, I.A. University Incubator Support and Entrepreneurial Intention among Tourism Graduates: Mediating Role of Personal Attitude. *Sustainability* **2022**, *14*, 16045. [CrossRef]
25. Elshaer, I.; Sobaih, A.E.E.; Alyahya, M.; Abu Elnasr, A. The Impact of Religiosity and Food Consumption Culture on Food Waste Intention in Saudi Arabia. *Sustainability* **2021**, *13*, 6473. [CrossRef]
26. Sobaih, A.E.E. Excessive Food Buying in Saudi Arabia Amid COVID-19: Examining the Effects of Perceived Severity, Religiosity, Consumption Culture and Attitude toward Behavior. *Int. J. Environ. Res. Public Health* **2023**, *20*, 3126. [CrossRef]
27. Sobaih, A.E.E.; Abdelaziz, A.S. The Impact of Nutrition Labelling on Customer Buying Intention and Behaviours in Fast Food Operations: Some Implications for Public Health. *Int. J. Environ. Res. Public Health* **2022**, *19*, 7122. [CrossRef]
28. Sobaih, A.E.E.; Algezawy, M.; Elshaer, I.A. Adopting an Extended Theory of Planned Behaviour to Examine Buying Intention and Behaviour of Nutrition-Labelled Menu for Healthy Food Choices in Quick Service Restaurants: Does the Culture of Consumers Really Matter? *Int. J. Environ. Res. Public Health* **2023**, *20*, 4498. [CrossRef]
29. Shin, Y.H.; Jung, S.E.; Im, J.; Severt, K. Applying an extended theory of planned behavior to examine state-branded food product purchase behavior: The moderating effect of gender. *J. Foodserv. Bus. Res.* **2020**, *23*, 358–375. [CrossRef]
30. Shin, Y.H.; Im, J.; Jung, S.E.; Severt, K. The theory of planned behavior and the norm activation model approach to consumer behavior regarding organic menus. *Int. J. Hosp. Manag.* **2018**, *69*, 21–29. [CrossRef]
31. Paul, J.; Modi, A.; Patel, J. Predicting green product consumption using theory of planned behavior and reasoned action. *J. Retail. Consum. Serv.* **2016**, *29*, 123–134. [CrossRef]
32. Chen, M.F.; Tung, P.J. Developing an extended theory of planned behavior model to predict consumers' intention to visit green hotels. *Int. J. Hosp. Manag.* **2014**, *36*, 221–230. [CrossRef]
33. Han, H.; Hsu, L.T.J.; Sheu, C. Application of the theory of planned behavior to green hotel choice: Testing the effect of environmental friendly activities. *Tour. Manag.* **2010**, *31*, 325–334. [CrossRef]
34. Moon, M.A.; Mohel, S.H.; Farooq, A. I green, you green, we all green: Testing the extended environmental theory of planned behavior among the university students of Pakistan. *Soc. Sci. J.* **2021**, *58*, 316–332. [CrossRef]
35. Judge, M.; Warren-Myers, G.; Paladino, A. Using the theory of planned behaviour to predict intentions to purchase sustainable housing. *J. Clean. Prod.* **2019**, *215*, 259–267. [CrossRef]
36. Chan, K.H.; Ng, T.H.; Fadi, A. What do undergraduates think about green investment? Empirical evidence from a developing nation. *Indian J. Public Health Res. Dev.* **2018**, *9*, 1627–1632. [CrossRef]
37. Gamel, J.; Bauer, A.; Decker, T.; Menrad, K. Financing wind energy projects: An extended theory of planned behavior approach to explain private households' wind energy investment intentions in Germany. *Renew. Energy* **2022**, *182*, 592–601. [CrossRef]
38. Yee, C.H.; Al-Mulali, U.; Ling, G.M. Intention towards renewable energy investments in Malaysia: Extending theory of planned behaviour. *Environ. Sci. Pollut. Res.* **2022**, *29*, 1021–1036. [CrossRef]

39. Chang, Y.C.; Chen, M.H.; Lin, Y.P.; Gao, Y.S. Measuring regional innovation and entrepreneurship capabilities. *J. Knowl. Econ.* **2012**, *3*, 90–108. [CrossRef]
40. Laroche, M.; Bergeron, J.; Barbaro-Forleo, G. Targeting consumers who are willing to pay more for environmentally friendly products. *J. Consum. Mark.* **2001**, *18*, 503–520. [CrossRef]
41. Cheah, I.; Phau, I. Attitudes towards environmentally friendly products: The influence of ecoliteracy, interpersonal influence and value orientation. *Mark. Intell. Plan.* **2011**, *29*, 452–472. [CrossRef]
42. Sobaih, A.E.E.; Elshaer, I.A. Risk-Taking, Financial Knowledge, and Risky Investment Intention: Expanding Theory of Planned Behavior Using a Moderating-Mediating Model. *Mathematics* **2023**, *11*, 453. [CrossRef]
43. Elshaer, I.A.; Sobaih, A.E.E. Antecedents of Risky Financial Investment Intention among Higher Education Students: A Mediating Moderating Model Using Structural Equation Modeling. *Mathematics* **2023**, *11*, 353. [CrossRef]
44. Delener, N. The Effects of Religious Factors on Perceived Risk in Durable Goods Purchase Decisions. *J. Consum. Mark.* **1990**, *7*, 27–38. [CrossRef]
45. Mokhlis, S. Relevancy and Measurement of Religiosity in Consumer Behavior Research. *Int. Bus. Res.* **2009**, *2*, 75. [CrossRef]
46. Ahmed, S. *What Is Islam?* Princeton University Press: Princeton, NJ, USA, 2015.
47. McDaniel, W.; Burnett, J. Consumer religiosity and retail store evaluative criteria. *J. Acad. Mark. Sci.* **1990**, *18*, 101–112. [CrossRef]
48. Vitell, S.; Ramos-Hidalgo, E.; Rodríguez-Rad, C. A Spanish perspective on the impact on religiosity and spirituality on consumer ethics. *Int. J. Consum. Stud.* **2018**, *42*, 675–686. [CrossRef]
49. Rakrachakarn, V.; Moschis, P.; Ong, S.; Shannon, R. Materialism and life satisfaction: The role of religion. *J. Relig. Health* **2015**, *54*, 413–426. [CrossRef]
50. Liobikiene, G.; Niaura, A.; Mandravickaite, J.; Vabuolas, Ž. Does religiosity influence environmental attitude and behaviour? The case of young Lithuanians. *Eur. J. Sci. Theol.* **2016**, *12*, 81–96.
51. Roshani Perera, C.; Rathnasiri Hewege, C. Religiosity and environmentally concerned consumer behaviour:'becoming one with god (nature)'through surrendering environmental identities. *Int. J. Consum. Stud.* **2018**, *42*, 627–638. [CrossRef]
52. Singh, G.; Sharma, S.; Sharma, R.; Dwivedi, Y.K. Investigating environmental sustainability in small family-owned businesses: Integration of religiosity, ethical judgment, and theory of planned behavior. *Technol. Forecast. Soc. Change* **2021**, *173*, 121094. [CrossRef]
53. Timberlake, D.S.; Rhee, S.H.; Haberstick, B.C.; Hopfer, C.; Ehringer, M.; Lessem, J.M.; Smolen, A.; Hewitt, J.K. The moderating effects of religiosity on the genetic and environmental determinants of smoking initiation. *Nicotine Tob. Res.* **2006**, *8*, 123–133. [CrossRef]
54. Bhuian, S.N.; Sharma, S.K.; Butt, I.; Ahmed, Z.U. Antecedents and pro-environmental consumer behavior (PECB): The moderating role of religiosity. *J. Consum. Mark.* **2018**, *35*, 287–299. [CrossRef]
55. Siyavooshi, M.; Foroozanfar, A.; Sharifi, Y. Effect of Islamic values on green purchasing behavior. *J. Islam. Mark.* **2019**, *10*, 125–137. [CrossRef]
56. Wang, S.; Wang, J.; Li, J.; Zhou, K. How and when does religiosity contribute to tourists' intention to behave pro-environmentally in hotels? *J. Sustain. Tour.* **2020**, *28*, 1120–1137. [CrossRef]
57. Chen, M.F. Consumer attitudes and purchase intentions in relation to organic foods in Taiwan: Moderating effects of food-related personality traits. *Food Qual. Prefer.* **2007**, *18*, 1008–1021. [CrossRef]
58. Mohd Suki, N. Green product purchase intention: Impact of green brands, attitude, and knowledge. *Br. Food J.* **2016**, *118*, 2893–2910. [CrossRef]
59. Gopi, M.; Ramayah, T. Applicability of theory of planned behavior in predicting intention to trade online: Some evidence from a developing country. *Int. J. Emerg. Mark.* **2007**, *2*, 348–360. [CrossRef]
60. Amin, H.; Rahim Abdul Rahman, A.; Abdul Razak, D. Consumer acceptance of Islamic home financing. *Int. J. Hous. Mark. Anal.* **2014**, *7*, 307–332. [CrossRef]
61. Jaffar, M.A.; Musa, R. Determinants of attitude towards Islamic financing among halal-certified micro and SMEs: A preliminary investigation. *Procedia-Soc. Behav. Sci.* **2014**, *130*, 135–144. [CrossRef]
62. Zeithaml, V.A.; Berry, L.L.; Parasuraman, A. The behavioral consequences of service quality. *J. Mark.* **1996**, *60*, 31–46. [CrossRef]
63. Nunnally, J.C. The Assessment of Reliability. In *Psychometric Theory*; McGraw-Hill: New York, NY, USA, 1994.
64. Podsakoff, P.M.; MacKenzie, S.B.; Lee, J.-Y.; Podsakoff, N.P. Common Method Biases in Behavioral Research: A Critical Review of the Literature and Recommended Remedies. *J. Appl. Psychol.* **2003**, *88*, 879. [CrossRef] [PubMed]
65. Hair, J.F., Jr.; Hult, G.T.M.; Ringle, C.M.; Sarstedt, M.; Danks, N.P.; Ray, S. *Partial Least Squares Structural Equation Modeling (PLS-SEM) Using R: A Workbook*; Springer Nature: Berlin/Heidelberg, Germany, 2021; ISBN 978-3-030-80519-7.
66. Hair, J.F., Jr.; Hult, G.T.M.; Ringle, C.M.; Sarsted, M. *A Primer on Partial Least Squares Strutural Equation Modeling (PLS-SEM)*; Sage Publications Inc.: New York, NY, USA, 2014.
67. Leguina, A. A Primer on Partial Least Squares Structural Equation Modeling (PLS-SEM). *Int. J. Res. Method Educ.* **2015**, *38*, 220–221. [CrossRef]
68. Kline, R.B. *Structural Equation Modeling*; Guilford: New York, NY, USA, 1998.
69. Henseler, J.; Ringle, C.M.; Sinkovics, R.R. The Use of Partial Least Squares Path Modeling in International Marketing. In *New Challenges to International Marketing*; Sinkovics, R.R., Ghauri, P.N., Eds.; Advances in International Marketing; Emerald Group Publishing Limited: Bingley, UK, 2009; Volume 20, pp. 277–319; ISBN 978-1-84855-469-6.

70. Chin, W.W. The Partial Least Squares Approach to Structural Equation Modeling. *Mod. Methods Bus. Res.* **1998**, *295*, 295–336.
71. Wetzels, M.; Odekerken-Schröder, G.; Van Oppen, C. Using PLS Path Modeling for Assessing Hierarchical Construct Models: Guidelines and Empirical Illustration. *MIS Q.* **2009**, *33*, 177–195. [CrossRef]
72. Cohen, J. *J 1923–1998. Statistical Power Analysis for the Behavioral Sciences*; Lawrence Erlbaum Associates: Mahwah, NJ, USA, 1988.
73. Hofstede, G. Culture and organizations. *Int. Stud. Manag. Organ.* **1980**, *10*, 15–41. [CrossRef]
74. Gharbi, H.; Sobaih, A.E.E.; Aliane, N.; Almubarak, A. The Role of Innovation Capacities in the Relationship between Green Human Resource Management and Competitive Advantage in the Saudi Food Industry: Does Gender of Entrepreneurs Really Matter? *Agriculture* **2022**, *12*, 857. [CrossRef]

agriculture

MDPI

Article

The Impact of Technical Training on Farmers Adopting Water-Saving Irrigation Technology: An Empirical Evidence from China

Ding Xiuling [1], Lu Qian [1,*], Li Lipeng [2,*] and Apurbo Sarkar [3]

1 College of Economics and Management, Northwest A&F University, Yangling, Xianyang 712100, China; dxl1227@nwafu.edu.cn
2 School of Economics and Management, Ningxia University, Yinchuan 750021, China
3 School of Agriculture and Food Science, The University of Queensland, Brisbane 4072, Australia
* Correspondence: luqian@nwafu.edu.cn (L.Q.); feixiang2022@nxu.edu.cn (L.L.)

Abstract: Farmers' adoption of water-saving irrigation technology (WSIT) is essential for achieving high-quality agricultural development. An in-depth analysis of the impact of risk aversion, technical training and their interaction on farmers' adoption of WSIT will help the government to promote WSIT to facilitate agricultural resource conservation and sustainable development. The study takes 707 farmers who grow watermelons and muskmelon in Yuncheng and Xian City of Shanxi and Shaanxi provinces as the research object to analyse the influence of risk aversion and technical training and their interaction terms on farmers' WSIT adoption behaviour. The study uses the Probit and moderating effect models to outline the findings. The empirical analysis reveals the following outcomes: (i) 27.44% of the sample farmers adopt water-saving irrigation technology, indicating that the current adoption rate and the enthusiasm for adoption are relatively low; (ii) risk aversion has a significant negative impact on farmers' adoption of WSIT; (iii) both online and offline technical training have a significant positive impact on farmers' adoption of WSIT; (iv) significant group differences exist in the effects of risk aversion, online technical training, offline technical training and interaction items on farmers' WSIT adoption behaviour. Therefore, the study proposes to strengthen the role of technical training in the diffusion of WSIT and implement differentiated technical training for different types of farmers to reduce the degree of risk aversion of farmers.

Keywords: risk aversion; offline technical training; online technical training; farmers' water-saving irrigation technology adoption behaviour; moderating effect

Citation: Xiuling, D.; Qian, L.; Lipeng, L.; Sarkar, A. The Impact of Technical Training on Farmers Adopting Water-Saving Irrigation Technology: An Empirical Evidence from China. *Agriculture* **2023**, *13*, 956. https://doi.org/10.3390/agriculture13050956

Academic Editors: Roberto Henke and Filiberto Altobelli

Received: 14 March 2023
Revised: 14 April 2023
Accepted: 18 April 2023
Published: 26 April 2023

1. Introduction

Water scarcity is now emerging as an underappreciated challenge to the integrity of China's comprehensive development goals [1]. China's per capita water resources are one-fourth of the world's average level [2], and day by day, it is decreasing at an alarming rate [3,4]. With the significant development of China's overall economy, and high water consumption trends by manufacturing industries, the contradiction between the supply and demand of water resources will further intensify [5,6]. As a dominant user of water resources, China's agricultural sector consumes a significant proportion of the existing water resources. According to the Bulletin of the Ministry of Water Resources of China, the agricultural water consumption in 2021 will be 235 billion m^3, accounting for 74% of the total water consumption [7]. However, the effective utilization coefficient of China's farmland irrigation water is only 0.6, which is still far behind the average level of 0.7–0.8 in developed countries [8,9], further exacerbating the contradiction between the supply and demand of water resources for agricultural production. Under the combined effects of the widespread shortages of water resources and poor agricultural irrigation conditions, China's high-quality agricultural development and green transformation face significant bottlenecks.

Moreover, the contradiction between the supply and demand of water resources has severely impacted farmers' agricultural production and livelihood and poses a severe threat to their future sustainable development [10,11]. Compared with traditional irrigation methods, high-efficiency water-saving irrigation technologies such as channel seepage, droppers, micro-sprinkler and drip irrigation systems can improve water resource utilization efficiency, effectively improve cultivated land quality and increase planting intensity [12]. Those systems can also foster the dual effects of improving the ecology and increasing agricultural income [13]. Therefore, supporting farmers to adopt high-efficiency water-saving irrigation technology and improving farmers' cognitive prospects to reduce water loss has become an inevitable choice to solve the above contradictions [14,15]. However, farmers' adoption of water-saving irrigation technology is not optimistic, and they are not very aware of water-saving irrigation technology and even less motivated to adopt it [16]; especially in developing countries, the situation is even worse [17,18]. Even rural farmers resist water-saving irrigation technology and technology adoption, and average household sizes are low [19,20]. Given this, how to effectively promote farmers to adopt water-saving irrigation technology has become a practical problem to be solved urgently by the government and academia. It is generally believed that farmers' production technology selection behaviour is closely related to internal factors and external factors [21]. Regarding internal factors, it mainly analyses the effects of family demographic characteristics, economic endowment characteristics, differences in socioeconomic status, social capital and production risks on farmers' adoption of water-saving irrigation technology [22,23]. Regarding the external environment, the impact of technical attributes, market environment, natural environment, policy environment and land property rights may be crucial to the adoption behaviour of farmers' water-saving irrigation technology [24].

According to the development economics theory, high-risk aversion is an essential characteristic of small farmers. Some studies (such as Grové et al. [25], Hu et al. [26] and Adere et al. [27]) indicate that higher risk aversion of farmers will lead to slow technology diffusion. Logically, it will make them less motivated to adopt water-saving irrigation technology. Therefore, it is necessary to pay attention to the impact of technical training in the external environment on farmers' water-saving irrigation technology adoption to solve the problem of low water-saving irrigation technology adoption due to farmers' risk aversion [28]. On the one hand, technical training is an essential channel for farmers to understand and adopt risk avoidance measures [29,30]. On the other hand, technical training can efficiently transmit information and increase the availability of new technologies [31]. It is helpful for farmers to learn agricultural knowledge and technology and use the experience to continuously accumulate and improve the structure of agricultural technology knowledge [32,33]. It also plays a leading role in technology demonstration to increase the diffusion speed of water-saving irrigation technology and to increase the speed of water-saving irrigation for farmers' technology adoption rate [34].

In the existing literature, risk aversion and the impact of technical training have been explored separately. Very limited literature has integrated these two into an integrated framework on farmers' water-saving irrigation technology adoption behaviour has not been analysed sufficiently. However, most studies only start from the perspective of offline technical training, such as government or cooperatives. Few studies include online technical training in analysing farmers' water-saving irrigation technology adoption behaviour. The study aims to analyse how risk avoidance and online and offline technical training work together for farmers adopting water-saving irrigation technology. To the best of our knowledge, the study will be one of the first attempts to explore the impact of online and offline training in farmers' water-saving irrigation technology adoption. Moreover, using a behavioural analysis framework, the study also provides a comparative analysis of the impact of online technical training on farmers' risk aversion behaviour, which is of prime significance to the study. The empirical setup of the study comprised a data set of 707 melon farmers in Yuncheng and Xian City of Shanxi and Shaanxi provinces, China. Specifically, the researchers utilized the Probit and moderating effect models to analyse the influence of risk aversion, technical training, and interaction items on farmers' water-saving irrigation technology adoption behaviour. The study will comprehensively explain the behavioural logic of Chinese farmers' water-saving irrigation technology adoption and provide a decision-making reference for relevant governmental departments to promote farmers' water-saving irrigation technology adoption and the sustainable use of water resources.

2. Theoretical Analysis and Research Hypotheses

2.1. Influence of Risk Aversion on Farmers' Adoption Behavior of Water-Saving Irrigation Technology

Technically, as a core economic entity, farmers are risk-sensitive and tend to avoid risk [34], especially in adopting novel approaches or tactics [35]. Moreover, various studies (such as Zhou et al. [36], Mooney et al. [37] and Ojo et al. [38]) outline that farmers have very limited risk-taking and coping capabilities in the process of agricultural production and operation. They must consider profit maximization and risk minimization when adopting production technology. Therefore, the motivation of farmers' risk preference is the key to slow technology diffusion [39], and risk aversion leads to a low adoption rate of farmers' water-saving irrigation technology, which inhibits farmers' enthusiasm for adoption [40]. This is because the higher the degree of risk aversion of farmers, the more generally they maintain a scrutinous and cautious attitude in the production process and rational thinking to avoid risks and shocks. In agriculture economics, water-saving irrigation technology can play an irreplaceable role in land–water resource utilization efficiency, improving cultivated land quality, increasing yield per hectare and promoting sustainable agricultural development [41]. Specifically in China, various factors influence farmer's behaviour in adopting water-saving irrigation technology, such as, for farmers, the risk of uncertain returns of agricultural products, the frequent fluctuations in the price of agricultural products, the weak bargaining power of marginal farmers, the poor market price information and the high requirements for input costs in the early stage of water-saving irrigation technology adoption [42]. However, there are still more significant risks in adopting water-saving irrigation technology. According to the study of Wang et al. [43], the cost invested in water-saving irrigation technology may not bring the corresponding expected return, inhibiting farmers' adoption of water-saving irrigation technology. Existing literature (such as Tian et al. [44], Yang et al. [45] and Bakhshi et al. [46]) highlighted the potential risk of the improper application of water-saving irrigation technology and found that there are dual attributes of knowledge-intensive and capital-intensive at play, which requires high-quality knowledge of the subject of technology adoption. When farmers have a relative deprivition of risk regarding the technical know-how of new technology, they are more likely to avoid it [47,48]. Therefore, it can be argued that the higher the degree of risk aversion of farmers, the lower the framer's enthusiasm for water-saving irrigation technology, which eventually hinders the adoption rate. Accordingly, this study proposes the first research hypothesis:

H1. *Risk aversion has a negative impact on farmers' adoption of water-saving irrigation technology.*

2.2. Effects of Technical Training on Farmers' Water-Saving Irrigation Technology Adoption Behavior

Agricultural technology training is a scientific, structured and promotional activity that takes farmers as the training objects [49,50] and may improve farmers' agricultural technology cognition, information acquisition ability and agricultural literacy through various channels including knowledge sharing and demonstration [51,52]. Water-saving irrigation technology is an exogenous technology with knowledge-intensive attributes [53]. Therefore, according to the existing literature, technical training mainly affects farmers' water-saving irrigation technology adoption through the following two channels: First, technical training helps farmers break down information barriers, increases farmers' information and understanding of water-saving irrigation technologies, improves farmers' agricultural technical literacy, and deepens their knowledge of water-saving irrigation technologies to improve the quality of cultivated land, increase crop yield and income per hectare and save energy [54,55]. The degree of awareness of water resources and other benefits makes them more active in adopting water-saving irrigation technology [56,57], which promotes the adopters to obtain relevant information actively and helps farmers break information barriers [58]. Second, technical training can help obtain product, market and policy information to promote the deepening and expansion of knowledge and experience of farmers and further rationally optimize the endowment of agricultural production resources [59,60]. Moreover, it may ease the factor endowment constraints of farmers adopting water-saving irrigation technology so that farmers can actively adopt it without significantly impacting the current family business situation [61,62]. With the promotion and use of infrastructure and digital technology, technical training can be divided into two categories: (i) offline and (ii) online technical training, according to different training forms [63,64].

Offline technical training refers to the publicity, promotion of technical knowledge and transfer of relevant information by agricultural technicians or experts by distributing agricultural information materials, broadcasting, classroom explanations and field demonstrations. Online technical training refers to relying on the internet and digital technology, using computers and smartphones as a platform for public accounts or web pages, and using short videos to help farmers obtain relevant technical knowledge and information. Online technical training can break time and regional boundaries and transmit information to more farmers at a lower cost, break down information barriers, and reduce farmers' information asymmetry [65]. Due to the different emphases of offline and online technical training, these two may impact the adoption behaviour of farmers' water-saving irrigation technology [66]. Compared with online technical training, offline technical training can alleviate the contradiction between the knowledge- and capital-intensive attributes of water-saving irrigation technology and farmers' technical cognition and application ability and help promote the adoption of water-saving irrigation technology [67]. Accordingly, the study proposes the second and third research hypotheses:

H2. *Participating in technical training positively impacts farmers' adoption of water-saving irrigation technologies.*

H3. *Unlike online technical training, offline technical training has a more substantial positive effect on farmers' adoption of water-saving irrigation technology.*

2.3. Mitigation Effect of Technical Training on Risk Aversion Inhibiting Farmers' Adoption of Water-Saving Irrigation Technology

Farmers' risk preference is the key to technology diffusion and can significantly alter the behavioural factors of farmers [68,69]. In order to solve the "dilemma" of farmers' risk aversion and technology adoption, technical training is regarded as promoting the adoption of production technology by farmers and saving agricultural production [70]. The impact of risk aversion on farmers' behaviour is not static but changes with the external environment [71,72]. As a typical form of the external environment, technical training can effectively change the endowment constraints of farmers and enhance farmers' confidence and skills in using technology effectively [73]. Therefore, it is essential to alleviate the inhibitory effect of risk aversion on farmers' adoption of water-saving irrigation technology. This is mainly reflected in two aspects: First, technical training can correct the information asymmetry between farmers' risk aversion and water-saving irrigation technology adoption by introducing external technical knowledge and confidence and prompt farmers to evaluate water-saving irrigation correctly and rationally [74]. The risks faced by technology adoption form a positive expected return and reduce the negative impact of risk aversion on farmers' water-saving irrigation technology adoption behaviour. Second, technical training can improve farmers' awareness of water-saving irrigation technology [75]. Farmers can learn standardized technical operation knowledge through online and offline technical training and are familiar with various water-saving irrigation technical facilities and their use and maintenance [76]. It can also enhance the confidence and application ability of water-saving irrigation technology adoption, reduce farmers' concerns about the risk of unsuitable technology or improper operation and effectively resolve the negative impact of risk aversion on farmers' water-saving irrigation technology adoption behaviour [77]. Accordingly, this paper proposes a fourth research hypothesis.

H4. *Technical training can alleviate the inhibitory effect of risk aversion on farmers' adoption of water-saving irrigation technology.*

3. Materials and Methods

3.1. Data Sources

The study's empirical data comprised a face-to-face survey of farmer's households in the central melon-producing region of Shanxi and Shaanxi Province, China, in December 2020. At the same time, the responses were recorded with a structured questionnaire covering the risk aversion test experiment, technical training situation, individual characteristics of the head of household, family situation, external environmental characteristics and water-saving irrigation technology adoption of farmers. The study adopted multistage sampling criteria to ensure the rationality of the selection of sample farmers, while the researchers adopted typical random sampling tactics to identify the potential respondents. First, the study consulted with the local agricultural extension officers to determine the major melon-producing region of the selected provinces and the associated characteristics of the farmer's water usage mechanism. Yuncheng City, Shanxi Province, and Xi'an City, Shaanxi Province, were selected based on the agriculture extension officers' inputs. The selected two cities belong to the Yellow River Irrigation Area and the Fen River Irrigation Area, respectively, where farmers usually use the traditional flood irrigation method. The method is considered water resources intensive and can lead to severe soil erosion and a sharp decline in soil quality. Therefore, the region was suitable for fulfilling the prime research objectives. Second, the researchers randomly selected Yanhu and Xia County from Yuncheng City, Shanxi Province, and Yanliang County from Xi'an City, Shanxi Province. In the third stage, 3–5 towns were randomly selected from each of the selected districts/counties, providing 19 towns. After that, the researchers randomly selected 2–5 villages from this township, comprising 35 villages. Finally, 19–25 farmers who grow watermelons and muskmelon were randomly selected in each sample village as the research objects.

Before conducting the formal survey, the study utilized a pilot test with randomly selected 20 farmers from four villages from two provinces to test the instrument, and according to the inputs, the study adjusted the instrument accordingly, which we believe improved the accuracy of the instruments. Moreover, we chose the respondent household head to be the priority (if not present, we choose the immediate farming decision maker), which we believe ensured the quality of the information we have gathered. During the final survey process, 731 farmers were consulted. Among them, 707 valid questionnaires were obtained for further analysis, and the effective rate of the survey was 96.715%. We eliminated 24 responses as they gave up midway or there was missing information regarding the core variables required for performing the analysis. As the prime respondents of the study are farmers, we acknowledge that potentially biased responses may occur. Therefore, the study adopted a two-stage strategy to reduce the potentially biased responses, as suggested by Podsakoff et al. [78]. First, before asking questions, the research team discussed all the variables and essential information with the respondent to reduce this issue. Second, the team ensured the questionnaire was well equipped with neutrally worded questions and answer options were not leading. Moreover, the study performs a robustness test to depict the reliability of the outcomes.

Table 1 summarizes the essential characteristics of the sample farmers. In terms of the age of household heads, 18–30, 31–45, 46–60 and over 61 accounted for 1%, 17.96%, 66.05%, and 14.99% of the respondents, respectively, indicating that the current rural households are relatively older, mainly middle-aged and older adults. Regarding the education level of the household heads, the proportions of households with an education of 6 years or less, 6–9 years, 9–12 years and more than 12 years were 34.80%, 53.32%, 11.46 and 0.42%, respectively. At present, the education level of farmers is generally low, and most of them are at the level of junior high school or below. Regarding cooperative participation, 230 farmer's households are participating in cooperatives, accounting for only 32.53% of the total sample, indicating that the current participation in cooperatives is low. In terms of the planting scale, farmers with less than 1 hectare, 1–2 hectare, 2–3 hectare and more than 3 hectares were 56.58%, 36.63%, 5.09% and 1.70% of the respondents, respectively, indicating that the surveyed farmers mainly focus on small-scale planting. Regarding the proportion of income from farming, farming income accounting for less than 10%, 10–30%, 30–50%, and 50–100% of farmers' income was reported by 2.97%, 22.49%, 24.75% and 49.79% of respondents, respectively, indicating that farming is an essential source of income for most farmers. Regarding market prospect expectations, 633 farmers (89.53%) were optimistic about the prospects of the melon and fruit market, indicating that most farmers are optimistic about the development of the melon and fruit industry.

Table 1. Descriptive analysis of sample farmers.

Feature	Options	No.	Proportion (%)	Feature	Options	Frequency	Proportion (%)
Age of head of the household	18–30 years old	7	1	Planting scale	Less than 1 hectare	400	56.58
	31–45 years old	127	17.96		1–2 hectare	259	36.63
	46–60 years old	467	66.05		2–3 hectare	36	5.09
	61 years old and above	106	14.99		More than 3 hectares	12	1.70
Head of the household education level	Under 6 years	246	34.80	The proportion of planting income	10% or less	twenty one	2.97
	6–9 years	377	53.32		10–30%	159	22.49
	9–12 years	81	11.46		30–50%	175	24.75
	12 years or more	3	0.42		50–100%	352	49.79
Cooperative participation	Participate/Not involved	230	32.53	Market outlook	Yes/No	633	89.53

3.2. *Variable Selection*

3.2.1. Explained Variables

The adoption behaviour of farmers' water-saving irrigation technology is the explanatory variable in the study. Drawing on existing literature (such as Ho et al. [79], Zhang et al. [13] and Mushtaq et al. [19]) and consulting with experts from the agricultural machinery industry, water-saving irrigation technologies such as a dropper, channel seepage irrigation, micro-sprinkler irrigation and film-covered irrigation were determined to be research objects. When farmers adopt any of them or when there are multiple technologies adopted, the value of farmers' water-saving irrigation technology adoption behaviour is 1; otherwise, this value is 0.

3.2.2. Core Explanatory Variables

Risk aversion and technical training refer to the study's core explanatory variables. The experimental economic method was chosen as the experimental measurement of risk aversion, as recommended by Qiu et al. [80] and Xu et al. [81]. The experiment was completed in three stages: first, the "lottery draw" game rules were introduced to the respondents, and the game plan was pre-tested. Second, ten sets of game questions were provided to the respondents, and each question included option A (low-risk option) and option B (high-risk option). Each option corresponds to a different cash reward to let the respondents know that the choice of the risk option will affect their final income. Finally, the respondents from the first questions were selected one by one. Only after the respondent had completed the selection of each question did the researcher allow the respondent to see the next question. In the step-by-step selection process of each question, as long as the respondent chose option B, he can no longer choose option A in subsequent games. This experiment links the final reward of the respondents with the experimental results to ensure that the acquisition of the degree of risk aversion of the respondents is authentic and reliable and to avoid data bias. Respondents can obtain a reward of 10 yuan within 20 min, which can stimulate the enthusiasm of respondents to participate in the lottery game. Table 2 shows the specific content of the experimental design.

Table 2. The experimental design and experimental results of the degree of risk aversion of farmers.

Question Number	Low-Risk Program (Option A)		High-Risk Program (Option B)		The Proportion of Choosing High-Risk Options
	30% Chance	70% Chance	10% Chance	90% Chance	
1	Exchange 200 yuan	Exchange 50 yuan	Exchange 3 00 yuan	Exchange 25 yuan	12.87
2	Exchange 200 yuan	Exchange 50 yuan	Exchange 330 yuan	Exchange 25 yuan	20.50
3	Exchange 200 yuan	Exchange 50 yuan	Exchange 370 yuan	Exchange 25 yuan	32.24
4	Exchange 200 yuan	Exchange 50 yuan	Exchange 420 yuan	Exchange 25 yuan	43.13
5	Exchange 200 yuan	Exchange 50 yuan	Exchange 480 yuan	Exchange 25 yuan	49.35
6	Exchange 200 yuan	Exchange 50 yuan	Exchange 580 yuan	Exchange 25 yuan	56.24
7	Exchange 200 yuan	Exchange 50 yuan	Exchange 700 yuan	Exchange 25 yuan	62.18
8	Exchange 200 yuan	Exchange 50 yuan	Exchange 900 yuan	Exchange 25 yuan	67.83
9	Exchange 200 yuan	Exchange 50 yuan	Exchange 1100 yuan	Exchange 25 yuan	72.07
10	Exchange 200 yuan	Exchange 50 yuan	Exchange 1400 yuan	Exchange 25 yuan	76.45

As the amount exchanged from the first question to the tenth question gradually increases, the possibility of respondents choosing high returns and risks also increases. Option A has a 30% possibility of a bonus of 200 yuan and a 70% possibility of a bonus of 50 yuan. Option B has a 10% possibility of a bonus of 300 yuan, and there is a 90% probability that the bonus will be reached by 25 yuan. According to the experimental results and referring to the risk aversion index formula of Xu et al. [81], the risk aversion degree of farmers can be calculated as the following: risk aversion index = 1(number of high-risk schemes/10). If the number of times the farmer chose high risk was 0, this is extreme risk aversion. On the contrary, if the number of times the farmer chose high risk was 10, this is extreme risk preference.

Offline technical training means agricultural technical experts or personnel disseminate knowledge and information about water-saving irrigation technology to farmers at a fixed time and place through conference lectures and on-site training, most of which take place via the face-to-face medium. If farmers participate in offline technical training, the value is 1; otherwise, they are assigned a value of 0. Online technical training means farmers use the internet, various online apps, WeChat and Weibo official accounts or web browsing, voice, video and other forms to obtain knowledge and information related to water-saving irrigation technology. If farmers use online technical training, the assigned value is 1; otherwise, the assigned value is 0.

3.2.3. Control Variables

The adoption behaviour of farmers' water-saving irrigation technology depends on individual or family internal factors and the external environment. Therefore, along with household head characteristics (age of household head and education level of household head), household management characteristics (family planting years, the proportion of planting income, planting area and number of family workers) and social capital (participation in cooperatives and social network), the study chose the external environment (market outlook and natural disasters) to be the core control variable. In addition, to ensure the estimation effect, regional variables were controlled. The specific meaning and assignment of each variable are shown in Table 3.

Table 3. Variable description and descriptive statistical analysis.

Variable Name		Variable Meaning and Assignment	Average	Standard Deviation
Explained variable				
Adoption behaviour of farmers' water-saving irrigation technology		Whether to use water-saving irrigation technology: adopted = 1, not adopted = 0	0.274	0.447
Core explanatory variable				
Risk Aversion		Risk aversion degree value (between 0 and 1): 0 means extreme risk preference type, 1 means extreme risk aversion type	0.508	0.365
Technical training	Offline technical training	Whether you have received offline technical training: Yes = 1, No = 0	0.226	0.419
	Online technical training	Whether you have received online technical training: Yes = 1, No = 0	0.495	0.501
Control variable				
Age of Head of household		Respondent's age (years)	52.301	8.866
Head of the household education level		Respondents' years of education (years)	7.854	2.720
Family Planting Years		Family planting years (years)	26.266	11.141
The proportion of planting income		The proportion of melon and fruit income in household income (%)	0.544	0.996
planting scale		Family watermelons and muskmelon planting area (hectare)	0.979	0.601
Land levelness		The flatness of the land where the crop is planted: very uneven = 1, uneven = 2, normal = 3, relatively flat = 4, very flat = 5	3.777	0.779
Number of migrant workers		Number of family workers (person)	1.147	0.995
cooperative participation		Whether to participate in cooperatives: yes = 1, no = 0	0.325	0.469
social network		Number of mobile phone contacts (number)	129.484	111.346
market outlook		Whether the respondents are optimistic about the prospects of the melon and fruit market: Yes = 1, No = 0	0.898	0.307
natural disaster situation		Number of natural disasters in the past three years (times)	1.164	1.203
Regional location		Location: Shanxi Province = 0, Shaanxi Province = 1	0.506	0.501

The 707 farmers of the sample were divided into two groups according to the age of the household head, planting scale and education level to grasp the adoption of water-saving irrigation technology more simply and accurately. The specific distribution is shown in Table 4. These groups of farmers included: (i) the first group, based on age (18–50 years old), was more substantial than that of the second one (50 years or more); (ii) regarding the farming scale of the farmers, the second group (more than 0.667 hectares) was more substantial than the first group (0.667 hectares and below); (iii) regarding the education level of the head of the household, the first group (9 years and below) was more substantial than the second group (over 9 years).

Table 4. Adoption of water-saving irrigation technology by sample farmers (%).

Adoption Behaviour	Age of Head of Household		Farming Scale		Head of the Household Education Level	
	18 to 50 (Group A)	Over 50 Years old (Group B)	0.667 Hectares and Below (Group A)	More than 0.667 Hectares (Group B)	9 Years and Below (Group A)	Over 9 Years (Group B)
Adopted	49.49%	50.51%	23.19%	76.81%	84.54%	15.46%
Not adopted	41.72%	58.28%	48.54%	51.46%	89.47%	10.53%

3.3. Model Building

3.3.1. Benchmark Regression Model

The explanatory variable in this paper is "the adoption behaviour of farmers' water-saving irrigation technology", which is a binary classification variable. Therefore, this paper uses the binary Probit model for empirical analysis. Specifically, the model of Formula (1) is as follows:

$$Z = \ln(\frac{p_i}{1-p_i}) = \beta_0 + \beta_1 RA_i + \beta_2 TT_i + \beta_3 MT_i + \sum_{k=1} \beta_{4k} C_i + D_i + \varepsilon_i \quad (1)$$

where P_i represents the probability of farmers adopting water-saving irrigation technology, $1 - P_i$ represents the probability that farmers do not adopt water-saving irrigation technology, $P_{1i}/1 - P_i$ is the probability ratio or relative risk, RA_{1i} represents the degree of risk aversion of farmers, T_i represents offline technical training, MT_i represents the line C_i represents the control variable, D_i represents the dummy variable in the area where farmer i is located, β_0 is the intercept item of the model, B_k is the regression coefficient corresponding to the independent variable and ε_i is the random disturbance item.

3.3.2. Modulation Effect Model

In order to explore the influence mechanism of risk aversion, technical training and farmers' water-saving irrigation technology adoption behaviour, drawing on the research of Wen et al. [82], the following adjustment effect model was constructed:

$$Z = \ln(\frac{p_i}{1-p_i}) = \beta_0 + \beta_1 RA_i + \beta_2 TT_i + \beta_1 RA \times \beta_2 TT_i + \sum_{k=1} \beta_{4k} C_i + D_i + \varepsilon_i \quad (2)$$

$$Z = \ln(\frac{p_i}{1-p_i}) = \beta_0 + \beta_1 RA_i + \beta_2 TT_i + \beta_1 RA \times \beta_3 MT_i + \sum_{k=1} \beta_{4k} C_i + D_i + \varepsilon_i \quad (3)$$

$$Z = \ln(\tfrac{p_i}{1-p_i}) = \beta_0 + \beta_1 RA_i + \beta_2 TT_i + \beta_3 MT_i + \beta_1 RA \times \beta_2 TT_i + \beta_1 RA \times \beta_3 MT_i + \sum_{k=1} \beta_{4k} C_i + D_i + \varepsilon_i \quad (4)$$

In (2), $\beta_1 RA \times \beta_2 TT_i$ represents the interaction term between risk aversion and offline technical training. Meanwhile, (3) represents the interaction term between risk aversion and online technical training; $\beta_1 RA \times \beta_3 MT_i C$ represents the interaction term between risk aversion and online technical training; C represents the control variable; D_i represents the dummy variable of the area where farmer i is located; β_0 is the intercepted item of the model; β_k is the regression coefficient corresponding to the independent variable; ε_i is the random disturbance item.

4. Results

The results of the multicollinearity diagnosis showed that the variance inflation factor (VIF) values of each variable were less than 2, indicating no multicollinearity problem among variables. Stata15.0 [William Gould, StataCorp, https://www.stata.com (accessed on 23 January 2023, Texas, United States] was used for regression, and the estimated results were as follows.

4.1. Benchmark Model Results and Analysis

Table 5 shows the regression results of the model with core explanatory variables introduced in turn. Model (1), model (2), model (3) and model (4) were all tested using the Wald test, and all of them reached a significance level of 1%, indicating that the overall fitting degree of the model is good. The following analysis is mainly based on the estimated results of model (4).

Table 5. Benchmark model regression results.

Variable		Model (1)	Model (2)	Model (3)	Model (4)
Risk Aversion		−0.3061 * (0.1630)			−0.4841 ** (0.1950)
Technical Training	Offline technical training		1.2784 *** (0.1525)		1.2821 *** (0.1635)
	Online technical training			1.1819 *** (0.1337)	1.2056 *** (0.1463)
age of Head of household		0.0155 * (0.0083)	0.0191 ** (0.0088)	0.0154 * (0.0087)	0.0178 * (0.0095)
Head of the household's education level		0.0487 ** (0.0237)	0.0419 * (0.0247)	0.0490 * (0.0254)	0.0312 (0.0274)
Planting years		−0.0156 ** (0.0064)	−0.0184 *** (0.0067)	−0.0127 * (0.0069)	−0.0159 ** (0.0074)
The proportion of farming income		1.6409 *** (0.2648)	1.3707 *** (0.2836)	1.5358 *** (0.2888)	1.2868 *** (0.3145)
Farming scale		0.4665 *** (0.1035)	0.4065 *** (0.1095)	0.3720 *** (0.1110)	0.3030 ** (0.1185)
Land levelness		0.3804 *** (0.0807)	0.1796 ** (0.0858)	0.3367 *** (0.0851)	0.2009 ** (0.0948)
Number of migrant workers		0.0837 (0.0754)	0.0462 (0.0792)	0.0506 (0.0831)	0.0370 (0.0896)
Cooperative participation		0.4696 *** (0.1338)	0.2454 * (0.1454)	0.5595 *** (0.1438)	0.3572 ** (0.1584)
social network		0.0009 * (0.0005)	0.0010 * (0.0006)	0.0007 (0.0006)	0.0008 (0.0006)
Market outlook		0.8430 *** (0.2286)	0.6627 *** (0.2357)	0.7529 *** (0.2399)	0.6411 ** (0.2571)
Natural disaster shock		−0.0668 (0.0496)	−0.0760 (0.0528)	−0.0804 (0.0539)	−0.0789 (0.0589)
Regional location		YES	YES	YES	YES
_cons		−5.3421 *** (0.6899)	−4.7211 *** (0.7283)	−5.8278 *** (0.7380)	−5.1322 *** (0.8055)
Pseudo R 2		0.2580 _	0.3418 _	0.3588 _	0.4420
L R chi 2		2 13.88	283.35	297.40	366.35
Prob > chi 2		0.0000 _	0.0000 _	0.0000 _	0.0000 _
Wald value		163.21 ***	204.56 ***	196.31 ***	208.94 ***

Note: * $p < 0.10$, ** $p < 0.05$, *** $p < 0.01$; the robust standard errors are in brackets.

4.1.1. Core Explanatory Variables

The estimated coefficient of risk aversion to farmers' water-saving irrigation technology adoption behaviour is negative and passed the 5% significance test. This indicates that risk aversion has a significant negative impact on farmers' water-saving irrigation technology adoption behaviour. The higher the degree of farmers' risk aversion, the less likely they are to adopt water-saving irrigation technology, because it is both knowledge-intensive and capital-intensive. For farmers, the cost and risk of adoption are high. Failure to do so will seriously affect the continuous operation of farmers' agricultural production. Therefore, when the degree of risk aversion of farmers is high, they are less willing to bear the potential risks and costs of adopting water-saving irrigation technology, so risk aversion has a significant negative impact on the adoption of water-saving irrigation technology by farmers. Based on this, Hypothesis H1 is verified.

Water irrigation technology adopts risk and costs to encourage farmers to adopt water-saving irrigation technology. In terms of technical training, the estimated coefficients of offline technical training and online technical training on the adoption behaviour of farmers' water-saving irrigation technologies were positive and passed the 1% significance test regarding the effect of offline technical training on farmers' adoption of water-saving irrigation technologies. The influence coefficient is more significant, indicating that obtaining online or offline technical training can significantly promote farmers' adoption of water-saving irrigation technology. The effect of obtaining offline technical training is relatively more substantial because technical training can save farmers' information search costs and improve their cognition level regarding water-saving irrigation technology. It also assists in alleviating the factor endowment constraints of farmers' adoption of water-saving irrigation technology, optimizing household resource allocation, and reducing energy-saving costs. In addition, compared with online technical training, the form and content of offline technical training are relatively more targeted, and the concepts, knowledge and technologies taught are easier to be understood and accepted by farmers. The positive effect of technology adoption behaviour is relatively more substantial. Therefore, Hypotheses H2 and H3 are verified.

4.1.2. Other Explanatory Variables

Among the household head characteristics, the age of the head is significant, at the 10% level, and the coefficient is positive, indicating that older farmers are more likely to adopt water-saving irrigation technology. The possible reason is that water-saving irrigation technology is a resource-saving technology. Older farmers pay more attention to resource conservation and the ecological environment than young farmers, so they tend to adopt water-saving irrigation technology. The effect of planting years on farmers' water-saving irrigation technology adoption is significant, at the 5% significance level, and the coefficient is positive, indicating that farmers with longer planting years are more likely to adopt water-saving irrigation technologies. The possible reason is that the longer the planting years, the more sufficient the agricultural production skills of farmers and the easier it is to adopt water-saving irrigation technologies. Therefore, the planting years significantly positively impact the adoption behaviour of farmers' water-saving irrigation technologies. Market prospect expectations positively impact farmers' adoption of water-saving irrigation technology at a significance level of 5%, indicating that farmers who are more optimistic about market prospects are more likely to adopt water-saving irrigation technology. This is because the more optimistic the market prospect is, the more optimistic the farmers are about adopting water-saving irrigation technology, thus encouraging farmers to adopt water-saving irrigation technology actively.

Regarding family characteristics, the impact of the planting income on farmers' adoption of water-saving irrigation technology is significant at the 1% level, and the coefficient is positive, indicating that the higher the proportion of planting income, the higher the contribution and importance of melon and fruit planting income to farmers' families. In this situation, farmers' adoption of water-saving irrigation technology can significantly improve production stability and obtain stronger income protection; planting income has a significant positive effect on farmers' adoption of water-saving irrigation technology. Participation in cooperatives has a significant positive impact on farmers' adoption of water-saving irrigation technologies at the 5% significance level, indicating that farmers participating in cooperatives are more inclined to adopt water-saving irrigation technologies. This is because the participation of cooperatives can significantly improve the degree of organization of farmers, help them obtain core agricultural information and improve the bargaining power of farmers.

Therefore, the participation of cooperatives can promote the adoption of water-saving irrigation technology by farmers. The planting scale significantly affects farmers' adoption of water-saving irrigation technology at the significance level of 1%, indicating that farmers with large planting scales are more likely to adopt water-saving irrigation technology. The possible reasons are as follows: on the one hand, the larger the planting scale, the less willing farmers are to bear the huge potential losses caused by drought and the more willing they are to adopt water-saving irrigation technology. On the other hand, the larger the planting scale, the lower the average cost of water-saving irrigation technology. Therefore, the planting scale can significantly promote farmers' adoption of water-saving irrigation technology. Land levelness has a significant positive impact on farmers' adoption of water-saving irrigation technology at a significance level of 5%, indicating that the more level the household land is, the more farmers will adopt water-saving irrigation technology. The high land level can reduce the labour and material cost of farmers' water-saving irrigation technology and encourage farmers to adopt water-saving irrigation technology actively.

Regarding external environmental characteristics, the estimated coefficient of natural disaster impact is negative but has not passed the significance test. The reason for this is that the impact of natural disasters in the study area is dominated by strong wind and hail, which will destroy water-saving irrigation facilities and inhibit farmers from adopting water-saving irrigation technology. The geographical location variable passed the significance test at the statistical level of 1%, indicating that farmers in Shaanxi Province are more inclined to adopt water-saving irrigation technology. The possible reason is that, compared with Yuncheng City in Shanxi Province, Xi'an City in Shaanxi Province has a faster economic development. The population is large and concentrated, and a better sales market encourages farmers to adopt water-saving irrigation technology.

4.2. Moderation Effect Results and Analysis

The interaction term between risk aversion and offline technical training and the interaction item between risk aversion and online technical training were added based on the benchmark model regression and the measurement to further analyse the moderating effect of technical training on risk aversion and inhibition of farmers' adoption of water-saving irrigation technology. Table 6 presents the regression results. From the estimated results of model (5) to model (7), it can be seen that the interaction term between offline technical training and risk aversion has a significant positive impact on the adoption behaviour of farmers' water-saving irrigation technology at the significance level of 1%. It indicates that offline technical training can effectively alleviate the negative effect of risk aversion on farmers' water-saving irrigation technology adoption.

Table 6. Test of the moderating effect.

Variable		Model (5)	Model (6)	Model (7)
Risk Aversion		−0.9001 *** (0.2187)	−1.8028 *** (0.4426)	−2.3222 *** (0.5555)
Technical Training	Offline technical training	0.4711 ** (0.2381)		0.6019 ** (0.2495)
	Online technical training		0.4754 ** (0.2111)	0.5761 *** (0.2209)
interaction term	Risk avoidance × offline technical training	1.7565 *** (0.4042)		1.5437 *** (0.4712)
	Risk Avoidance × Online Technical Training		1.9048 *** (0.4859)	1.8018 *** (0.5763)
control variable		YES	YES	YES
Constant term		YES	YES	YES
Pseudo R 2		0.3720	0.3876	0.4702
L R chi 2		308.32	321.32	389.75
Prob > chi 2		0.0000	0.0000	0.0000
Wald value		2 02.87 ***	1 67.81 ***	1 70.83 ***

Note: ** $p < 0.05$, *** $p < 0.01$; the robust standard errors are in brackets.

The interaction term between online technical training and risk aversion has a significant positive impact on the adoption behaviour of farmers' water-saving irrigation technologies at the 1% significance level, indicating that online technical training can also effectively alleviate the impact of risk aversion on farmers' adoption of water-saving irrigation technologies. As the two primary components of technical training, whether offline or online, it can quickly and effectively transmit relevant technical knowledge and information on water-saving irrigation technology to reduce the degree of information asymmetry in farmers' water-saving irrigation. It can also reduce farmers' relative knowledge deprivation about the uncertainty and risk of the water-saving irrigation technology adoption process, thereby increasing the possibility of farmers adopting water-saving irrigation technology. At the same time, technical training can improve farmers' agricultural production resource and risk management levels and optimize family income. Finally, the allocation of agricultural production resources increases farmers' water-saving irrigation technology adoption rate. Therefore, Hypothesis H4 is verified.

4.3. Robustness Test

The method of replacing the core model is used for verification to test the robustness of the baseline regression results and the moderation effect. In this study, model (8) and model (9) were used for regression in the binary Logit model, and the results are shown in Table 7. It can be seen that risk aversion has a negative impact on farmers' water-saving irrigation technology adoption behaviour 5% level of significance. Offline and online technical training and their interaction items significantly impact farmers' water-saving irrigation technology. A positive impact was that the estimated results are consistent with the regression results of the binary Probit model, in terms of significance and impact direction, proving that the regression and the moderating effect test are relatively robust.

Table 7. Robustness test of the regression results.

Variable Name		Model (8)	Model (9)
Risk Aversion		−0.8157 ** (0.3473)	−4.2843 *** (1.0080)
Technical Training	Offline technical training	2.2113 *** (0.2913)	0.9965 ** (0.4366)
	Online technical training	2.1584 *** (0.2753)	0.9648 ** (0.3894)
interaction term	Risk avoidance × offline technical training		2.7235 *** (0.8510)
	Risk Avoidance × Online Technical Training		3.4010 *** (1.0389)
Age of Head of household		0.0354 ** (0.0177)	0.0325 * (0.0178)
Head of the household education level		0.0614 (0.0497)	0.0492 (0.0500)
Family Planting Years		−0.0313 ** (0.0135)	−0.0296 ** (0.0135)
The proportion of planting income		2.2535 *** (0.5699)	2.2300 *** (0.5781)
Planting scale		0.5895 *** (0.2145)	0.5475 ** (0.2175)
Land levelness		0.3590 ** (0.1671)	0.3794 ** (0.1700)
Number of migrant workers		0.0423 (0.1635)	0.0573 (0.1678)
Cooperative participation		0.6242 ** (0.2852)	0.6281 ** (0.2895)
Social network		0.0014 (0.0011)	0.0015 (0.0011)
Market outlook		1.2104 ** (0.4743)	1.2056 ** (0.4811)
Natural disaster shock		−0.1559 (0.1040)	−0.1597 (0.1082)
Regional location		YES	YES
_cons		−9.3351 *** (1.4979)	−7.9467 *** (1.5166)
Pseudo R 2		0.4414	0.4691
L R chi 2		365.92	388.82
Prob > chi 2		0.0000	0.0000

Note: * $p < 0.10$, ** $p < 0.05$, *** $p < 0.01$; the robust standard errors are in brackets.

5. Further Socio-Demographics

Differential analysis was carried out on the age, education level, and planting scales to further clarify the mechanism of risk aversion and technical training on farmers' water-saving irrigation technology adoption. The main reason for considering these three aspects is that age and years of education can reflect farmers' views and understanding of new technologies to a certain extent. Farmers of different age groups and educational levels will have specific differences in their risk aversion and information acquisition capabilities, so the influence of risk aversion and technical training on the adoption behaviour of water-saving irrigation technology of farmers of different ages and educational levels may be different. The planting scale represents the endowment of land resources of farmers, and the more land resources there are, the stronger farmers' dependence on the land. Similarly, risk aversion and technical training may have differential effects on farmers with different planting scales' adoption behaviour of water-saving irrigation technology. The specific results are shown in Table 8.

Table 8. Regression results of the impact of risk aversion and technical training on different types of farmers' adoption of water-saving irrigation technology.

Variable Name		Age		Education Level		Business Scale	
		50 and under	Over 50 years old	9 years and below	Over 9 years	0.667 hectare and below	More than 0.667 hectare
Risk Aversion		−2.0699 *** (0.6037)	−6.0150 ** (3.0233)	−6.7048 ** (2.9787)	−1.7876 *** (0.5767)	−3.6654 ** (1.4970)	−1.9302 *** (0.6515)
Technical Training	Offline technical training	0.6411 ** (0.2796)	0.5380 (0.7399)	1.1508 ** (0.5015)	0.3616 (0.3156)	0.7269 (0.6421)	0.6748 ** (0.2932)
	Online technical training	0.7524 *** (0.2555)	0.7124 (0.6516)	0.5691 (0.4850)	0.6632 ** (0.2699)	0.5044 (0.4832)	0.7391 *** (0.2693)
Interaction term	Risk avoidance × offline technical training	1.5630 *** (0.5168)	2.5312 (1.8631)	1.1955 (1.0048)	1.8408 *** (0.5711)	0.9511 (1.2405)	1.6222 *** (0.5613)
	Risk Avoidance × Online Technical Training	1.3315 ** (0.6287)	5.6601 * (3.0643)	6.3843 ** (2.9987)	1.0701 * (0.6268)	2.4685 (1.5404)	1.5155 ** (0.6699)
Control variable		YES	YES	YES	YES	YES	YES
Constant term		YES	YES	YES	YES	YES	YES
Pseudo R 2		0.4757	0.6198	0.5542	0.4745	0.3349	0.5207
L R chi 2		323.36	91.64	159.50	255.07	52.50	330.67
Prob > chi 2		0.0000	0.0000	0.0000	0.0000	0.0000	0.0000

Note: * $p < 0.10$, ** $p < 0.05$, *** $p < 0.01$; the robust standard errors are in brackets.

5.1. The Influence of Risk Aversion and Technical Training on the Adoption Behaviour of Farmers' Water-Saving Irrigation Technology under the Age Difference

It can be seen from Table 8 that risk aversion has a significant negative effect on the adoption of water-saving irrigation technology by farmers of different age groups. The absolute value of the coefficient of the group over 50 years old is greater than that of the group under 50 years old, indicating that risk aversion significantly impacts farmers' adoption of water-saving irrigation technology. The inhibitory effect of technology adoption behaviour increases with the age of farmers. This is because, with increasing age, farmers are more cautious about investment in agricultural production, avoid investment risks and adopt prudent management strategies to obtain a safer investment return to stabilize their livelihoods. Therefore, as age increases, the inhibitory effect of risk aversion on farmers' water-saving irrigation technology adoption becomes stronger.

In technical training, both online and offline technical training have a positive impact on the adoption of water-saving irrigation technology by farmers of different age groups, and only the coefficient of the group aged 50 and below passed the significance test, indicating that technical training has a positive effect on the adoption of water-saving irrigation technology of farmers aged 50 and below. This is because, on the one hand, farmers aged 50 and below are less dependent on traditional technologies and have more channels to acquire new agricultural technologies, which are more modern. It is easy to break the path dependence of technology diffusion and obtain water-saving irrigation technologies at the minimum cost. Information prompts farmers to adopt water-saving irrigation technology; on the other hand, farmers aged 50 and below are more adventurous and innovative and are more willing to take risks and adopt water-saving irrigation technology with innovative attributes; therefore, as age increases, the promotion effect of technical training on the adoption of water-saving irrigation technology by farmers is gradually weakened. The interaction term of risk aversion and offline technical training positively impact the adoption behaviour of water-saving irrigation technology for farmers in different age groups, and only the group aged 50 and below passed the significance test, indicating the interaction between risk aversion and offline technical training. The item has a stronger effect on promoting the adoption of water-saving irrigation technology by farmers aged 50 and below. The interaction term of risk aversion and online technical training has a significant positive impact on the adoption behaviour of water-saving irrigation technology by farmers in different age groups. The absolute value of the coefficient of the group over 50 years old is greater than the absolute value of the coefficient of the group under 50 years old, indicating that, regarding risk aversion, compared with online technical training, the effect of promoting the adoption of water-saving irrigation technology by farmers aged 50 and below is stronger.

5.2. The Impact of Risk Aversion and Technical Training on the Adoption Behaviour of Farmers' Water-Saving Irrigation Technology under the Difference of Educational Level

It can be seen in Table 8 that risk aversion has a negative and significant effect on the water-saving irrigation technology adoption behaviour of farmers with different educational levels, and the absolute value of the coefficient of the group with an education level of 9 years or less is greater than the absolute value of the group with an education level of 9 years or more. This indicates that risk aversion positively affects the inhibitory effect of irrigation technology adoption weakens with farmers' educational level improvement. Farmers' cognitive ability and cognitive level are significantly improved with the increase in education level. They can more objectively evaluate technical risks and personal technical capabilities and reduce unnecessary worries and concerns in adopting water-saving irrigation technologies. The higher the degree, the weaker the inhibitory effect of risk aversion on farmers' adoption of water-saving irrigation technology. In technical training, offline technical training has a positive impact on the adoption of water-saving irrigation technology by farmers with different levels of education, and only the coefficient of the group of 9 years of education and below passed the significance test, indicating that offline technical training has a positive impact on the adoption of water-saving irrigation technology by farmers in the group of 9 years of education and below. The promotion effect of water-saving irrigation technology adoption is relatively substantial. Online technical training positively impacts the adoption of water-saving irrigation technology by farmers with different educational levels, and only the coefficient of the group of more than 9 years of education passed the significance test. This indicates that the promotion effect on the adoption of water-saving irrigation technology by farmers in the group is relatively more robust.

The study also found that a higher education level significantly enhances farmers' preferences and dependence on technical training. More specifically, the farmers with 9 years of education and below depend more on specific, visual, and face-to-face offline training, while those with more than 9 years of education depend more on multiple forms and content. The interaction term of risk aversion and offline technical training positively impacted the adoption of water-saving irrigation technology by farmers with different educational levels, and only the group with more than 9 years of education passed the significance test. This indicates that the interaction term of risk aversion and offline technical training positively impacts the promotion effect of the water-saving irrigation technology adoption behaviour of farmers in the group (group B). Interestingly, the interaction term of risk aversion and online technical training has a significant positive impact within group A (9 years of education or less). The absolute value of the coefficient of group A is more significant than Group B, indicating that the relationship between risk aversion and online technical training has a more substantial effect on promoting the adoption of water-saving irrigation technology by farmers in group A.

5.3. The Impact of Risk Aversion and Technical Training on the Adoption Behaviour of Farmers' Water-Saving Irrigation Technology under the Difference of Planting Scale

It can be seen from Table 8 that risk aversion has a significant negative effect on the adoption of water-saving irrigation technology by farmers with different planting scales. The absolute coefficient value of group A is greater than that of group B, indicating that the inhibitory effect of technology adoption weakens with farmers' planting scale expansion. The possible reason for this is that farmers' management ability, technical cognition level and management confidence have significantly improved with the increase in planting scale. They can objectively evaluate technical risks and reduce unnecessary worries and concerns while adopting water-saving irrigation technologies. Therefore, the larger the scale, the weaker the inhibitory effect of risk aversion on farmers' adoption of water-saving irrigation technology. Both offline and online technical training positively affects farmers' adoption of water-saving irrigation technology with different planting scales. The promotion effect of adopting water-saving irrigation technology by farmers in the above group is relatively more substantial. With the expansion of planting scale, the average cost of technology adoption by farmers is decreasing, and the economies of scale in adopting water-saving irrigation technologies are gradually emerging. Therefore, the impact of offline and online technical training on the adoption behaviour of farmers' water-saving irrigation technology is gradually increasing if the planting scale is expanded and increased.

6. Conclusions

Based on the empirical data of 707 watermelon and muskmelon farmers in Shanxi and Shaanxi provinces, this study analyses the effects of risk aversion, technical training (online and offline) and their interaction on farmers' water-saving technology adoption behaviour. We further conducted a robustness test and provide an in-depth comparison between the two forms of technical training. Based on the findings, the following conclusions were made: (i) The impact of risk aversion on farmers' adoption of water-saving irrigation technology is significant at the 5% statistical level, and the estimated coefficient is positive. The estimated coefficients of offline and online technical training are positive and significant at the 1% statistical level. (ii) Offline and online technical training have a positive regulatory effect between risk aversion and farmers' water-saving irrigation technology adoption behaviour, which can alleviate the inhibitory effect of risk aversion on water-saving irrigation technology adoption behaviour. (iii) The effects of risk aversion, technical training and interaction items on farmers' water-saving irrigation technology adoption behaviours have noticeable inter-group differences regarding age, education level and planting scale.

Based on the above conclusions, the following specific policy recommendations are drawn: (i) Alleviate farmers' degree of risk aversion and actively promote farmers to adopt water-saving irrigation technology: In this notion, the interaction of various risk-sharing networks and organizations should be strengthened. (ii) The government should facilitate innovative water-saving irrigation technology with easy conditions and relatively lower costs. Moreover, financial and technical support should also be strengthened. Optimize the agricultural technology training system and improve farmers' ability to acquire and apply technical information. (iii) Agricultural technology demonstration bodies and extension offices should act more responsibly to disseminate up-to-date knowledge and technical know-how by implementing "learning by seeing" and "learning by doing" prospectives. (iv) Awareness-building campaigns and technical dissemination platforms should be strengthened to enhance the farmers' cognitive level. A well-structured "water-saving irrigation model" should be established at the national level to effectively guide farmers to confidently use the water-saving irrigation technology and alleviate farmers' negative concerns about the risks of adopting water-saving irrigation technology. (v) Private and public partnerships and agricultural cooperatives should also be more responsible and enhance the social promotion system. Increase investment in agricultural offline technical training, optimize the content of water-saving irrigation technology-related training and expand the coverage of water-saving irrigation technology-related training. (vi) Moreover, farmers should be guided to use modern agricultural digital media such as websites and mobile apps to receive online technical training and improve their ability to obtain technical information. This has great potential to provide farmers with timely information on water-saving irrigation technologies. Innovate the form of agricultural technology training, combine online and offline technical training, actively expand the channels for farmers to receive technical training, provide farmers with more credible and more innovative training methods and improve the effectiveness of technical training. (vii) Implement differentiated guidance methods to meet the needs of different types of farmers. Different types of farmers have different objectives in pursuit of agricultural production management, so they also have different focus points in adopting water-saving irrigation technology. Therefore, training methods with different emphases can be adopted according to different types of farmers.

This study has some limitations. First, using cross-sectional data, this paper cannot analyse the dynamic impact of risk aversion and technical training on farmers' water-saving irrigation technology adoption behaviour. Second, this article only considers water-saving irrigation technology in agricultural resource conservation technology. Further research is needed to evaluate the impact of risk avoidance and technical training on farmers' adoption behaviour of different types of resource conservation technologies. Finally, for farmers, the cost of technology adoption is one of the critical factors affecting the adoption behaviour of water-saving irrigation technology. This article considered the possibility of potential measurement errors and did not include them in the model analysis. Whether this impacts the estimation results of this article still needs further testing.

Author Contributions: Conceptualization, D.X. and A.S.; methodology, D.X. and A.S.; software, A.S. and L.Q.; validation, D.X., A.S. and L.L.; formal analysis, D.X. and L.Q.; investigation, L.L. and A.S.; resources, A.S. and D.X.; data curation, D.X. and A.S.; writing—original draft preparation, D.X., L.L. and A.S.; writing—review and editing, D.X., A.S. and L.L.; visualization, D.X. and L.L.; supervision, L.Q. and A.S.; project administration, D.X., L.Q. and A.S.; funding acquisition, L.L. and L.Q. All authors have read and agreed to the published version of the manuscript.

Funding: This research was funded by the Humanities and Social Sciences Youth Program of the Ministry of Education of China (22YJC790057), China National Natural Science Foundation Project (71973105) and Ningxia University Communist Youth League Project (2022).

Institutional Review Board Statement: As the study does not involve any personal data and the respondents were well aware that they could opt out anytime during the data collection phase, any written institutional review board statement was not required.

Data Availability Statement: The associated dataset of the study is available upon request to the corresponding author.

Conflicts of Interest: The authors declare no conflict of interest.

References

1. Chai, Q.; Gan, Y.; Turner, N.C.; Zhang, R.-Z.; Yang, C.; Niu, Y.; Siddique, K.H.M. Chapter Two—Water-Saving Innovations in Chinese Agriculture. In *Advances in Agronomy*; Sparks, D.L., Ed.; Academic Press: Cambridge, MA, USA, 2014; Volume 126, pp. 149–201.
2. Wang, H.; Huang, J.; Zhou, H.; Deng, C.; Fang, C. Analysis of Sustainable Utilization of Water Resources Based on the Improved Water Resources Ecological Footprint Model: A Case Study of Hubei Province, China. *J. Environ. Manag.* **2020**, *262*, 110331. [CrossRef] [PubMed]

58. Sun, L.; Yang, Y.; Hu, J.; Porter, D.; Marek, T.; Hillyer, C. Reinforcement Learning Control for Water-Efficient Agricultural Irrigation. In Proceedings of the 2017 IEEE International Symposium on Parallel and Distributed Processing with Applications and 2017 IEEE International Conference on Ubiquitous Computing and Communications (ISPA/IUCC), Guangzhou, China, 12–15 December 2017; pp. 1334–1341.
59. Lynne, G.D.; Franklin Casey, C.; Hodges, A.; Rahmani, M. Conservation Technology Adoption Decisions and the Theory of Planned Behavior. *J. Econ. Psychol.* **1995**, *16*, 581–598. [CrossRef]
60. Ngwira, A.; Johnsen, F.H.; Aune, J.B.; Mekuria, M.; Thierfelder, C. Adoption and Extent of Conservation Agriculture Practices among Smallholder Farmers in Malawi. *J. Soil Water Conserv.* **2014**, *69*, 107–119. [CrossRef]
61. Lu, C.; Ji, W.; Hou, M.; Ma, T.; Mao, J. Evaluation of Efficiency and Resilience of Agricultural Water Resources System in the Yellow River Basin, China. *Agric. Water Manag.* **2022**, *266*, 107605. [CrossRef]
62. Sharma, B.C.; Kumar, R.; Slathia, P.S.; Puniya, R.; Vaid, A. Evaluation of Refresher Training Programme on Conservation Agriculture Practices. *Indian J. Ext. Educ.* **2022**, *58*, 49–52. [CrossRef]
63. Selim, T.; Eltarabily, M.G. Impact of COVID-19 Lockdown on Small-Scale Farming in Northeastern Nile Delta of Egypt and Learned Lessons for Water Conservation Potentials. *Ain Shams Eng. J.* **2022**, *13*, 101649. [CrossRef]
64. Zheng, Y.; Zhu, T.; Jia, W. Does Internet Use Promote the Adoption of Agricultural Technology? Evidence from 1 449 Farm Households in 14 Chinese Provinces. *J. Integr. Agric.* **2022**, *21*, 282–292. [CrossRef]
65. Enkvetchakul, P.; Surinta, O. Effective Data Augmentation and Training Techniques for Improving Deep Learning in Plant Leaf Disease Recognition. *Appl. Sci. Eng. Prog.* **2022**, *15*, 3810. [CrossRef]
66. Zhu, X.; Xiang, C.; Wang, M.; Luo, Z.; Xiang, X.; Li, H. Research on Business Model Construction and Innovation of Smartphone Training for Farmers in Rural Areas. In Proceedings of the Fifth International Conference on Economic and Business Management (FEBM 2020), Online, 10–17 October 2020; Atlantis Press: Amsterdam, The Netherlands, 2020; pp. 176–184.
67. Van Mele, P.; Okry, F.; Wanvoeke, J.; Fousseni Barres, N.; Malone, P.; Rodgers, J.; Rahman, E.; Salahuddin, A. Quality Farmer Training Videos to Support South–South Learning. *CSIT* **2018**, *6*, 245–255. [CrossRef]
68. Eitzinger, A.; Binder, C.R.; Meyer, M.A. Risk Perception and Decision-Making: Do Farmers Consider Risks from Climate Change? *Clim. Chang.* **2018**, *151*, 507–524. [CrossRef]
69. Sabbagh, M.; Gutierrez, L. Micro-Irrigation Technology Adoption in the Bekaa Valley of Lebanon: A Behavioural Model. *Sustainability* **2022**, *14*, 7685. [CrossRef]
70. Senapati, AK Evaluation of Risk Preferences and Coping Strategies to Manage with Various Agricultural Risks: Evidence from India. *Heliyon* **2020**, *6*, e03503. [CrossRef]
71. Zeweld, W.; Van Huylenbroeck, G.; Tesfay, G.; Speelman, S. Impacts of Socio-Psychological Factors on Smallholder Farmers' Risk Attitudes: Empirical Evidence and Implications. *Agrekon* **2019**, *58*, 253–279. [CrossRef]
72. Neisi, M.; Bijani, M.; Abbasi, E.; Mahmoudi, H.; Azadi, H. Analyzing Farmers' Drought Risk Management Behavior: Evidence from Iran. *J. Hydrol.* **2020**, *590*, 125243. [CrossRef]
73. Bagheri, A.; Emami, N.; Damalas, C.A. Farmers' Behavior towards Safe Pesticide Handling: An Analysis with the Theory of Planned Behavior. *Sci. Total Environ.* **2021**, *751*, 141709. [CrossRef]
74. Mariano, M.J.; Villano, R.; Fleming, E. Factors Influencing Farmers' Adoption of Modern Rice Technologies and Good Management Practices in the Philippines. *Agric. Syst.* **2012**, *110*, 41–53. [CrossRef]
75. Lamm, A.J.; Warner, L.A.; Martin, E.T.; White, S.A.; Fisher, P. Enhancing Extension Programs by Discussing Water Conservation Technology Adoption with Growers. *J. Agric. Educ.* **2017**, *58*, 251–266. [CrossRef]
76. Jha, S.; Kaechele, H.; Sieber, S. Factors Influencing the Adoption of Water Conservation Technologies by Smallholder Farmer Households in Tanzania. *Water* **2019**, *11*, 2640. [CrossRef]
77. Hoang-Thi, H.; Fahad, S.; Ahmad Shah, A.; Nguyen-Huu-Minh, T.; Nguyen-Anh, T.; Nguyen-Van, S.; To-The, N.; Nguyen-Thi-Lan, H. Evaluating the Farmers' Adoption Behavior of Water Conservation in Mountainous Region Vietnam: Extrinsic and Intrinsic Determinants. *Nat. Hazards* **2023**, *115*, 1313–1330. [CrossRef]
78. Podsakoff, P.M.; MacKenzie, S.B.; Podsakoff, N.P. Sources of Method Bias in Social Science Research and Recommendations on How to Control It. *Annu. Rev. Psychol.* **2012**, *63*, 539–569. [CrossRef]
79. Ho, T.Q.; Hoang, V.-N.; Wilson, C. Sustainability Certification and Water Efficiency in Coffee Farming: The Role of Irrigation Technologies. *Resour. Conserv. Recycl.* **2022**, *180*, 106175. [CrossRef]
80. Qiu, H.G.; Su, L.; Zhang, Y.; Tang, J. Risk Preference, Risk Perception and Farmers' Adoption of Conservation Tillage. *Chin. Rural Econ.* **2020**, *7*, 59–79.
81. Xu, L.; Jin, W.; Chen, K. Risk Preference, Information Acquisition and Pig-Breeding Pollution Exposure Behaviors: An Empirical Analysis Based on 1489 Pig Breeders Survey Data. *China Rural Obs.* **2020**, *4*, 129–144.
82. Wen, Z.L.; Hou, J.T.; Zhang, L. Comparison and Application of Moderating Effect and Mediating Effect. *Acta Psychol. Sin.* **2005**, *37*, 268–274.

agriculture

Article

Measuring Short Food Supply Chain Sustainability: A Selection of Attributes and Indicators through a Qualitative Approach

Patrizia Borsotto [1], Roberto Cagliero [1], Francesca Giarè [1,*], Giorgia Giordani [2], Rita Iacono [1], Ilenia Manetti [1] and Roberta Sardone [1]

1 Council for Agricultural Research and Economics (CREA) Centre for Agricultural Policies and Bioeconomy, Via Barberini 36, 00187 Rome, Italy
2 Council for Agricultural Research and Economics (CREA) Centre for Food and Nutrition, Via Ardeatina 546, 00178 Rome, Italy
* Correspondence: francesca.giare@crea.gov.it

Abstract: Short food supply chains (SFSCs) are one of the most direct approaches to more directly connecting consumers with producers. The scaling-up of SFSCs is often challenged by critical issues which can be overcome with identification of the most sustainable, replicable schemes. This paper presents the results of a participatory analysis conducted within the agroBRIDGES H2020 project, with the aim of defining a list of economic, social, and environmental attributes and indicators to assess the sustainability of SFSCs and set up a decision-making tool to support producers in self-assessing their sustainability level and choosing the most appropriate business model (BM) from those identified within the project. The proposed framework was based on a literature review and validated using co-creation exercises (Delphi rounds and focus groups) with relevant European stakeholders. A final set of 47 indicators was identified, and their potential for use in assessing the sustainability level of various BMs was also validated. Early results highlighted three main issues: indicator calculation feasibility, business model categorization, and the simplicity of the framework for sustainability self-assessment. Some recommendations are made, including the importance of using a participatory process in building an evaluation framework on SFSC sustainability and the necessity of its adaptation to territorial contexts and needs.

Keywords: short supply chains; producers; sustainability; participatory methods; co-creation exercise; farmers

Citation: Borsotto, P.; Cagliero, R.; Giarè, F.; Giordani, G.; Iacono, R.; Manetti, I.; Sardone, R. Measuring Short Food Supply Chain Sustainability: A Selection of Attributes and Indicators through a Qualitative Approach. *Agriculture* **2023**, *13*, 646. https://doi.org/10.3390/agriculture13030646

Academic Editor: Giuseppe Timpanaro

Received: 31 January 2023
Revised: 16 February 2023
Accepted: 19 February 2023
Published: 9 March 2023

1. Introduction

Short food supply chains (SFSCs) can be understood as supply chains with a minimized number of intermediaries. Although they have been proven to bring economic, social, and environmental benefits, they represent a niche phenomenon in the agri-food market [1–5]. The so-called "gold standard" for SFSCs would be direct contact between the producer and the consumer in terms of maximizing revenue and income for farmers and producers [6,7].

SFSCs are also considered in the Farm to Fork strategy as a useful way of improving the resilience of regional and local food systems, considering their production, processing, and selling processes, as an alternative to conventional longer chains [6,8].

The literature presents and catalogues many different types of SFSCs; these are classified according to different criteria, including the geographical distance between the production and sales points and the chain's organizational aspects [5,9–11].

Nowadays, the spread of SFSCs is enhanced by many factors, including an interest in and awareness of the consumption of local and secure products, as well as a willingness to establish direct contact with the producers or a level of trust regarding the origin and traceability of a product [7,9,12,13]. Despite this growing trend, some barriers limiting the scaling-up of SFSCs still exist, including a lack of information (e.g., unclear labels or difficulties in communicating the added value of products), weak cooperation between producers, a generational gap, and infrastructural deficiencies (e.g., critically minimal Internet connection in rural areas, and logistics and distribution issues). Several studies, such as that of Hyland et al., conducted in 2021 [14] within the framework of the H2020 agroBRIDGES project, highlight that major consumer motivations to purchase within short chains include product quality in terms of taste and freshness, food safety issues, support for the local economy, and trust in SFSC producers.

This study is also part of the same H2020 research project, which is largely based on the development of an agri-food multi-actor framework and a set of practical support tools (called the agroBRIDGES toolbox) which can be used to connect producers with consumers in new SFSC business and marketing models (BMs), a term which is further defined later in this work. Among the different tools available, a key role is played by the development of a producer decision-support tool (DST) to facilitate the identification of the most sustainable business model to be adopted by each producer. The tool is based on a multi-criteria decision analysis (MCDA) that compares BMs. In MCDA approaches, it is assumed that the relative attractiveness of the sales channels depends on a set of indicators [15]. The indicators are intended to measure the economic, environmental, and social sustainability of the sales channels.

The purpose of this study is to show how a participatory process succeeded in the identification of a set of attributes and indicators that constitute the basis of the DST, supporting farmers in self-measuring their production sustainability.

To achieve this goal, suitable indicators to compare the business models within an SFSC were identified through a two-stage qualitative analysis. After a first stage of in-depth analysis of the literature, experts and stakeholders from different European countries were included in two Delphi rounds and two focus groups (FGs) to discuss a primary list of indicators and choose those considered more suitable from the point of view of producers. A final set of 47 indicators, listed as 14 attributes corresponding to 3 sustainability dimensions, was created.

Although, in the literature, studies investigating the sustainability of supply chains are often focused on consumer benefits or preferences and the analysis of market demand [16–19], the novelty of this study lies in its consideration of the production side and the position of farmers in the supply and value chains. Moreover, the sustainability assessment was conducted using the direct engagement of producer and sector representatives, who were involved in participatory exercises.

This paper is divided into six sections: an introduction, presenting the general context of the research and objectives; a literature review on SFSC classification approaches; a methodology section illustrating the indicator definition process following the agroBRIDGES participatory approach, with a focus on how a qualitative approach has been effective during the definition process and considering the inclusiveness and interactions occurring in the focus groups and Delphi rounds; results and discussion sections on the concrete use of such evidences; and a conclusions section regarding the limits of the research and future steps.

2. Literature Review

Traditionally, SFSCs allowed producers to have a strong position in the food chain, but their role decreased with Europe's industrialization and the rise of long-distance transportation, urbanization, and technological advances [20]. Mass distribution rose spectacularly in the 1960s with the import of the American model of supermarkets to Western Europe, unbalancing the producer's position in the agri-food supply chain and decreasing their income. During the 1990s, many small farms disappeared, and local open-air markets were often dominated by retailers who procured from wholesalers and large chain suppliers. Nowadays, renewed consumer interest in direct purchasing, in relation to the demand for more secure products, has boosted the resurgence of SFSCs and of new and innovative business models [5,7,21]. In 2015, 15% of farmers sold half of their products via short chains [22].

On the policy side, several EU member states have developed legal frameworks and incentives to support short agri-food chains. At the EU level, support for short supply chain initiatives is provided by rural development policies. Within the "CAP towards 2020" proposals, the European Commission (EC) has also proposed that SFSCs may be subject to themed sub-programs within the oncoming Common Agricultural Policy (CAP) strategic plans. This is what occurred with the Farm to Fork strategy, in which the promotion of SFSCs had a central role.

SFSCs are attracting more and more attention in research on food systems, partly as a result of their growing popularity among consumers, producers, and policy makers. Longitudinal interdisciplinary assessments of different types of SFSCs could also be useful for identifying levers and barriers to sustainable production and consumption, as well as for assessing their role in improving the agro-industrial scheme based on intensive production and long chains [23]. Their potential input in the transition towards a more sustainable food system [24] offers many research insights.

Concerning SFSCs, different approaches are used to classify their models, including innovation, the interpretation of local concept, proximity, organizational issues, and trust (Figure 1).

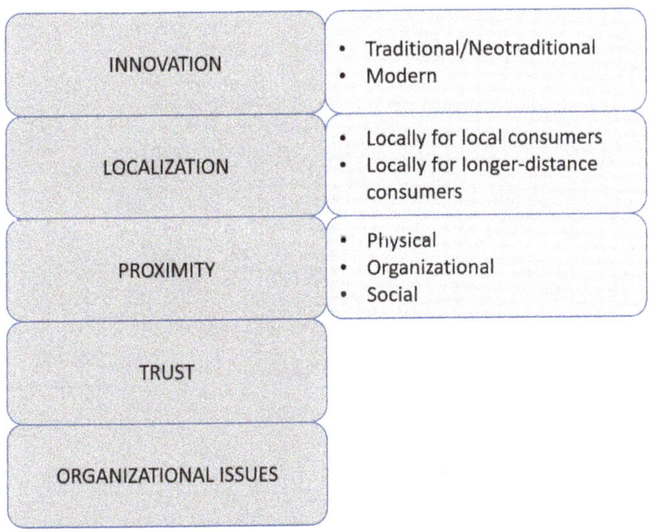

Figure 1. Approaches to classifying SFSC models.

One vision of SFSCs is based on the concept of innovation, dividing SFSCs into two overarching clusters: traditional or neo-traditional [6] and modern [25]. Within the SmartChain project, Sebök et al. (2022) [26] identified technological and non-technological innovations that can be applied in short food chains to increase their attractiveness for consumers and to improve the ability of SFSCs to deliver products and services reliably and consistently. The largest number of innovations identified concerns the issue of "logistics, product accessibility and short food chain channels", followed by food preservation and other processing technologies (i.e., preservation of freshness; nutritional value; packaging).

Some authors have focused on the context-based understanding of the concept of local food, distinguishing between "locally produced food for local consumers" and "locally produced food for longer-distance consumers" [27]. These terminological clarifications stress the complexity of SFSCs, their link with food, their local context, and the role of knowledge-based relations between local actors [27]. Thomé et al. (2021) [28] grouped chain models by convergence of interests and the need to add value criteria and described the conceptual coexistence framework of the food supply chains and SFSCs, this being at odds with the current bias of the literature.

Otherwise, according to Malak-Rawlikowska et al. (2019) [5], SFSCs can be categorized in terms of proximity between producers and consumers from three points of view: physical, organizational, and social. Contrary to what common sense might suggest, the application of physical proximity could lead to an unclear assessment of environmental impact [7]. For example, frequent deliveries of small quantities of products, even with very few displacements, may lead to negative impacts on environmental sustainability [29,30]. Majewski et al. (2020) [10] highlighted how SFSC models are characterized by the highest level of emissions, as they entail the use of personal cars (i.e., pick-your-own and on-farm sales). Eco-efficiency indicators display wide variability across the different types of SFSCs because the distribution process depends on numerous factors, including not only geographical proximity, but also supply chain infrastructure and logistics. On the other hand, organizational and social proximity generate social benefits [12], increasing consumer confidence in producers and bringing economic benefits to the local economy [4,31] while allowing producers to strengthen their position [5,7]. Different types of SFSCs also have different outcomes: for instance, farmers' markets may create stronger producer–consumer interactions, while direct sales are generally more efficient in terms of demand stability and economic return for producers [32].

Petropoulou et al. (2022) [33] highlight trust as the single most important determinant of success in SFSCs: "Without trust, any collective endeavor is doomed to fail. At the same time, trust is both an input and an outcome in SFSCs, where trust leads to more trust and vice versa". In general, organizational issues, i.e., the way initiatives are organized using traditional or new methods, seem to be an important factor in how the social, economic, and environmental sustainability of an SFSC is perceived, even in spite of geographical differences [12].

3. Materials and Methods

The focus of this study was to identify a group of attributes and indicators to assess the overall sustainability of SFSCs for the three main dimensions considered in the literature (economy, environment, social issues) [5,9,10,34]. The logical framework underlying the study is as follows:

Dimensions of sustainability → identification of attributes → selection of feasible indicators.

Within this framework, the three sustainability dimensions are composed of a set of identified attributes, each of them described by a selected set of different indicators. As suggested by Pyke et al. (2002) [35], we use the term attribute to describe a component of an SFSC that cannot be directly measured, and so a set of observable and measurable indicators can be used as a proxy. In this framework, the study aimed to determine a set of attributes and indicators to assess different SFSC business models through a multicriteria dashboard, as well as the development of the DST.

The research path was based on a two-stage qualitative analysis: (i) a literature review with the aim of proposing an initial list of issues and variables to be considered; (ii) a process of discussion and validation of the results of the first stage through a participatory process based on two Delphi rounds and two focus groups. The result of the second stage was the identification of a final list of feasible attributes and indicators by the so-called SFSC business models (BMs), as categorized in previous phases of agroBRIDGES [11]. The proposed categorization of SFCS BMs was firstly based on different types of relationships between producers and consumers, as shown in Table 1.

Table 1. agroBRIDGES Business Models.

agroBRIDGES Business Model Category	SFSC Business Model Type
Face-to-Face Trade	On-farm sales—Farm shops On-farm sales—Pick your own On-farm sales—Farm-based hospitality
Local Food Trade	Off-farm sales—Sales to retailers who source from local farmers and who make farmer identities clear Off-farm sales—Commercial Sector—Farmers' markets and other markets Off-farm Sales—Farm shops Off-farm sales—Sales directly to consumer co-operatives
Online Food Trade	Farm-direct deliveries—Internet sales Off-farm sales—Internet sales
Improved Logistics	Farm-direct deliveries—Delivery schemes Farm-direct deliveries—Specialty retailers
Other models	CSA

Source: Adaptation from AgroBRIDGES project D2.1.

In stage 1, the analytical framework following the economic, social, and environmental dimensions of sustainability was composed. To this aim, a first set of indicators was compiled, starting with a deep literature review of roughly 80 scientific references and other similar international project results, including SmartChain [26,33,35–39] and Strength2Food [12,40–42], focused on methodical models and instruments to assess SFSC sustainability [7,43]. This initial list included over one hundred indicators. Such a large set of indicators cannot be managed effectively, especially by farmers. Additionally, overlaps between indicators and poor practical feasibility of populating them were observed in many cases. Often, only theoretical hypotheses were proposed, while studies that had concretely measured the dimensions of sustainability were found to be in the minority. Consequently, applying criteria of non-redundancy and practical feasibility, the team selected a second list of indicators from the first list (63), aggregated into different attribute groups by sustainability dimensions, to be proposed operationally in the project and then aggregated.

In stage 2, an online participatory process using qualitative methods was proposed to assess and validate the attributes and indicators from the first stage. This approach—which could be considered a co-creation exercise—was chosen as the most appropriate method of validating the designed set, following RACER criteria (Relevant, Acceptable, Credible, Easy, and Robust). For the further phases of the agroBRIDGES project, a final and validated set of attributes and indicators was used to design the DST. The tool was then tested and validated by producers involved in the project around Europe, who implemented it in their business practices.

The following part of the paragraph outlines the participatory process carried out by the research team during the period of November 2021–January 2022 to formulate the final set of attributes and indicators, detailing the two qualitative methods chosen: Delphi rounds and focus groups.

The Delphi method is a structured methodological communication process that conveys competent opinions on specific questions, using questionnaires to reach shared conclusions that are as clear as possible based on consensus and stability [44–46]. For the purpose of this study, the Delphi technique was applied to explore the consensus of the participants on attributes and indicators chosen for the SFSC sustainability measurement. The activity was structured into two rounds.

For the first round (December 2021), a panel of experts from the European academic sector was invited by e-mail to complete an online questionnaire composed of 10 closed (Likert scale) and open-ended questions based on open-access software (i.e., the Google module). The purpose of the questionnaire was to analyze the level of agreement and disagreement regarding:

1. The different attributes presented for the measurement of the economic, environmental, and social sustainability of SFSCs;
2. The relationship between each economic, environmental, and social sustainability attribute and the five different business models identified (shown in Table 1) during the previous phases of the project.

In total, 19 experts from Finland, the United Kingdom, Greece, Belgium, Hungary, France and Italy completed the questionnaire; in order to reach a general consensus, the answers were analyzed following a feedback process [44].

The responding experts from the first round were invited to fill out the second-round questionnaire (January 2022). This second DELPHI round was composed of 9 closed-ended questions based on the results of the data collected in the first-round. Attributes that were either judged too negatively or met with a near-unanimous positive consensus were not reintroduced in round two, while attributes that did not meet with a sufficiently polarized and clear judgment were the subject of round two. The questionnaire was divided into different parts relating to the following themes:

1. A re-evaluation of the indicators that did not reach a sufficiently shared assessment in the first round;
2. The degree of expert agreement or disagreement with the attributes provided for the sustainability assessment of the five identified business models.

The focus group technique [47] is a research method that allows for the collection of qualitative data through a group discussion. It enables the gathering of potentially hidden information through the interactions between participants. For this study, two online meetings were organized. The two focus groups were set during the period of December 2021–January 2022.

The first FG involved members of the agroBRIDGES Stakeholder Reference Group (composed of one representative per each state participant as well as other stakeholders from the Experts Advisory Board and from the EU agri-food community), while the second involved experts from sister European projects on SFSCs (COACH, FOODRUS, and COCOREADO). The FGs aimed to present a list of indicators and select the most suitable and feasible ones to design an assessment model according to RACER criteria (Relevant, Acceptable, Credible, Easy, and Robust). The two events were conducted online and moderated by CREA researchers. The application Mentimeter (www.mentimeter.com) was used by the coordinators and experts to share comments, ideas, and opinions on-screen for each group of indicators presented and discussed, allowing for smoother and easier interactions among participants.

The discussions were based on analyzing the feasibility of the following attributes for measuring the SFSC sustainability in its three dimensions (economic, environmental, and social) and of their associated indicators (detailed indicators are shown in Tables 2–4):

- Attributes of the economic dimension: price, value chain, local producer sustainability, on-farm impact, bargaining power, regional economic impact;
- Attributes of the environmental dimension: food miles, energy consumption, type of process/production/packaging, food loss and waste;
- Attributes of the social dimension: labor/employment, human capital, social capital, food and nutrition, governance.

Table 2. Initial results of economic attributes and indicators.

Attribute	Indicator	DELPHI	FG 1	FG 2	Final Assessment
Price	Price different from farmgate (EUR)	A			A
	Premium price (%)	A			A
	Selling price is cheaper (%)	R			R
Value distribution	Chain value added (EUR)	A	+		A
	Chain value added (%)	A	+		A
	Reduction in production costs (%)	A	+	-	A
	Reduction in supply cost (%)	A	+	-	A
	Generated value is more equally distributed	A	+		A
Farm economic results	Turnover	A			A
	Financial support	A			A
	GVA	A	-	-	R
	Production costs	A			A
	Distribution costs	A			A
	Access to credit	A			A
Regional economic impact	Number of employees	A		-	A
	Number of producers involved	A		-	A
	Geographic scale including hectares farmed	V		-	R
	Sells to local customers	V		-	A
	Local supply	A		-	A
Bargaining power	Relationship with customers	A	+	+	A
	Relationship with suppliers	V	+	+	A
	Quantity of product sold	R	+	+	R
	Bargaining power self-assessment	A	+	+	A

Source: authors' own elaboration. Note: Columns 4 and 5 in the table refer to focus group evidence. The sign "+" indicates a positive and the sign "-" a negative judgment. Column 6 combines the results of the DELPHI and focus groups to build the final assessment: A = accepted; R = rejected; V = revised.

Table 3. Initial results of environmental attributes and indicators.

Attributes	Indicators	DELPHI	Focus Group 1	Focus Group 2	Final Assessment
Food Miles	Total food miles	R	-	-	R
	Carbon footprint related to food miles	A	-		A
	Reduced food miles—km/kg production	A	+		A
	Reduced food miles—km/kg distribution	A	+	+	A
	Use of fuel	A			A
Energy Consumption	% of clean energy from renewable sources	R			R
	Reduced % of energy consumption	A			A
	Increased energy efficiency measures	A			A

Table 3. *Cont.*

Attributes	Indicators	DELPHI	Focus Group 1	Focus Group 2	Final Assessment
Type of production process	% of organic products	A	+		A
	% of local/traditional products	A		+	A
	Certification	R	+	-	R
	Less packaging is used	R			R
	Increased ecofriendly packaging	A			A
Food loss and waste	Reduced kg of food loss and waste	A			A
	Increased circular economy initiatives	A			A

Source: authors' own elaboration. Note: Columns 4 and 5 in the table refer to focus group evidence, in particular: the sign "+" indicates a positive and the sign "-" a negative judgement. Column 6 combines the results of the DELPHI and focus groups to build the final assessment: A = accepted; R = rejected.

Table 4. Initial results of social attributes and indicators.

Attributes	Indicators	DELPHI	Focus Group 1	Focus Group 2	Final Assessment
Labor/Employment	Labor to production ratio (hours)	A			A
	Labor to production ratio (AWU)	R			R
	Presence of corporate welfare	V			A
	Increased resilience of employment	A			A
	Inclusion of disadvantaged people (%)	A			A
	Reduced wage difference (%)	V			R
Human Capital	Generational change	A			A
	Educational attainment	A			A
	Gender equality (%)	A		+	A
	No unequal treatment for same roles	A			A
	Smaller gender gap (%)	A			A
Social Capital	Influence by SFSC	A			A
	New local networks (formal orinformal)	A	+		A
	Customer and producer participation	A			A
	Stakeholder involvement	A	+		A
	Increased customer trust	A			A
Food and Nutrition	Increased access to food via SFSC	A			A
	Standards for food safety	A			A
	Certification	V			R
	Increased customer awareness	A		+	A
Governance	Coopetition index	A			A
	SFSC actor proactive involvement	A	+		A
	Typology of governance is moreinformal	R			R
	Collective investments	V		-	R
	Type and number of actors	R			R

Source: authors' own elaboration. Note: Columns 4 and 5 in the table refer to focus group evidence, in particular: the sign "+" indicates a positive and the sign "-" a negative judgement. Column 6 combines the results of the DELPHI and focus groups to build the final assessment: A = accepted; R = rejected; V= revised.

An analysis of the content and topics was used to study the results of this activity.

Finally, from the subsequent comparison of the results between the Delphi rounds and the focus groups, a final assessment of the feasibility and consistency of the attributes and indicators was composed. The evaluation of inclusion or exclusion was subsequently complemented by an analysis of the attributes and indicators for each business model, as identified by the agroBRIDGES project. Each of them could be useful for the assessment of a specific SFCS model, although it could be indifferent to another, or could represent a positive factor for one BM and a negative for another one. In addition, for each attribute, one lead indicator was identified as a milestone in the assessment, whereas a certain degree of freedom was possible in the collection and systemization process for the other indicators.

4. Results

The two qualitative methods were successfully used to discuss the results of the literature review from the first stage, and the final proposal of a list of attributes and indicators useful for assessing the sustainability of SFSCs was created. The outcomes, presented in the figures below, show a high level of agreement among the opinions on the proposed attributes and indicators, with only some minor exceptions.

The following Tables 2–4 represent the validation process from Delphi and focus groups for the three dimensions of sustainability. For each sustainability dimension, the first column lists the attributes, the second the indicators, and the next three columns report the assessments from the Delphi and focus groups. It must be remembered that these steps are linked; an issue that did not achieve a sufficiently clear evaluation in one step is not considered in the subsequent ones. The last column shows the final evaluation of possible use, indicating which attributes and indicators were feasible and which were not.

The use of a two-stage research method required an interpretative and mediating analysis to make the information consistent. In some cases, there was also a reversal of the results obtained, thus shaping the final assessment and the relative inclusion or exclusion of indicators in the final set. The gross value added indicator in Table 2 serves as an example. Although it advanced through the Delphi phase, it was deemed useful for operational practicality to discard the indicator due to the objections received during the two focus groups, as indicated by the "-" sign in the respective cells. The final assessment is, therefore, the result of the overall reading of the information obtained from the two phases of discussion involving experts.

Table 2 relates to the measurement of the economic sustainability of SFSCs through the following proposed attributes: price, value distribution, farm economic results, regional economic impact, and bargaining power.

We observed a high degree of agreement, although fewer disagreements and neutral values were registered for the attributes price, bargaining power, and regional economic impact, which were re-evaluated in the second Delphi round. More specifically, a critical reflection on the feasibility of the regional economic impact attribute emerged, as well as the need to clarify the definition of bargaining power. Additional comments concerned the reduction in indicators related to costs in order to avoid overlapping and redundancy. Some other suggestions aimed at simplifying the collection of indicators, especially at the level of on-farm impact (e.g., turnover could be an easier indicator for the farmers' self-assessment than gross added value).

Table 3 concerns the measurement of the environmental sustainability of SFSCs described by the proposed attributes: food miles, energy consumption, type of production process, and food loss and waste. There was quite a high level of agreement among the respondents. However, the presence of some neutral opinions about food loss and waste, a mixture of disagreements and neutral opinions about type of production process (particularly concerning the certification indicator), and, finally, a disagreement regarding the attribute food miles must be underlined. The attribute food miles proved controversial, as it could be interpreted from very different points of view. Additionally, the certification indicator was a topic of lengthy discussion due to ambiguity in its assessment, definition, identification, and use.

Table 4 concerns SFSC social sustainability: labor/employment, human capital, social capital, food and nutrition, and food system governance. Despite the commonly acknowledged difficulties in the actual measurement of this dimension overall, the experts underlined the usefulness of assessing the impacts of short supply chains by means of the attributes. They also suggested a simplification of the attributes proposed using simpler and more direct indicators that can be better understood by producers. The need to define the importance of SFSCs for human health emerged during the focus group sessions, while at the same time, the difficulty of finding suitable measurement indicators was discussed. The need to think "beyond the gates of the farm" was also highlighted, including aspects such as measuring the ability to build networks and to involve stakeholders in local/regional networks. In terms of employment, the issues of labor quality and seasonality were discussed, with the aim of including them in the evaluation set. Finally, the governance attribute needed to be simplified.

Overall, from the initial list of 63 indicators (23 economic, 15 environmental, and 25 social) from the literature review (stage 1) which were assessed by the Delphi and focus groups, 47 were confirmed, aggregated, and ranked into 14 attributes in three sustainability dimensions, and 14 lead indicators were identified as milestones. Table 5 shows the final set of attributes and indicators. For the economic dimension, the lead indicators identified are price difference from farmgate, equity in the generated value chain, turnover, number of producers involved, and bargaining power self-assessment. For the environmental dimension, these indicators were reduced food miles, reduced energy consumption, access to agri-environmental scheme support, and reduced food loss and waste. Finally, for the social dimension, the lead indicators were higher resilience of employment, educational attainment, stakeholder involvement, increased customer awareness, and coopetition index.

Table 5. Final set of SFSC sustainability attributes and indicators.

	Selected Attribute	Selected Lead Indicator	Other Selected Indicators
ECO	PRICE	Price difference from farmgate	Premium price
	VALUE DISTRIBUTION	Equity in the value chain generated	Chain value added
			Reduction in production costs
			Reduction in supply cost
	FARM ECONOMIC RESULTS	Turnover	Financial support
			Production costs
			Distribution costs
			Access to credit
	REGIONAL ECONOMIC IMPACT	Number of producers involved	Number of employees
			Sells to local customers
			Local supply
	BARGAINING POWER	Bargaining power self-assessment	Relationship with customers
			Relationship with suppliers
ENV	FOOD MILES	Reduced food miles (production and distribution)	Carbon footprint related to food miles
			Use of fuel
	ENERGY CONSUMPTION	Reduced energy consumption	Increased energy efficiency measures
	TYPE OF PRODUCTION PROCESS	Access to agri-environmental scheme support	% of organic products
			% of local/traditional products
			Increased eco-friendly packaging
	FOOD LOSS AND WASTE	Reduced food loss and waste	Reduced food loss and waste
			Increased circular economy initiatives

Table 5. *Cont.*

	Selected Attribute	Selected Lead Indicator	Other Selected Indicators
SOC	LABOR/EMPLOYMENT	Higher resilience of employment	Labor to production ratio
			Presence of corporate welfare
			Inclusion of disadvantaged people
	HUMAN CAPITAL	Educational attainment	Generational change
			Gender equality
			No unequal treatment for the same roles
	SOCIAL CAPITAL	Stakeholder involvement	Influence of SFSC
			New local networks
			Customer and producer participation
	FOOD AND NUTRITION	Increased customer awareness	Increased access to food via SFSC
			Standards for food safety
	GOVERNANCE	Coopetition index	SFSC actor proactive involvement
			Presence of corporate welfare

Source: authors' own elaboration.

Furthermore, through this participatory process, the feasibility of the identified attributes was also assessed in order to measure the sustainability of the five BMs identified within agroBRIDGES: community-supported agriculture (CSA), face-to-face, local food, online food, and improved logistics.

The overall process enabled the initial screening and final selection of the most suitable attributes for measuring the sustainability of each proposed BM. As can be seen in Table 6, the findings were summarized in a so-called performance matrix to connect and compare each BM to the most appropriate attributes. The connection is indicated with an "x". Instead, an empty cell is used where a specific attribute is irrelevant to a specific BM.

Table 6. Business models' performance matrices.

Attributes	CSA	FACE-TO-FACE	LOCAL FOOD	ONLINE FOOD	IMPROVED LOGISTICS
Price	x	x	x	x	x
Value Distribution	x	x	x	x	x
Farm Economic Results	x	x	x	x	x
Regional Economic Impact	x		x		x
Bargaining power	x	x	x		x
Food Miles	x		x		
Energy Consumption	x	x	x	x	x
Type of Production Process	x	x	x	x	x
Food loss and waste	x	x	x	x	x
Labor/Employment	x	x	x	x	x
Human Capital	x	x	x	x	x
Social Capital	x	x	x		x
Food and Nutrition		x			x
Governance	x		x	x	x

By observing the table, it can be seen that 3 out of 5 models are described by almost all of the identified attributes (13 out of 14).

Going through the individual models in detail, it can be assumed that the lack of certain attributes is due to the fact that the indicators of which they are composed measure factors that are not applicable to the SFSC model in question.

For example, the "food and nutrition" attribute is missing in the CSA, local food, and online food models because it is composed of several indicators that are not useful for measuring the sustainability of these typologies, including the consumer's full awareness regarding the product they buy and the consumer's perception regarding a farm's commitment to promoting short supply chains.

Similarly, for the local food and online food models, the attribute "regional economic impact" is missing because it is composed of indicators not relevant to the measurement of such models, such as the number of producers involved in representing the size structure of an SFSC on their territory or the number of employees.

5. Discussion

The issue of farm sustainability is at the heart of the forthcoming CAP, not only in economic and environmental terms, but, above all, within the boundaries of social analysis in relation to the newly introduced social cross-compliance. However, even though the CAP 2023-27 objectives require an important reorientation of the subsidy scheme, the reform proposal does not provide the instruments needed to address them [48].

It is, therefore, rational to analyze producers' views on the sustainability of short supply chains, often considered sustainable by definition, without using solid supporting indicators. At the same time, it is also important to empower farmers to engage with other actors to determine whether the identified criteria are understandable and useful. Agricultural knowledge and innovation system interventions to promote and share knowledge and innovation (AKIS) could be crucial for this purpose.

The two-stage path of this study was used for the identification of attributes and indicators that can be easily used by farmers to evaluate which business model matches their needs. For this reason, lead indicators have been identified to aid farmers in the collection of data needed for measurement.

In this light, it is important to consider environmental, economic, and social benefits in relation to the specific type of SFSCs to assess the potential sustainability of the different SFSC business models, taking into consideration the different actors involved in SFSCs themselves (farmers, consumers, processors, etc.), and not in a general rational context.

The co-creation approach used herein was particularly suitable for validating the list of attributes and indicators, because it was based on the direct involvement of experts with different levels of competence and experience arising from their participation in other sister projects on SFSCs. They were asked to assess the feasibility of the chosen attributes and indicators.

The early results highlight at least three elements for discussion that are relevant in theoretical and operational terms: feasibility, categorization (business models), and simplicity.

The main constraint remains the feasibility of obtaining basic information, and, therefore, the definitive possibility of collecting data for calculating the indicators. These are the basic building blocks for the assessment of the various attributes and, thus, allow for an assessment of the three main dimensions.

Despite a copious amount of literature on the topic, only very few references offer indications for an actual evaluation of SFCSs in operational terms [5,10,32]. Very often, only theoretical reflections are reported. Moreover, real attempts to quantify these dimensions are rare. Where quantifications have been possible, information gaps are observed in a spatial or temporal sense, which in turn invalidate the models. In this sense, the quality of data must also be evaluated in terms of robustness. A trade-off between cost/difficulty of collection and feasibility can be predicted. In many cases, the literature refers to self-assessment procedures to ensure the feasibility of collection, leading, however, to a reduction in the reliability of the information.

Basically, this research path made it possible to highlight that the initial set of attributes and indicators from the literature was adequate and feasible. Through the results of the Delphi and focus groups, there a high level of consensus was reached regarding the selections made. This is undoubtedly the most important result.

However, the results also revealed several critical points which were addressed and overcome. The first was the need to provide clear and agreed-upon definitions. For example, the bargaining power indicator proved to be difficult to handle for some participants and was debated at length during the discussions in the focus groups. A second consideration concerned the positive or negative connotation that an indicator should have. As an example, the meaning of the food miles indicator seemed to be clear and shared in the literature, but it was highly debated and reinterpreted in the FGs. Moreover, the meanings of some indicators and their collection modalities were discussed. Due to the multiplicity of existing types of certifications, the certification indicator was also discussed at length. To address it, it was considered appropriate to refer to specific adopted rural development measures.

Finally, participants agreed with the use of self-declaration for the ease of calculation and the level of reliability of the information, as was often proposed in the literature [9,34].

The second element of discussion concerns the identification of the most appropriate and explanatory attributes and indicators for the different BMs, which was based on the results of the participatory exercises and examined through the performance matrix. This highlights how this approach allows for categorization due to the flexibility of the grid based on attributes and indicators. The consensus on the proposed models resulting from the literature review in an earlier step of the project was the first interesting thing to be noted. Equally remarkable is that consistent assessments of the explanatory capacity of the selected indicators for each SFSC model were gathered. As was already noted from the results of the first Delphi Round, the polarization between adherent and non-adherent opinions on the indicators for each model was very clear. In fact, only a limited number of them required a second evaluation round. The overall rating of feasibility in relation to the SFCS models was also homogeneous.

The analysis conducted taking into account the different models of SFSC highlights some important differences concerning the practices adopted by farmers, and helps to overcome an excessively abstract vision of these distribution methods. Indeed, we often talk about short-chain models as if there were a single modality for reducing the number of intermediaries, yet the different practices considered herein affect sustainability dimensions and influence both prices and growth in different ways.

The evaluation model proposed by the research was positively perceived by the participants. This result is important because it reflects both the adequacy of the proposals and the concrete applicability of the performance matrix to the DST. In this sense, the trade-off between simplicity and explicative capacity seems to be considered adequate, and the course of action consistent and robust. This could be especially interesting for the purpose of reducing the distance between research and practices when considering agricultural sustainability. In fact, while the research generally uses complex models and procedures to measure sustainability and applies them with the aim of generalizing the results and confirming the validity of approaches, farmers need to have simple tools to understand immediately whether the choice to be made is in line with their vision and strategy.

6. Conclusions

The qualitative research approach was found to be suitable for validating the proposed framework of indicators through the direct involvement and inclusion of different representatives of the sector in co-creation activities (focus groups and Delphi rounds). In this sense, the research also takes into account producers' points of view, albeit indirectly, and reflects the main goal of the project, which is to balance the role of agricultural producers as active players for a more sustainable agri-food sector. This study lays an additional brick in the research framework regarding the measurement of short supply chain sustainability and the methodology of participatory approaches by considering the needs and views of the production sector instead of consumers only. However, it is also necessary to emphasize some limits of the research which could be attributed to methodological issues and the real engagement of farmers in the research pathway.

The novelty of this study is the determination of a process to handle the three main critical factors that the research results highlighted: feasibility, relationship with business models, and simplicity. This objective is even more challenging in this work because it must lead to set operational tools in the field and not only to a proposition of a methodological exercise, as can often be observed in the literature. The use of co-creation based on the literature review and qualitative techniques proved to be appropriate. This mixed and participatory approach provided both insights and operability to the process through the Delphi and focus groups. In relation to the three critical issues highlighted, the study provided concrete solutions: (i) it addressed the critical issues in terms of feasibility, building a common understanding of the indicators; (ii) it ensured the necessary flexibility of the indicator system in relation to the specificities of the different business models; and (iii) it simplified the process of gathering information from producers through self-declaration.

From a methodological point of view, the business models were identified within the countries of the project partners, starting from the most common practices. However, other practices and models could be widespread in Europe, and it would be useful to check whether there are other attributes and indicators identified as valid for them, and/or if there are any others to be used.

In fact, the participatory research involved different experts in SFSC, not all of whom were farmers, with potentially different opinions on appropriate, explanatory, and simple attributes and indicators for specific practices in their contexts. However, another stage of the project will specifically be aimed at verifying the results of this exercise.

These constraints could be addressed and overcome by exploring and analyzing specific geographical contexts, adapting the framework to local or national territorial needs, and collecting more feedback, mainly from farmers. Moreover, this research could add significant elements to other studies dealing with the application of tools that producers can adopt on their own to improve their business choices in terms of sustainability.

Author Contributions: Conceptualization, R.C., P.B., F.G. and R.S.; methodology, R.C. and R.S.; resources, R.C., R.S., P.B. and F.G.; software, R.I., I.M. and G.G.; validation, P.B., R.C., R.S. and F.G.; formal analysis, P.B., R.C., F.G., G.G., I.M. and R.S.; investigation, R.C.; data curation, R.C., R.S., I.M., G.G. and R.I.; writing—original draft preparation, P.B., R.C., F.G., G.G., I.M., R.I. and R.S.; writing—review and editing, P.B., F.G., I.M., G.G. and R.I.; visualization, G.G., I.M. and R.I.; supervision, F.G., R.S. and P.B. All authors have read and agreed to the published version of the manuscript.

Funding: This research was funded by the Project AgroBRIDGES, financed by the European Union's Horizon 2020 Research and Innovation Programme, grant number 101000788.

Data Availability Statement: Data are available upon request from the authors.

Conflicts of Interest: The authors declare no conflict of interest.

References

1. Italian NRN. *Contribution to the "Short Supply Chain" Workshop*; EC: Bad Schandau, Germany, 2011.
2. Pretty, J. Some Benefits and Drawbacks of Local Food Systems. Bioregional Connections to Sustainable Foodsheds. Briefing Note for TVU/Sustain AgriFood Network. 2001; pp. 1–11. Available online: https://www.sustainweb.org/pdf/afn_m1_p2.pdf (accessed on 16 December 2022).
3. Shindelar, R. The Ecological Sustainability of Local Food Systems. In *Think Global, Eat Local: Exploring Foodways*; Pimbert, M., Shindelar, R., Schösler, H., Eds.; Rachel Carson Center: Munich, Germany, 2015; pp. 19–23.
4. Galli, F.; Brunori, G. *Short Food Supply Chains as Drivers of Sustainable Development Evidence Document*; Laboratorio di studi rurali Sismondi: Pisa, Italy, 2013.
5. Malak-Rawlikowska, A.; Majewski, E.; Wąs, A.; Borgen, S.O.; Csillag, P.; Donati, M.; Freeman, R.; Hoàng, V.; Lecoeur, J.-L.; Mancini, M.C.; et al. Measuring the Economic, Environmental, and Social Sustainability of Short Food Supply Chains. *Sustainability* **2019**, *11*, 4004. [CrossRef]

6. Augere-Granier, M.L. *Short Food Supply Chains and Local Food Systems in the EU*; European Parliamentary Research Service (EPRS): Brussels, Belgium, 2016.
7. Kneafsey, M.; Venn, L.; Schmutz, U.; Balázs, B.; Trenchard, L.; Eyden-Wood, T.; Bos, E.; Foster, G.; Blackett, M. *Short Food Supply Chains and Local Food Systems in the EU. A State of Play of Their Socio-Economic Characteristics*; Joint Research Centre (JRC): Luxembourg, 2013.
8. Communication from the Commission. *A Farm to Fork Strategy for a Fair, Healthy and Environmentally-Friendly Food System*; DG SANTE/Unit Food Information and Composition, Food Waste: Brussels, Belgium, 2020; pp. 4–14.
9. Todorovic, V.; Maslaric, M.; Bojic, S.; Jokic, M.; Mircetic, D.; Nikolicic, S. Solutions for More Sustainable Distribution in the Short Food Supply Chains. *Sustainability* **2018**, *10*, 3481. [CrossRef]
10. Majewski, E.; Komerska, A.; Kwiatkowski, J.; Malak-Rawlikowsk, A.; Wąs, A.; Sulewski, P.; Gołaś, M.; Pogodzińska, K.; Lecoeur, J.-L.; Tocco, B.; et al. Are Short Food Supply Chains More Environmentally Sustainable than Long Chains? A Life Cycle Assessment (LCA) of the Eco-Efficiency of Food Chains in Selected EU Countries. *Energies* **2020**, *13*, 4853. [CrossRef]
11. Ariño, P.; Borgogno, C.; Serrano, J. AgroBRIDGES D2.1 SFSCs Business and Marketing Models Categorization. 2021. Available online: https://zenodo.org/record/6562875#.ZAm1pHbMK5c (accessed on 5 December 2022).
12. Vittersø, G.; Torjusen, H.; Laitala, K.; Tocco, B.; Biasini, B.; Csillag, P.; de Labarre, M.D.; Lecoeur, J.-L.; Maj, A.; Majewski, E.; et al. Short Food Supply Chains and Their Contributions to Sus-tainability: Participants' Views and Perceptions from 12 European Cases. *Sustainability* **2019**, *11*, 4800. [CrossRef]
13. Elghannam, A.; Mesias, F.J.; Escribano, M.; Fouad, L.; Horrillo, A.; Escribano, A.J. Consumers' Perspectives on Alternative Short Food Supply Chains Based on Social Media: A Focus Group Study in Spain. *Foods* **2020**, *9*, 22. [CrossRef]
14. Hyland, J.; Henchion, M.; Olomo, O. AgroBRIDGES D1.2 Report on European and Regional Analysis of Producer and Consumers Needs and Barriers towards SFSCs Implementation. Available online: https://cordis.europa.eu/project/id/101000788/results/it (accessed on 10 December 2022).
15. Azzabi, L.; Azzabi, D.; Kobi, A. *The Multi-Criteria Approach for Decision Support. An Introduction with practical Application*; Springer: Cham, Switzerland, 2020; Volume 300.
16. Cecchini, L.; Torquati, B.; Chiorri, M. Sustainable agri-food products: A review of consumer preference studies through experimental economics. *Agric. Econ.* **2018**, *64*, 554–565. [CrossRef]
17. Feldmann, C.; Hamm, U. Consumers' perceptions and preferences for local food: A review. *Food Qual. Prefer.* **2015**, *40*, 152–164. [CrossRef]
18. Giampietri, E.; Finco, A.; Del Giudice, T. Exploring consumers' behaviour towards short food supply chains. *Br. Food J.* **2016**, *118*, 618–631. [CrossRef]
19. Wang, M.; Kumar, V.; Ruan, X.; Saad, M.; Garza-Reyes, J.A.; Kumar, A. Sustainability concerns on consumers' attitude towards short food supply chains: An empirical investigation. *Oper. Manag. Res.* **2022**, *15*, 76–92. [CrossRef]
20. Chiffoleau, Y.; Millet-Amrani, S.; Canard, A. From Short Food Supply Chains to Sustainable Agriculture in Urban Food Systems: Food Democracy as a Vector of Transition. *Agriculture* **2016**, *6*, 57. [CrossRef]
21. Malak-Rawlikowska, A.; Majewski, E.; Wąs, A.G.; Gołaś, M.; Kłoczko-Gajewska, A.; Borge, S.O.; Coppola, E.; Csillag, P.; Duboys de Labarre, M.; Freeman, R.; et al. *Quantitative Assessment of Economic, Social and Environmental Sustainability of Short Food Supply Chains and Impact on Rural Territories; Deliverable 7.2.*; Strength2Food Project no.678024; Strength2Food: Krakow, Poland, 2019.
22. EIP AGRI. *Focus Group on Innovative Short Food Supply Chain Management*; Final Report; EIP AGRI: Brussels, Belgium, 2015.
23. Chiffoleau, Y.; Dourian, T. Sustainable Food Supply Chains: Is Shortening the Answer? A Literature Review for a Research and Innovation Agenda. Sustainability **2020**, *12*, 9831.
24. Southerton, D.; Welch, D. After Paris: Transitions for Sustainable Consumption. Sustain. *Sci. Pract. Policy* **2019**, *15*, 31–44.
25. Schweitzer, J.-P.; Gionfra, S.; Pantzar, M.; Mottershead, D.; Watkins, E.; Petsinaris, F.; ten Brink, P.; Ptak, E.; Lacey, C.; Janssens, C. *Unwrapped: How Throwaway Plastic Is Failing to Solve Europe's Food WASTE Problem (and What We Need to Do Instead)*; Institute for European Environmental Policy (IEEP): Brussels, Belgium, 2018.
26. Sebök, A.; Varsányi, K.; Kujáni, K.; Xhakollari, V.; Fricz Á, S.; Castellini, A.; Di Gioia, D.; Gaggia, F.; Canavari, M. Value propositions for improving the competitiveness of short food supply chains built on technological and non-technological innovations. *Int. J. Food Stud.* **2022**, *11*, SI161–SI181. [CrossRef]
27. Fonte, M. Knowledge, food and place. A way of producing, a way of knowing. *Sociol. Rural.* **2008**, *48*, 200–222. [CrossRef]
28. Thomé, K.M.; Cappellesso, G.; Ramos, E.L.A.; de Lima Duarte, S.C. Food Supply Chains and Short Food Supply Chains: Coexistence conceptual framework. *J. Clean. Prod.* **2021**, *278*, 123207. [CrossRef]
29. Gonçalves, A.; Zeroual, T. Logistic Issues and Impacts of Short Food Supply Chains: Case Studies in Nord—Pas de Calais, France. In *Toward Sustainable Relations between Agriculture and the City*; Soulard, C.-T., Perrin, C., Valette, E., Eds.; Springer: Berlin/Heidelberg, Germany, 2017; pp. 33–49.
30. Mancini, M.; Menozzi, D.; Donati, M.; Biasini, B.; Veneziani, M.; Arfini, F. Producers' and Consumers' Perception of the Sustainability of Short Food Supply Chains: The Case of Parmigiano Reggiano PDO. *Sustainability* **2019**, *11*, 721. [CrossRef]
31. Mundler, P.; Laughrea, S. The contributions of short food supply chains to territorial development: A study of three Quebec territories. *J. Rural Stud.* **2016**, *45*, 218–229. [CrossRef]
32. Mastronardi, L.; Marino, D.; Giaccio, V.; Giannelli, A.; Palmieri, M.; Mazzocchi, G. Analyzing alternative food networks sustainability in Italy: A proposal for an assessment framework. *Agric. Food Econ.* **2019**, *7*, 21. [CrossRef]

33. Petropoulou, E.; Benos, T.; Theodorakopoulou, I.; Iliopoulos, C.; Castellini, A.; Xhakollari, V.; Canavari, M.; Antonelli, A.; Pe-truzzella, D. Understanding social innovation in short food supply chains: An exploratory analysis. *Int. J. Food Stud.* **2022**, *11*, 182–195. [CrossRef]
34. Antonelli, A.; Petruzzela, D. SMARTCHAIN Deliverable D3.4: Social Innovation Assessment Template Work Package No.3. 2020. Available online: https://www.smartchain-platform.eu/sites/default/files/publication-files/Smartchain_D3.4_FINAL110521.pdf (accessed on 30 December 2022).
35. Pyke, D.A.; Herrick, J.E.; Shaver, P.; Pellant, M. Rangeland health attributes and indicators for qualitative assess-ment. *J. Range Manag.* **2002**, *55*, 584–597. [CrossRef]
36. Parrag, V.; Fricz Szegedyné, Á.; Sebők, A. Application of digital solutions to improve the operation of short food supply chains. *Int. J. Food Stud.* **2022**, *11*, SI151–SI160. [CrossRef]
37. Chrysanthopoulou, F.; Lamerisb, M.; Greil, G.; Vudragovic, D.; Flynn, K. An online innovation platform to promote collaboration and sustainability in short food supply chains. *Int. J. Food Stud.* **2022**, *11*, SI232–SI247. [CrossRef]
38. Živković, L.; Pešić, M.B.; Schebesta, H.; Nedović, V.A. Exploring regulatory obstacles to the development of short food supply chains: Empirical evidence from selected european countries. *Int. J. Food Stud.* **2022**, *11*, SI138–SI150. [CrossRef]
39. Aouinait, C.; Christen, D.; Carlen, C.; Mehauden, L.; Mora, P.; Massar, B.; Frederiks, M. Motivations and barriers for engagement in short food supply chains: Insights from european focus groups. *Int. J. Food Stud.* **2022**, *11*, SI219–SI231. [CrossRef]
40. Bellassen, V.; Drut, M.; Hilal, M.; Bodini, A.; Donati, M.; Duboys de Labarre, M.; Filipovic, J.; Gauvrit, L.; Gil, J.M.; Hoang, V.; et al. The economic, environmental and social performance of European certified food. *Ecol. Econ.* **2022**, *191*, 107244. [CrossRef]
41. Arfini, F.; Bellassen, V. *Sustainability of European Food Quality Schemes: Multi-Performance, Structure, and Governance of Pdo, Pgi, and Organic Agri-Food Systems*; Springer International Publishing: Cham, Switzerland, 2019.
42. Mattas, K.; Tsakiridou, E.; Karelakis, C.; Gorton, M.; Filipovic, J.; Hubbard, C.; Saidi, M.; Stojkovic, D.; Tocco, B.; Tregear, A.; et al. Strengthening the sustainability of European food chains through quality and procurement policies. *Trends Food Sci. Technol.* **2021**, *120*, 248–253. [CrossRef]
43. Bellassen, V.; Arfini, F.; Amilien, V.; Antonioli, F.; Bodini, A.; Boehm, M.; Brečić, R.; Chiussi, S.; Csillag, P.; Diallo, A.; et al. Strength2Food. Report on Assessment of the Social, Environmental and Economic Sustainability of Food Quality Schemes. Ph.D. Dissertation, Università degli Studi di Parma, Parma, Italy, 2019.
44. Marbach, G. Il punto sul metodo DELPHI. In *Le previsioni: Fondamenti Logici e Basi Statistiche*; Marbach, G., Mazziotta, C., Rizzi, A., Eds.; ETASLIBRI: Milano, Italy, 1991; p. 47.
45. Hsu, C.-C.; Sandford, B.A. The DELPHI Technique: Making Sense of Consensus. *Pract. Assess. Res. Eval.* **2007**, *12*, 10.
46. Avella, J.R. DELPHI panels: Research design, procedures, advantages, and challenges. *Int. J. Dr. Stud.* **2016**, *11*, 305–321. [CrossRef]
47. Morgan, D.L. *Basic and Advanced Focus Groups*; Sage Publications: Thousand Oaks, CA, USA, 2018.
48. Heyl, K.; Döring, T.; Garske, B.; Stubenrauch, J.; Ekardt, F. The Common Agricultural Policy beyond 2020: A critical review in light of global environmental goals. *Rev. Eur. Comp. Int. Environ. Law* **2021**, *30*, 95–106. [CrossRef]

agriculture

MDPI

Review

Can Agroforestry Contribute to Food and Livelihood Security for Indonesia's Smallholders in the Climate Change Era?

Aris Sudomo [1,*], Budi Leksono [1], Hesti Lestari Tata [2], Anita Apriliani Dwi Rahayu [1], Aziz Umroni [1], Heny Rianawati [1], Asmaliyah [1], Krisnawati [1], Ali Setyayudi [1], Marcellinus Mandira Budi Utomo [3], Levina Augusta Geraldine Pieter [3], Arini Wresta [4], Yonky Indrajaya [2], Syed Ajijur Rahman [5] and Himlal Baral [6]

1 Research Center for Plant Conservation, Botanic Gardens, and Forestry, National Research and Innovation Agency (BRIN), Jalan Raya Jakarta-Bogor Km 46, Cibinong 16911, Indonesia; budi.leksono@brin.go.id (B.L.); anit015@brin.go.id (A.A.D.R.); aziz.umroni@brin.go.id (A.U.); heny.rianawati@brin.go.id (H.R.); asmaliyah@brin.go.id (A.); krisnawati@brin.go.id (K.); ali.setyayudi@brin.go.id (A.S.)
2 Research Center for Ecology and Ethnobiology, National Research and Innovation Agency (BRIN), Jalan Raya Jakarta-Bogor Km 46, Cibinong 16911, Indonesia; made009@brin.go.id (H.L.T.); yonky.indrajaya@brin.go.id (Y.I.)
3 Research Center for Society and Culture, National Research and Innovation Agency (BRIN), Jl. Gatot Subroto No. Kav 10, Jakarta 12710, Indonesia; marcellinus.mandira.budi.utomo@brin.go.id (M.M.B.U.); levi001@brin.go.id (L.A.G.P.)
4 Research Center for Environmental and Clean Technology, National Research and Innovation Agency (BRIN), Bandung 40135, Indonesia; arini.wresta@brin.go.id
5 Institute of Ecology and Key Laboratory for Earth Surface Processes of the Ministry of Education, College of Urban and Environmental Science, Peking University, Beijing 100871, China; sumonsociology@yahoo.com
6 Center for International Forestry Research and World Agroforestry (CIFOR-ICRAF), Bogor 16115, Indonesia; h.baral@cgiar.org
* Correspondence: aris.sudomo@brin.go.id

Citation: Sudomo, A.; Leksono, B.; Tata, H.L.; Rahayu, A.A.D.; Umroni, A.; Rianawati, H.; Asmaliyah; Krisnawati; Setyayudi, A.; Utomo, M.M.B.; et al. Can Agroforestry Contribute to Food and Livelihood Security for Indonesia's Smallholders in the Climate Change Era? *Agriculture* **2023**, *13*, 1896. https://doi.org/10.3390/agriculture13101896

Academic Editors: Roberto Henke and Filiberto Altobelli

Received: 9 August 2023
Revised: 6 September 2023
Accepted: 11 September 2023
Published: 27 September 2023

Abstract: In Indonesia, smallholders have historically practiced agroforestry, which warrants examination in terms of food and livelihood security within sustainable community forest frameworks. Based on a literature review, we analyzed these two forms of security related to smallholder agroforestry practices. Findings indicate diverse agroforestry systems, with 88% focusing on non-timber forest products (NTFPs) and 12% on timber. While 42% prioritize direct food supply, 58% emphasize income generation through product sales. However, agroforestry that does not produce food for direct consumption by smallholders generates revenue for purchasing food necessities. Agroforestry supports both food needs (46–61%) and income (51–54%) for smallholders, surpassing traditional agriculture (13%). Semi-commercial agroforestry (57%) is a predominant livelihood prospect. The remaining 27% are purely subsistence, and 15% are purely commercial. However, the commercialization of agroforestry that focuses only on high-value commodities results in a negative impact on biodiversity. There is a concomitant decrease in environmental services for climate change mitigation and adaptation. Biodiversity remains crucial for climate resilience, health care, and food security in rural communities. Semi-commercial agroforestry is a midpoint for achieving multifunctional agriculture (biodiversity, soil and water conservation, food security, and income) in the climate change era. The research directly related to food security and ecosystem services quantification remains limited, necessitating further investigation. Policy support and incentives are essential for smallholders practicing complex agroforestry for climate adaptation and mitigation.

Keywords: agroforestry; sustainability; subsistence; commercialization; community forestry

1. Introduction

Growing populations and the depletion of agricultural land are creating enormous challenges for the sustainability of food production and supply systems [1]. The declining quantity and quality of agricultural land, together with water scarcity and climate variability, is not only threatening global food security but also overall rural livelihoods that are vastly dependent on agricultural production [2,3]. Approximately 80% of fresh water is used for agricultural activities to support food production, while only about 10% of irrigated water in developing countries comes from reused wastewater [4,5]. In this context, crops in agroforestry systems require less water due to their efficient use of available soil water content ('green water') than many monoculture systems [5].

Foods such as grains, vegetables, fruits, dairy, and meat are important to human nutrition [6]; the production of such foods in conventional agriculture has contributed to negative environmental and social effects, e.g., climate change, biodiversity, ecosystem and land degradation, water scarcity, and stressed social structures [7,8]. This is also because the intensification of conventional agriculture mostly relies on chemical fertilizers, excessive water use, mechanization, and hybrid genetic engineering [9,10]. Across the landscape, the conversion of forest land to agricultural land has had an impact on water availability, thus causing the loss of hydrological functions associated with infiltration [11] as well as climate change on a local to global scale [12], trigging a shift to sustainable and multifunctional agriculture [13]. Sustainable agricultural supply chains generate greater production and/or higher agricultural productivity while at the same time achieving enhanced environmental, economic, and social outcomes [14].

Multifunctional agriculture can improve food production and positively impact social and environmental aspects, contributing to sustainable development [8,15]. Combining the Sustainable Development Goals (SDGs) of the forestry sector (SDG 15) and agriculture (SDG 2) in integrated land management has the potential to achieve three broad groups of SDGs [16]. Agroforestry as regenerative farming by food producers suits the SDGs program [8,17]. Hence, it can potentially be the future of agriculture [18]. Conversely, unsustainable agriculture practices [13] can be transformed into environmentally friendly agroforestry [19]. Agroforestry systems create multifunctional landscapes for income diversity and environmental services (clean water, biodiversity, carbon sequestration, and cultural conservation) [20,21]. Land use is often characterized by an integration of forests, agroforestry, agriculture, and settlement areas, which have to meet many interests and conflicting needs to produce products and services that support the SDG agenda [19].

In general, land covered by trees with the dual functions of producing food and environmental services is called agroforestry [22]. Agroforestry must be developed to integrate forestry and agriculture [23]. The forms of an agroforestry system can be a combination of commodities, such as agri-silviculture, silvofishery, silvopasture, agrosilvofishery, agro-silvopasture, and apiculture. The main requirement of an agroforestry system is that there are tree stands as the main component. A 'forest', according to the Food Agriculture Organization of Global Forest Resource Assessment (FAO FRA) 2000 program, has a tree canopy cover of >10% and an area > 0.5 ha or 10–30% of the tree canopy area and conservation of tree diversity [16] for climate change mitigation [24] as well as social, market, goods, and ecosystem services [25]. In Indonesia, the forest represents land with a minimum area of 0.25 hectares, and that contains trees with a canopy cover of at least 30%, capable of reaching a minimum height of 5 m at maturity [26]. Agroforestry can provide forest ecosystem functions as well as food and other products. It is a system involving the use of natural resources based on ecology through the combination of trees and crops, with various kinds and benefits of products (social, economic, and environmental), in a sustainable manner [19,21,26].

Indonesia's agroforestry practices have been developed since ancient times [27,28]. Even in Southeast Asian countries, there is a trend to develop a relationship between agroforestry and food security. However, studies on agroforestry in Indonesia that focus on food security are still limited [29]. The assumption is that most agroforestry systems are still based on traditional subsistence practices [27], in which management intensification is needed to increase business prospects and sustainable food security [30]. The economic and environmental impacts of multifunctional agroforestry farms are influenced by several factors, including scale, regional conditions, management practices adopted and landscape design [31]. However, no study classifies smallholder agroforestry practices in Indonesia in detail into three levels of business (subsistence, semi-commercial and commercial) as indicators of their contribution to smallholders' livelihood in the climate change era. Therefore, it is necessary to analyze the prospective business characteristics of smallholder agroforestry practices that have long been applied and their contribution to smallholder food security in rural Indonesia. This paper aims to evaluate agroforestry's contribution to food and livelihood security for smallholders in the climate change era.

Country Context

In Indonesia, smallholders' land in rural areas is typically spread among the home garden, fields (*tegalan*) and community forests with agroforestry practices. In its development, silviculture is defined as establishing and maintaining tree communities that produce tangible or intangible value to human beings, such as timber, non-timber forest products (NTFPs), food, and conservation and ecosystem value [32]. Forest and tree crop products can be a basis for income and food security [21]. Interrelating the concepts of food security, forests, agroforestry, environmental services, and sustainable development is still a challenge for Indonesia [22].

Indonesia's overall food production failed to meet the demand of its 270.20 million people in 2020 [33]. In 2023, Indonesia's population reached 277.7 million [34]; however, the overall farming area has decreased by 12.9% annually [35]. Conversion of rice fields to non-agricultural land occurred at a rate of around 80,000 ha per year [36], in the context in which the rate of expansion of such rice fields was around 20,000–30,000 ha per year [37]. The decline in rice production owing to a decrease in paddy fields was by up to 2.4 million tons over five years [38]. Moreover, climate change causes a decrease in rice production by 1.37% of the total production per year [39]. Therefore, the data show that rice imports reached 407,741.40 tons in 2021 [36].

The percentage of the population classified as 'food insecure' in Indonesia was 7.9% in 2019 [40]. Food must be accessible to communities in remote areas. In Indonesia, the population is spread over 83,931 villages, of which 3.3% are in forest areas (2768 villages), 22.18% are on the edge of forest areas (18,617 villages), and 74.52% are outside forest areas (62,546 villages). Among those, 99% of the villages located inside the forest depend on agricultural production as their main source of income, which is 93.8% and 79.4% for the villages located at the edge of the forest and outside the forest, respectively [33]. The ability to access food must be accompanied by ensuring viability, stability, and sustainability. Indonesia needs to develop widespread, sustainable, regenerative agriculture [10].

Indonesia's land area (191.1 million ha) features potential dry land of 144.5 million ha (76%), 42.7 million ha (22.4%) of wetlands, and 4.6 million ha for other uses. Dry land with a wet climate is spread across Sumatra, Kalimantan, and Java (133.7 million ha), and dry land with a dry climate is spread across Eastern Indonesia (10.7 ha) [41]. Nevertheless, based on the FAO FRA 2000 program, the rate of deforestation in Indonesia reached 0.78 million ha/year during 2010–2020. The latest data for 2020–2021 show that deforestation in Indonesia was 113,534.3 ha [42]. The agricultural land expansion rate for rice fields was around 20,000–30,000 ha annually [37].

Indonesia has already committed to promoting sustainable agriculture to achieve food security through more diverse food production systems using local resources [40,43]. Indonesia has 77 types of carbohydrate sources, 26 types of nuts, 389 types of fruits, 228 types of vegetables, and 110 types of spices and seasonings [39]. Local food consumed by a community can be sustainable [10] if it contributes positively to three aspects: (1) the environment (reducing greenhouse gas (GHG) emissions and sustaining the supply chain, e.g., reducing food loss and reducing packaging; (2) socio-economic (a local identity that can unite communities and provide income); and (3) health (providing healthy, diverse foods and reducing loss and waste [10]. People's reliance on rice in Indonesia as a source of carbohydrates can be reduced through food diversification, e.g., by consuming tubers, sago, breadfruit, and sugar palm. Several types of tubers as alternative food sources include cassava, sweet potato, *suweg* (*Amorphophallus paeniifolius*), arrowroot (*Maranta arundinaceae*), taro (*Colocasia esculenta*), *kimpul* (*Xanthosoma sagottifolium*), *gembili* (*Dioscorea esculenta*), canna (*Canna edulis*), and *porang* (*Amorphophallus muelleri*) [39,44]. These food commodities are produced from agroforestry to various types of land use systems, cultivated by smallholder farmers.

In the era of climate change, one of the growing obstacles to sustainable agriculture is the availability of water resources. Hence, utilizing dry land for food production is a strategic challenge. The Government of Indonesia issued Law no. 16/2014 and Presidential Decree 61/2011 regarding Planning, Implementation and MRV Systems to Achieve Emission Reduction (*Rencana, Implementasi dan Sistem MRV untuk Mencapai Penurunan Emisi* or RAN-GRK) as part of the national commitment to the Paris Agreement, which includes the development of environmentally friendly (low carbon) agriculture. The Indonesian government issued Minister of Environment and Forestry Regulation No. 9/2021 concerning Social Forestry Management to provide opportunities for communities to gain access to and benefits from forest management. Ministry of Environment and Forestry Regulation 8 Number 2021 concerning Forest Management and Preparation of Forest Management Plans and Forest Utilization in Protected Forests and Production Forests encourages increased productivity of forest land by applying agroforestry and multi-business forestry. Agroforestry practices can also be applied to dry land areas because they are relatively efficient with water resources. Therefore, the option of promoting or linking agroforestry, tree crops, and household food security should be considered a critical theory (sustainable agriculture in climate change) [45] and climate-smart agriculture [46].

2. Materials and Methods

This study is based on a literature review of both peer-reviewed and grey literature [47]. The review mainly focused on six scientific areas of interest in English and Indonesian—agroforestry, *hutan rakyat* (community forest), smallholders, food security, business prospects, and Indonesia—through an intensive search of online publications that primarily appeared from 2000 onwards. A preliminary scoping study was conducted based on a Google Scholar search targeted at finalizing keywords and search phrases and contributing to the framing of the manuscript. After finalizing keywords and phrases, relevant literature was gathered using scientific research search sites, i.e., Google Scholar, Mendeley, Scopus, and Web of Science. The selection of papers in reputable journals and several proceedings was carried out to identify research directly related to the contribution of agroforestry to smallholder food security and business prospects in Indonesia. Papers about agroforestry related to climate change adaptation and mitigation were also examined. Previous studies have assumed that agroforestry is an appropriate practice in the era of climate change for adaptation and mitigation, so it still contributes to food security and livelihood [8,21,48]. After removing any duplicates and considering the timeframe for the study, we selected 38 documents for thorough review by considering their relevance. The stages of searching and screening the publications are shown in Figure 1.

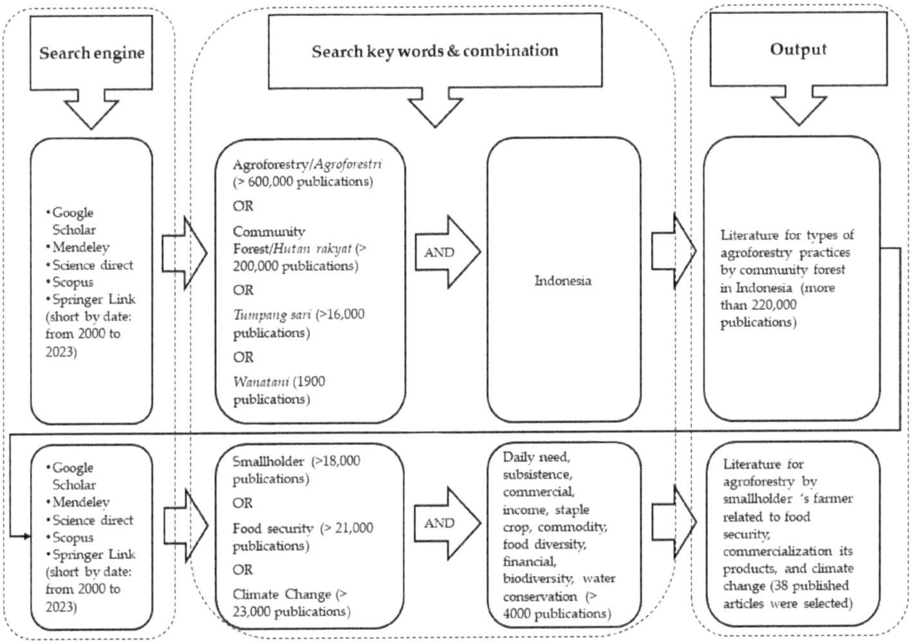

Figure 1. Stages of literature selection from various web-based databases. The process was adopted from Paudyal et al. [49].

The review was carried out by reading the content of the literature in detail. Relevant information was carefully compiled point by point, and scientific interpretations were made by using narrative qualitative and narrative compare analysis methods, including tables and figures [50]. The analysis process classified business and food security of smallholder agroforestry practices in Indonesia. Agroforestry businesses were classified as (1) subsistence, (2) semi-commercial, and (3) commercial [51]. 'Subsistence' agroforestry is typically applied to small areas with various crops in random planting arrangements and less intensive maintenance. Farmers manage subsistence agroforestry to meet their daily needs through day-to-day activities mainly related to providing family food. Any timber plantations are considered savings and a source of income when smallholders need money, with the timber being cut and sold. 'Semi-commercial' is an intermediate or transitional form of agroforestry, from subsistence to commercial, characterized by cultivating semi-commercial plant species with products that the household can consume and sell locally [51]. 'Commercial' agroforestry consists of two to three combinations of plant species, one of which may be the staple or main commodity developed on a broad scale with adequate technological input [51]. Commercial agroforestry requires professional management and a well-organized supply chain.

Food security consists of supply adequacy, simple physical or economic access, utilization, stability, and sovereignty [52]. Based on the World Food Program [53], food security is classified into six priority groups: 1 to 2: most vulnerable; 3 to 4: moderately vulnerable; and 5 to 6: food secure. The criteria used are food access, dietary diversity, nutritional security, and income [53]. Based on this method, related to food security, we found vulnerable food prospects on all islands in Indonesia: Sumatra (priorities 2–6); Kalimantan (priorities 3–6); Sulawesi (priorities 2–6); Java (priorities 2–6); and West Nusa Tenggara, East Nusa Tenggara, Maluku, and Papua (priorities 1–4) [29,53]. The logical framework of the literature study was qualitatively described, analyzed, and synthesized (Figure 2).

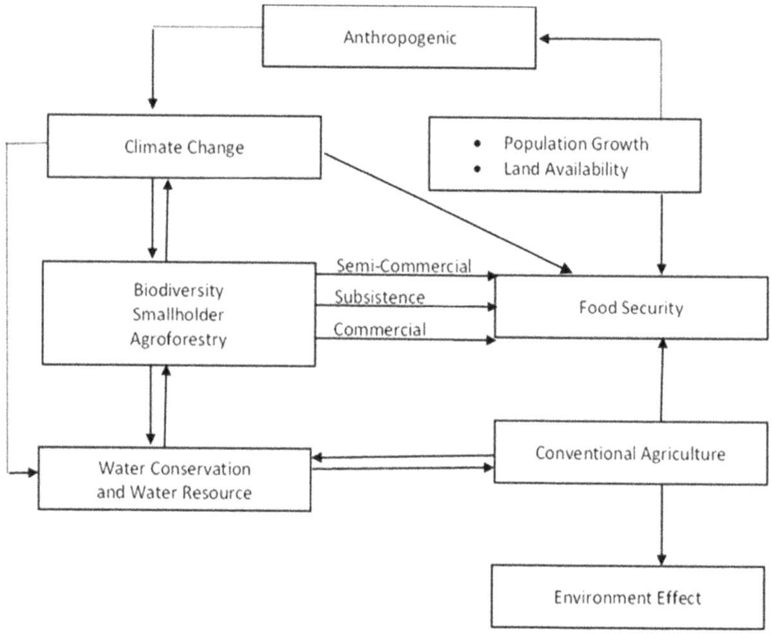

Figure 2. The logical framework of the literature study.

The criteria of the food security level in this study are determined by the level of food availability and dietary diversity. 'High' indicates an adequate staple food quantity that meets farmers' daily needs, with the example of agroforestry in which diverse food sources—such as staples, fruits and vegetables—are produced. 'Medium' indicates food produced only during particular seasons; typically, the diversity of crops is lower than the high level. 'Low' indicates restricted food availability and dietary diversity, even with no food crops in the agroforestry system. The level of food security, which reflects the level of crop diversity, is related to the resilience to climate change. The higher the contribution to food security and the higher the biodiversity from agroforestry practices, the higher the level of resilience and mitigation in the climate change era.

3. Results

Based on the representativeness of smallholder agroforestry practices in Indonesia (Figure 3), we found 33 references to be grouped into three business levels as indicators of their contribution to smallholders' livelihoods and local community food security. In total, 33 peer-reviewed studies were grouped into business types and their prospects for food security. In total, 8 peer-reviewed studies on subsistence agroforestry practices, 19 studies on semi-commercial agroforestry practices and 5 studies on commercial practices explicitly fit the definitions of the three business types.

Agroforestry practices are related to access to food, food diversity, nutritional security, and income of smallholder farmers in rural areas. Forty-two percent (42%) of the peer-reviewed studies were directly related to contributions to a community's food needs. At the same time, 19 (58%) peer-reviewed studies were related to the contributions of increasing income, which means an indirect contribution to meet a community's food needs. From 14 reviews of direct contributions to domestic food consumption, households were grouped into low (6 cases or 43%), medium (7 cases or 50%), and high (1 case or 7%) contributions.

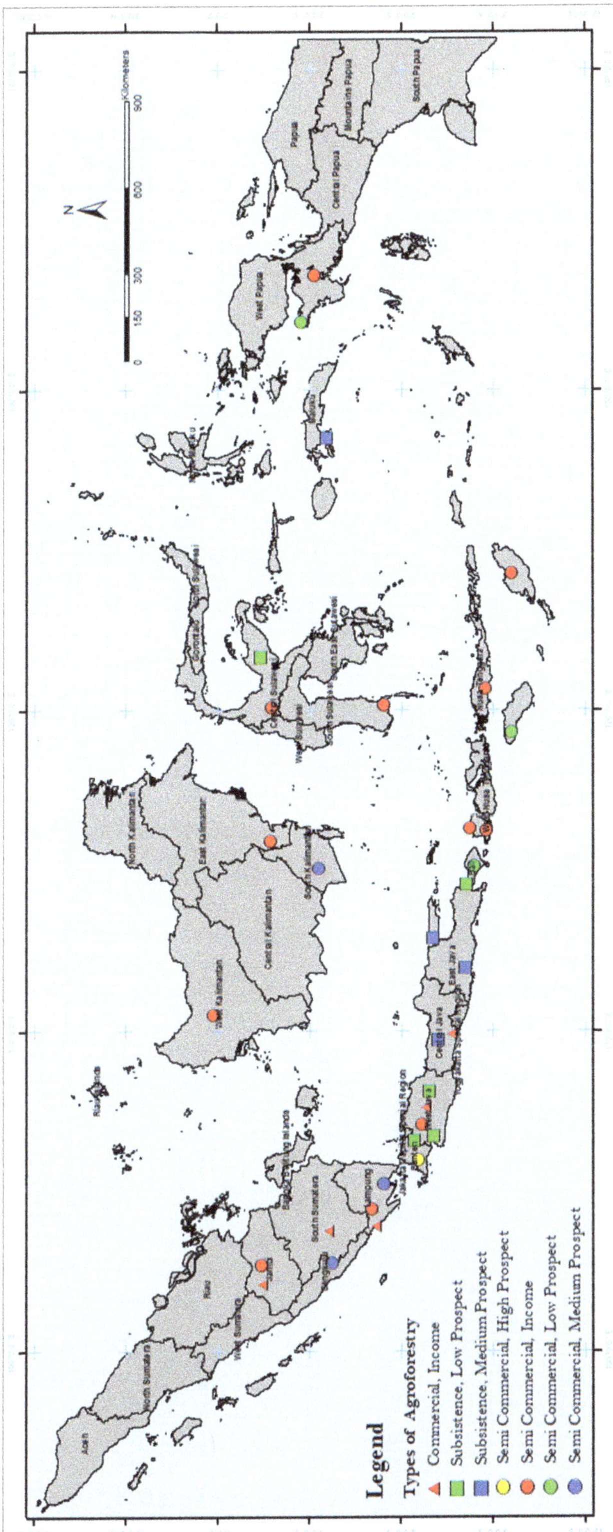

Figure 3. The prospect level of business and food security from different types of agroforestry practices in Indonesia.

Of the 33 peer-reviewed studies, we found that most smallholders practiced NTFP-based agriculture (29 studies), and only four cases practiced timber-based agrisilviculture. The results of the review indicate that the composition of plants in agroforestry was impacted by several factors, such as the width of the agroforestry area, the location, and the farmer's economic background, culture, and beliefs. Agroforestry plants were often cultivated in the form of (1) timber species, such as teak, *sengon* (*Falcataria falcata* (synonyms: *Albizia falcata*, *Falcataria moluccana* and *Paraserianthes falcataria*) and mahogany (*Swietenia* spp.); (2) multipurpose species, such as mango, durian, coconut and sugar palm; (3) food crops, such as upland rice, maize, cassava, taro, and sweet potatoes; (4) high-value commodity crops, such as cocoa, clove, nutmeg and coffee; (5) spices and medicinal plants, such as chili, ginger, turmeric, and galangal; and (6) fodder, such as *Leucaena leucocephala*, *Gliricidia* sp., and *Erythrina* sp. In some areas, the greater the area of land for agroforestry, the greater the proportion of trees compared with annual crops [54].

Moreover, agroforestry's contribution to the total revenue varies greatly depending on plant composition and land size [46,54–58]. Communities frequently practice agroforestry on small plots of land and prioritize supplying their basic needs. Therefore, the business classification of agroforestry in Indonesia is often 'subsistence' or 'semi-commercial'. Only 7 peer-reviewed studies provided detailed financial analysis, and most of the studies (26 articles) only stated the contribution to food production and the percentage of contribution to the total income of smallholders, which was related to complicated commodities, most of which were side businesses, meaning that such smallholder production was generally not recorded with detailed production inputs and outputs. From the seven peer-reviewed articles, six studies were classified as 'semi-commercial' agroforestry, and only one study was 'commercial' agroforestry. Agroforestry practices, which consist of high-value commodities—such as coffee, cocoa, rubber, and nutmeg—provide higher contributions to smallholders' incomes. The average NPV from this agroforestry system was more than IDR 4 million ($\approx$USD 256 as of July 2023) per year. The Internal Rate of Return (IRR) was higher than the interest rate; the ratio of benefit and cost (BCR) was more than 1. The main contribution of this agroforestry system was income for smallholders, but a lack of food production is an issue, as illustrated in West Java, West Timor, West Nusa Tenggara, West Papua, and Central Java.

3.1. Subsistence Smallholder Agroforestry Practices

A 'subsistence-scale' business means that most products are consumed directly to meet the domestic needs of the household, and only a small portion, if any, is sold to the market. 'Subsistence-scale management' means cultivating various plants with non-intensive maintenance in the yard or field around a smallholder's house. This is similar to the type of complex agroforests, tree-dominated home gardens, or smallholder tree crop plantation agriculture that rural households have established to obtain short-, medium- and long-term income [59]. As an example, in West Bandung, most smallholders used their farm products for domestic consumption: only 3.03% sold their products [60].

Based on nine references (Table 1), the cases of intercropping crops under teak in West Java [61] and alley cropping in Bali [62]; *dusung* traditional agroforestry in Maluku [63], mixed cropping in Central Java [64] and Central Sulawesi [57]; home garden agroforestry in Central Java [65], West Java [66], and Madura [67]; and agroforestry farms (teak and fruits) in Gunung Salak, Bogor, West Java [68]; were grouped as 'subsistence' businesses (consisting of NTFP-based agrisilviculture (eight cases) and timber-based agrisilviculture (one case). Most of the studies focused on densely populated areas in Java, Madura, and Bali, with limited community forest land areas. The contribution prospects to food security were five cases (low), which included two cases of alley cropping, two cases of home garden and one case of farm agroforestry with mixed agroforestry. Only two cases had a medium contribution to food security.

The other characteristic of subsistence agroforestry is the variation in food crops. While the other studies had fewer direct links to food availability, some contained elements of relevance to food access. The peer-reviewed studies showed that there were two cases of subsistence alley cropping of food crops (chili, rice, maize, peanut, cassava, and medicinal plants) among woody stands [61,62]. The tree species that made up this system were also relatively diverse, such as teak, *Azadiractha indica*, *Leucaena glauca*, *Swietenia macrophylla*, *Albizia falcataria* and *Dalbergia latifolia*. In the mixed cropping pattern (five cases), most were a combination of timber and fruit trees, annual food crops, *mpon-mpon* (medicinal plants) and fodder. This mixed cropping pattern was usually applied to fields and private forests owned by smallholders. The diversity of plant species in the mixed cropping pattern was relatively high, making it more multifunctional in its ability to meet various needs and positive adaptation to climate change. Home garden agroforestry featured food-oriented commodities for direct domestic consumption, such as chili peppers, tomatoes, spinach, long beans, and fruits and decorative and medicinal plants (two cases) [68,69].

3.2. *Semi-Commercial Smallholder Agroforestry Practices*

'Semi-commercial' is a transition from subsistence to commercial: the business may still be a mix of crops for subsistence and commodities for local-scale commercial sale. Our results show that applied agroforestry typically involves several crops, and each plays a subsistence and commercial function. Thus, the combination of crops produces multi-function agroforestry (semi-commercial). Table 2 features examples of semi-commercial agroforestry, including a combination of subsistence and commercial integration in mixed cropping in Lampung [70], cacao and coffee agroforestry in Central Sulawesi [71], rubber and fruit agroforestry in Jambi [72], fallow agroforestry system in East Kalimantan [73], coffee agroforestry in Lampung [74] and South Kalimantan [75], private forest agroforestry in Bogor, West Java [76], *huma* traditional agroforestry in West Java [77], *tembawang* traditional agroforestry in West Kalimantan [78], mixed garden in South Sumatra [79], *mamar* traditional agroforestry in West Timor [80], intercropping of agarwood in Flores [81], home garden agroforestry in South Sulawesi [54], *dusung* traditional agroforestry in Maluku [55], home garden with *trigona* [82] and mixed planting in West Nusa Tenggara [58], shifting cultivation garden [46] and yard agroforestry in West Papua [83] and mixed garden in Bali [84]. Nineteen agroforestry units are included in NTFP-focused agrisilviculture and only one is timber-focused agrisilviculture.

Agroforestry business contributes to food security in medium prospects (three cases), both a high contribution to food security (one case) and a low contribution (one case), and contributes to household income that can be used to buy staple foods (14 cases). MPTs and food crops in agroforestry play an important role in daily food availability for rural communities, although most only contribute to medium levels of food security. This is because agroforestry is only a side job for some rural communities. Semi-commercial agroforestry can be in the form of commodity crops and forestry plants (timber and NTFPs), including coffee, cocoa, candlenut, coconut, and rubber [51]. A semi-commercial type of business features various commodities that have subsistence and commercial functions. This is a form of balance between the interests of protecting biodiversity and commercialization, which, nevertheless, tends to reduce biodiversity.

Table 1. Subsistence business and food security of smallholder agroforestry practices in Indonesia.

No.	Type of Agroforestry	Agroforestry Commodity	Food Security	Business Prospect	Location and Source
1	Intercropping agricultural crops under teak	Teak (main species) and food crops (**)	L	S	West Java [61]
2	*Dusung* (traditional agroforestry)	Timber species, food crops, multipurpose tree species (MPTs), and high-value plantation crops (clove and nutmeg) (*)	M	S	Maluku [63]
3	Mixed cropping system agroforestry	Timber species, food crops, medicinal plants, and MPTs (*)	M + I	S	Central Java [64]
4	Home garden agroforestry	Food crops, spices, and MPTs (*)	M	S	Central Java [65]
5	Agroforestry farms (teak and fruits)	Teak (main species), other timber species, food crops, and MPTs (*)	L + I	S	West Java [68]
6	Home garden agroforestry	Food crops and medicinal plants (*)	L	S	West Java [66]
7	Mixed agroforestry	Timber species, high-value commodity crops (cacao), MPTs, fodder, spices, and medicinal plants (*)	L	S	Central Sulawesi [57]
8	*Taneyan lajang* (shared home garden)	Food crops, MPTs, fodder, spices, and medicinal plants (*)	M + I	S	Madura [67]
9	Alley cropping	Timber species (as main species), food crops, spices, and medicinal plants (**)	L	S	Bali [62]

Note: * NTFP-focused agrisilviculture; ** timber-focused agrisilviculture; S (subsistence); I (income); L (low-prospect food secure); M (medium-prospect food secure).

Table 2. Semi-commercial business and food security of smallholder agroforestry practices in Indonesia.

No.	Type of Agroforestry	Agroforestry Commodity	Food Security	Business Prospect	Source
1	Mixed cropping pattern/mixed garden	Timber species, MPTs, and high-value commodity crops (coffee, cocoa, and rubber) (*)	M (NPV: IDR 4,168,660.18/ha–IDR 4,589,627.36/ha)	SC	Lampung [70]
2	Cacao traditional agroforestry and coffee agroforestry	High-value commodity crops (cacao and coffee as main species) and fodder (*)	I	SC	Central Sulawesi [71]
3	Rubber agroforestry with fruit trees	Food crops, MPTs, and high-value plantation crops (rubber as a main species) (*)	I	SC	Jambi [72]
4	Fallow agroforestry system (oil palm, rubber, fruits, and rattan)	Food crops, MPTs, and high-value commodity crops (oil palm, rubber, and rattan as main species) (*)	I	SC	East Kalimantan [73]

Table 2. *Cont.*

No.	Type of Agroforestry	Agroforestry Commodity	Food Security	Business Prospect	Source
5	Protected forest (coffee agroforestry)	*Dalbergia latifolia*, high-value commodity crops (coffee as a main species), MPTs, and fodder (*)	I	SC	Lampung, Sumatra [74]
6	Private agroforestry	*Albizia moluccana* (main species), food crops, high-value commodity crop (clove) and MPTs (**)	I (NPV: IDR 64,197,125 (6 years); IRR 15%; BCR 2)	SC	West Java [76]
7	Coffee agroforestry with mixed garden and home garden/coffee and fruits)	High-value commodity crops (coffee as the main species) and MPTs (*)	M + I	SC	South Kalimantan [75]
8	*Huma* agroforestry	Food crops (rice as the main species), MPTs, and medicinal plants (*)	H	SC	Banten Province, Java [77]
9	*Tembawang* agroforestry	MPTs (main species), food crops, and medicinal plants (*)	I	SC	West Kalimantan [78]
10	Mixed garden and *talang*/*umo* agroforestry (traditional agroforestry)	MPTs, high-value plantation crops (coffee), and fodder (*)	M + I	SC	South Sumatra [79]
11	*Mamar* agroforestry	Timber species, food crops, MPTs, high-value commodity crops (cashew, cocoa, and coffee), and fodder (*)	I (NPV: IDR 19,862,245 (5 years); IRR 18%; BCR 1.63)	SC	Timor [56,80,85]
12	Intercropping of agarwood	Agarwood (as main species), high-value commodity crops (cacao, clove, coffee, and candlenut), MPTs (*)	I	SC	Flores [81]
13	Home garden	High-value commodity crops (cocoa, coffee, and cloves), MPTs, food crops, and fodder (*)	I	SC	South Sulawesi [54]
14	*Kaliwu*	Timber species, high-value commodity crop (coffee), and MPTs (*)	L + I (Income: IDR 103,100/month or IDR 4,520,863/year)	SC	Sumba, East Nusa Tenggara [86–88]
15	Home garden with *Trigona* sp.	MPTs and flowering plants (*)	L + I	SC	Lombok, West Nusa Tenggara [82]
16	Mixed planting	Mahogany, *Falcataria moluccana*, high-value commodity crops (cacao, coffee, and candlenut), and MPTs (*)	I (Income: IDR 14,942,031–38,547,093/year/ha	SC	Sesaot, West Nusa Tenggara [58]

Table 2. *Cont.*

No.	Type of Agroforestry	Agroforestry Commodity	Food Security	Business Prospect	Source
17	Shifting cultivation garden	*Instia bijuga*, high-value commodity crops (*Cryptocarya massoia*), MPTs, food crops, fodder, spices, and medicinal plants (*)	L + I (NPV: USD 10,965 or around IDR 163,334,000/ha (15 years))	SC	Teluk Patipi, Papua [46]
18	*Telanjakan*, *abian*, *kebon* mixed garden	Timber species, MPTs, food crops, fodder, spices, and medicinal plants (*)	L	SC	Bali [84,89]
19	Yard	Food crops, MPTs, high-value commodity crops (nutmeg), and spices (*)	I (Income: IDR 7,693,000/year)	SC	West Papua [83]

Note: * NTFP-focused agrisilviculture; ** Timber-focused agrisilviculture; SC (semi-commercial); I (income); L (low-prospect food secure); M (medium-prospect food secure); H (high-prospect food secure).

3.3. Commercial Smallholder Agroforestry Practices

Commercial agroforestry tends to be in the form of simple agroforestry with two or three combinations of plant species, one of which is a main commodity that is developed more intensively. Based on the reference (Table 3), our results show that this form of applied agroforestry features a few commodities, and each commodity plays a commercial function. Although this system is often a combination of commodities, it is typically only focused on two or three species for commercialization. So, the diversity of commercial agroforestry is significantly less compared to semi-commercial or subsistence agroforestry. This system cannot contribute directly to domestic needs (food) but obtains income, which increases consumption power to buy food. Commercial agroforestry will increase income but lower biodiversity. As an example, commercial vegetable agroforestry was found in home garden agroforestry systems in upstream West Java [90], dragon blood fruit agroforestry in South Sumatra [91], *repong damar* agroforestry in Lampung [92], oil-palm agroforestry in Jambi [93] and cardamom agroforestry in private forest in Central Java [94]. These cases were dominated by NTFP-focused agrisilviculture, with only one case of timber-focused agrisilviculture. All of them aimed at earning income and indirectly contributing to the fulfillment of domestic food needs of small holders. The income could be used for purchasing staple foods and meeting daily needs. Because the contribution of the commercial type of agroforestry is to obtain income, the possibility of contributing to food security, although not directly, is relatively high.

Table 3. Commercial business and food security of smallholder agroforestry practices in Indonesia.

No.	Type of Agroforestry	Agroforestry Commodity	Food Security	Business Prospect	Source
1	'Tradition to commerce' home garden agroforestry	MPTs, high-value commodity crops (onion, carrot, and cabbage) (*)	I	C	Upstream, West Java [90]
2	*Jernang* (dragon blood, *Daemonorps draco*) fruit agroforestry	Dragon blood (as the main species) and rattan (*)	I	C	Jambi and South Sumatra [91]
3	*Shorea javanica* agroforestry gardens (*repong damar*)	*Shorea javanica* (as the main species), other high-value commodity crops, food crops, or MPTs (*)	I	C	West Lampung, Lampung [92]
4	Oil-palm agroforestry	Oil palm and rubber trees (*)	I	C	Jambi [93]
5	Private forest agroforestry	*Falcataria moluccana* and cardamom (**)	I (NPV: IDR 33,599,884–IDR 112,039,098; IRR 13–35%; BCR 1.58–2.32 and revenue IDR 5,672,957–IDR 18,916,524/year	C	Central Java [94]

Note: * NTFP-focused agrisilviculture; ** Timber-focused agrisilviculture; C (commercial); I (income).

4. Discussion

Agroforestry in Indonesia is complex, as can be seen in the development of agroforestry as a form of community thought in the Indonesian archipelago. It shows the perseverance of agrarian communities in cultivating land that eventually becomes a distinctive hereditary culture in certain niches, thus creating diverse forms of agroforestry in Indonesia. Cultural background, preferences, and needs determine the peculiarities of agroforestry in each area. The influence of the surrounding environment and the information will also affect a person's decisions [95], including smallholders' decisions to plant species commodities on their land.

Agroforestry systems increased food production and improved environmental conditions, depending on the land management practices and tree management [96]. Silvicultural practices, such as planting, assisted migration, thinning, or natural regeneration, can cost-efficiently help reduce the impact of climate change on forest structure, composition, and function [97]. Smallholders with agroforestry systems are often aware of functionality in broad contexts, including different product uses, different tree characteristics (e.g., differences in phenology), or risk management options. For example, smallholders manage different species for different purposes, contributing to their livelihoods, addressing competition between species, and assisting ecosystem processes [98]. The changing of a natural ecosystem to agroforestry practices and other uses has implications for livelihoods and ecosystem services [16,99].

Benefits from the impact of tree cover on climate at local, regional, and continental scales require broader acknowledgment [12]. Agroforestry provides smallholders resiliency to dryland conditions and climate change for accessing food, income, health, and ecosystem or environmental stability [29,53].

4.1. Food Security and Agroforestry Systems in Indonesia

The Indonesian smallholder has been developing agroforestry community forests as a source of food, NTFPs and timber since ancient times, demonstrating that community forests can be managed to meet the food needs of smallholders and contribute to national food security. As an illustration, community forests have been shown to contribute 61.34% of the daily food needs of communities, which were produced from 23 types of food crops [100]. In addition, agroforestry practices in community forests have been shown to produce 46.01% of food commodities, consisting of 12 types of food, for example, sugar palm (*Arenga pinnata*), cocoa (*Theobroma cacao*) and mango (*Mangifera indica*) [101]. Agroforestry practices in community forests in Lampung contributed to farmers' income by 53% and were sufficient for household food security [102]. However, communities with limited landholdings and a homogeneous local culture with traditional rice farming systems produce seasonal employment opportunities, reducing the risk of long-term tree cultivation with limited resources [19,96,103].

Complex agroforests with various commodity crops are the foundation of many businesses, which can be categorized as 'subsistence' for certain products and 'commercial' for others. Hence, it is not easy to determine whether an agroforest falls into 'subsistence' or 'commercial' prospects. For instance, agroforestry systems range from traditional to commercial in Bandung, West Java [90], rubber and fruit tree agroforestry in Jambi [72], fallow agroforestry (oil palm, rubber, rattan, and fruits) in East Kalimantan [104], and mixed garden (coffee and fruits) in South Kalimantan [75]. These four agroforests are forms of NTFP-focused agrisilviculture, which contribute to income and indirectly contribute to family food security. There is a mixed garden of coffee and fruit trees in South Kalimantan where fruits are consumed by the family, even though they are on a subsistence scale, and coffee is a commercial product. However, coffee agroforestry in Sumatra contributed 54% of household income compared to 12.5% from traditional agricultural components [74], and *damar* agroforestry contributed up to 51% of household income [92]. Smallholders' involvement in coffee agroforestry for the rehabilitation of degraded land resulted in greater social stability with an increase in incomes and greater access to agricultural land [34]. Several studies also highlighted a 'risk reduction' (specifically concerning lack of food from crop failure and income volatility) as an outcome of greater diversification through agroforestry [105–107].

The business feasibility of agroforestry is illustrated in the net present value (NPV), internal rate of return (IRR), and benefit/cost ratio (BCR) in some cases globally. The cases of agroforestry in Indonesia indicate that agroforestry is worthy of being cultivated as a business that achieves profits for smallholders. Income has been derived from timber species or high-value commodity crops such as coffee, cacao, cloves, and candlenuts. However, the commercial value of timber species and high-value commodity crops in agroforestry systems could decrease the availability or diversity of food crops. Smallholders could only cultivate food crops (corn, upland rice, peanut, banana, cassava, etc.) in the initial three years before the canopy of timber species or high-value crops closed [108].

Subsistence, semi-commercial, and commercial categories cannot always be a firm predictor of the prospect of income and welfare for landowners. Many factors make landowners choose this or that model of agroforestry. These factors include (1) the land area, which affects the prospect of land productivity; (2) the need for crops and the financial situation of the landowner; (3) the type and quality of commodities that can be cultivated; and (4) market conditions (which are often influenced by access and transportation, which ultimately affect the demand and selling price of a commodity). The ability of the land to produce economic value can also be different in each area, depending on the tenacity of the cultivator, property security, type of commodity, smallholder's capital ability, market access, crop quality, and seasonal suitability. All of the above factors are intertwined and shape the pattern of agroforestry in Indonesia today. For example, the results of research in East Priangan [109] show that owing to lost demand during the peak of the COVID-19 pandemic, agroforestry products, which were originally for semi-commercial purposes, became full subsistence, with many crops not absorbed by the market because of poor road access, which further dropped commodity prices.

Several agroforestry practices were described as 'living savings accounts' by others, and they primarily became a significant source of revenue [90,110]. Agroforestry was a method to increase food production from the forest and a new source of income for smallholders, with increases ranging from 41.32–68.67% of total income [111,112]. Nonetheless, about 60.97% of smallholders in Lampung were classified as 'poor' [112]. This demonstrates that agroforestry, in this case, is a small business that can only fulfill basic livelihood needs by either using agroforestry products for domestic consumption or sale. Although the latter indirectly contributes to food production (particularly in commercial agroforestry), the income may also play a role in increasing the system's stability and resilience. Dependency on rice as a staple food will be reduced by diversifying agricultural commodities for subsistence prospects of semi-commercial systems.

On the other hand, the commercialization of agroforestry in rubber, coffee, cacao, and vegetables in uplands and palm-oil commodities elsewhere decreases plant diversity and smallholders' preference to grow food, mostly in semi-commercial and subsistence prospects [113]. This occurred, for example, in the Upper Citarum Watershed, West Java, where commercialization of short-term perennials with high international demand (e.g., cacao, coffee, and pepper) reduced the planting area for other fruit trees and food crops. In the same case, the commercialization of vegetables decreased the diversity of agroforestry [90]. Further, owing to a 20% decline in species diversity caused by the transition from subsistence to commercial agroforestry, some ecological and sociocultural functions were dramatically reduced [90]. Whereas commercial home gardens throughout the region directly increased food availability and utilization through income generation, the outputs were recognized as less diverse than traditional home gardens [29]. In Java, agroforestry practices in smallholder systems reduced access to food subsistence, and negative ecological effects could be reduced by planting fruit trees [68]. The commercialization of agroforestry tends to apply simple agroforestry (a mix of perennial and annual crops) with one tree species and one-to-a-few annual crops, which are ecologically not as good as complex agroforestry (a complex vegetation structure that looks like a forest) [59].

4.2. Food Security and Water Conservation in Agroforestry

In Indonesia, agroforestry has been applied predominantly to dry or marginal land in yards, fields, private forests, buffer lands of forest areas, and community forests. These lands are usually without irrigation, so they are very efficient in using water and relatively resistant to climate change. Although the increase in water productivity due to the microclimatic modification by tree crops tends to be limited [20,114], the existence of forests and trees as an agroforestry system plays a role in soil and water conservation and water use efficiency. Although the greater productivity of agroforestry is mainly due to the higher amount of water used [20,114], an agroforestry system is more resilient in the face of climate change and various drought and flood disasters [21,99,115–117]. The water function of natural forests with high biodiversity is impossible to reproduce if changed to oil-palm monoculture with shallow roots [118]. This can be anticipated by planting fruit or nut trees (such as stinky bean and jackfruit) using an agroforestry system, which promotes the development of multifunctional landscapes to conserve or increase the quality of catchment areas [17,119].

Agroforestry increases water productivity in two ways: (1) tree transpiration and (2) the resulting tree biomass [120]. Trees can link local to regional and global water cycles through modification of infiltration, water use, hydraulic redistribution of groundwater, and their role in rainfall recycling. Agroforestry contributes to improving ecosystem services for water, including transmission, buffering peak flows, increased infiltration, water quality, slope and riparian stability, reduced erosion, modified microclimate, coastal protection, and rainfall triggering [17]. Trees also affect a soil's ability to capture, store, and release water. The presence of litter (organic matter) assists soil in retaining water and improving soil structure and porosity [121]. In areas with limited water resources, the presence of trees as shade for coffee agroforestry systems can protect agricultural crops by reducing soil evaporation and coffee transpiration. Shade trees affect the microclimate (light, temperature, water saturation, vapor pressure deficit) and radiant energy in a system [122]. Tree canopy cover, understorey vegetation, and litter necromass are strong indicators of watershed health in terms of low run-off and high soil infiltration [11].

An illustration of quality enhancement in an agroforestry system is the shifting of cacao monoculture to cacao agroforestry, which increased the soil organic carbon by 1 g kg^{-1} (0.1%) and soil water capacity by 6% [123]. The application of agroforestry affected watershed quality indicators, such as controlling 97% of erosion, increasing 100% of water retention, and CN (curve number) below 80 [124]. In an upstream watershed, >55% of tree canopy cover was associated with infiltration rates, while in the midstream, >80% of tree canopy cover was qualified as an 'infiltration-friendly' land use [11,124]. In another example, in Palu Watershed, Central Sulawesi, various agroforestry patterns (alley cropping, alternate rows, random mixture, and trees along borders) resulted in low erosion rates: 5.17 g ha^{-1}; 4.93 g ha^{-1}; 0.78 g ha^{-1}; and 0.47 g ha^{-1}, respectively [125]. Additionally, agroforestry in mangrove forests with a silvofishery pattern enhanced water quality owing to the removal of nutritional pollutants (absorb nitrate, ammonia, and nitrite), oxygen level and pH improvement, and nitrate fixation, as well as shrimp culture production improvement [126–128]. In addition, agroforestry considerably improves smallholders' access to food in the face of potential severe famine crises triggered by climate change [129,130]. It also boosts livelihood resilience by reducing the reliance on remote product markets [68]. Smallholder agroforestry, practiced in a home garden, *tegalan*/dry land, or community forest, needs to be preserved as a form of natural ecosystem service for the adaptation and mitigation of climate change by providing biodiversity and soil and water conservation for the sustainability of future results [19,103].

4.3. *Food Security and Biodiversity of Smallholder Agroforestry in Climate Change*

Food security is one of the various ecosystem services that can be provided by agroecosystems (Figure 4). If an agroforestry area is focused on obtaining financial profit, then the form of agroforestry only tends to become more monocultural and will lose its multifunctionality [131]. In the era of climate change, the more biodiverse the agroforestry, the higher the carbon absorption and productivity, benefitting the livelihoods of rural communities and protecting nature [103,132]. In contrast to conventional agriculture, this smallholder agroforestry practice is clearly a form of smallholder resilience in obtaining food and livelihoods in the modern era, which positively affects the environment. Therefore, the practice of diverse trees in agroforestry needs to be maintained in order to provide ecosystem services and increase production for climate change adaptation and mitigation.

Figure 4. Three ecosystem services from agroforestry systems, adapted from Palacios and Bokelmann, 2017 [133].

There are two issues related to biodiversity in the implementation of agroforestry. First, the biodiversity constituents of agroforestry land: to ensure that agroforestry land has a variety of functions—provisioning, regulating, and cultural ecosystem services [95]—the biodiversity of species will contribute positively [134]. Biodiversity will help to reach ecological equilibrium at the household farm prospect, combat climate change, achieve food security, and expand market opportunities for communities [135–138]. In addition, Garí [139] concludes that for indigenous people, the preservation of biodiversity is crucial for ecological resilience, health care, and food security.

Second, the diversity of animal biodiversity: agroforestry is very likely to provide habitat but will not be able to replace the role of natural forests. Compared to old-growth forests, agroforests supported around 23% fewer species and 47% fewer endemic species [140]. It should also be noted that even in natural forests in Indonesia, wildlife is extremely difficult to find owing to poaching [141], especially on agroforestry land with easier access and where protection of animal biodiversity is not the core business of agroforestry farmers. There have been many studies that explain that wildlife and human conflicts often occur owing to the destruction of habitats whereby wildlife food security is affected by the establishment of agricultural (including agroforestry) businesses [107,142–146]. Forestry practices that frequently result in habitat homogenization, habitat heterogeneity, and forest biodiversity reduction are tightly related [147]. Ecosystems subjected to intensive management experience a decline in biodiversity as well as services, and forests that are subjected to intensive management show significantly reduced multifunctionality, which recovers more slowly the longer the practice is continued [148]. In general, expecting agroforestry land to be able to provide maximum environmental benefits and building intact forest–agroforestry gardens seem to be the best approaches, even though these require larger areas of land so that there may still be high prospects of both commercial value and food security.

Tree biodiversity resulting from agroforestry practices is a form of adaptation and mitigation of climate change. Climate change will affect the economic and environmental productivity of agroforestry practices. The higher the biodiversity, the higher the agroforestry productivity in the climate change era [51,138]. Increasing human populations will further increase anthropogenic activities that affect climate change [52,149]. Conventional agriculture's approach to increasing food production has been proven to have a negative impact on the environment, while agroforestry has been proven to be good from an environmental perspective (water and soil conservation, climate change adaptation, and biodiversity enhancement) [20,150]. Agroforestry businesses have built rural livelihoods that have contributed to food security, biodiversity, and environmental services to realize sustainable development [21]. Indirectly, agroforestry practices are related to the protection of rural ecosystems in providing income and food for the community in climate change conditions.

The ambition to strengthen food security in Indonesia is still great owing to the vast land resources. The Global Forest Watch summarized research results from Potapov et al. [151] and concluded that 24.1 million hectares of Indonesian forests were 'disturbed'. These degraded forests require reforestation, which can be an opportunity for society, biodiversity, and the climate to win [103]. Smallholders and biodiversity will both benefit from the restoration of degraded land [140]. The most recent project for reforestation is through the social forestry program, which covers 12.7 million hectares [152] and in which agroforestry has become the major activity. Having successful social forestry and reforestation programs will certainly help to improve food security in Indonesia.

Until recently, agroforestry in Indonesia has been challenged to achieve sustainability, and its adoption of characteristics of success at a broader scale is required [153]. There are technical, financial, market, and social constraints to agroforestry development, especially for smallholders [154]. This sector requires policy support from the government [155], indicated by currently unfavorable inter-sectoral policies, viz. legal frameworks and coordination between different government mandates, such as agriculture, forestry, rural development, environment, and trade [116]. In rural development planning, agroforestry is a strategically advantageous land use if the inherent complexities are considered in policy measures [156]. Successful agroforestry systems are characterized by well-functioning institutions, management, capacity building, and infrastructure [45].

5. Conclusions

Agroforestry in Indonesia is a crucial form of land utilization. The practice of agroforestry, particularly among small landholders, demonstrates adaptability to land conditions, household needs, and market opportunities, resulting in a diversity of agroforestry implementations. Most smallholders use agroforestry systems to earn income from product sales, timber as 'savings accounts', and food for daily consumption. Some smallholders do not produce food for domestic consumption from their agroforestry practices; however, they generate income from selling products that increase their purchasing power for food needs (57% of peer-reviewed case studies). Two studies show that agroforestry can contribute to smallholder's food needs by 46% to 61% and three studies state that agroforestry contributes to small farmers' income by 51–54%. The contribution of agroforestry to the income of smallholders is greater than that of traditional agriculture.

Furthermore, agroforestry contributes to SDGs through climate change mitigation (SDG 13) since unsuitable agricultural land can still substantially contribute to food provision (SDGs 2 and 15). Traditional subsistence agroforestry practices show high diversity in producing food, medicine, NTFPs, and timber (27% of peer-reviewed case studies). The commercialization of agroforestry with the intensification of several high-value plant commodities (oil palm, cardamom, vegetables, and dragon fruit) has reduced the environmental services produced in climate change adaptation and mitigation (15% of peer-reviewed case studies). The limited land availability for small landholders poses a constraint to meeting their food and income.

Nevertheless, the role of agroforestry remains highly significant, especially in the context of semi-commercial agroforestry, as it provides irreplaceable social security for small landholders in Indonesia. Semi-commercial agroforestry with a mixture of trees—commercial and subsistence species—has higher diversity, a form of 'local wisdom' in maintaining community forest sustainability and contributing to family income and food needs. Smallholder agroforestry practices in Indonesia are shifting from traditional subsistence agroforestry to semi-commercial agroforestry. It is necessary to promote complex semi-commercial agroforestry to maintain productivity in times of climate change with compensation or incentives for smallholders. Agroforestry practices can maintain landscape ecosystems with soil and water conservation and biodiversity to continue sustainably producing food and income for rural communities in an era of climate change. Therefore, it is not surprising that agroforestry practices are at the core of various forestry programs, including peatland, mangrove, critical land rehabilitation, and social forestry. Agroforestry plays a key role in environmental improvement efforts while delivering direct and indirect economic benefits to communities. This study confirms these findings. The weakness of this review is the limited quantification of agroforestry ecosystem services and the direct link between agroforestry and food security, so further research is needed. It is necessary to measure food production and livelihoods from the agroforestry sector outside the conventional agricultural sector by the Central Bureau of Statistics.

Author Contributions: Main authors, A.S. (Aris Sudomo), B.L., H.L.T., M.M.B.U., A.A.D.R., A.U., L.A.G.P., Y.I., S.A.R. and H.B.; conceptualization, A.S. (Aris Sudomo), B.L., H.L.T., M.M.B.U., L.A.G.P., Y.I., S.A.R. and H.B.; methodology, A.S. (Aris Sudomo), A.A.D.R., H.R., K., A., A.U., B.L., H.B., A.S. (Ali Setyayudi) and S.A.R.; investigation, A.S. (Aris Sudomo), A.U., B.L., S.A.R. and H.B.; writing—original draft preparation, A.S. (Aris Sudomo), B.L., M.M.B.U., A.A.D.R. and L.A.G.P.; writing—review and editing, A.S. (Aris Sudomo), Y.I., A.A.D.R., A.W., B.L., H.R., H.L.T., S.A.R., L.A.G.P., M.M.B.U., K., A.W. and H.B.; English editing, A.S. (Aris Sudomo), A.A.D.R., A.W., H.L.T., K., S.A.R., Y.I., H.B., L.A.G.P. and M.M.B.U.; supervision, A.S. (Aris Sudomo), A.U., B.L., H.L.T., Y.I., S.A.R. and H.B.; project administration, A.U., A.A.D.R., A.S. (Ali Setyayudi), H.R., A., A.W. and K., A.S. (Aris Sudomo) and B.L. have created the flow and contents of the paper. B.L. contributed to deepening the Results and Discussion sections and some parts of the research method. All authors have read and agreed to the published version of the manuscript.

Funding: This research paper was funded by the Center for International Forestry Research and World Agroforestry (CIFOR-ICRAF) Indonesia.

Institutional Review Board Statement: Not applicable.

Data Availability Statement: Not applicable.

Acknowledgments: The authors highly appreciate CIFOR-ICRAF Indonesia for providing financial support (Article Processing Charge) on the publication of the manuscript. The authors thank the Biofuel and Biopharmaceutical Breeding Research Group for jointly following up on the Minimum Performance Output regulations. We also thank Fuad Hasyim from the Geospatial Information Agency for supplying a map in this manuscript.

Conflicts of Interest: The authors declare no conflict of interest.

References

1. Karabulut, A.A.; Crenna, E.; Sala, S.; Udias, A. A Proposal for Integration of the Ecosystem-Water-Food-Land-Energy (EWFLE.) Nexus Concept into Life Cycle Assessment: A Synthesis Matrix System for Food Security. *J. Clean. Prod.* **2018**, *172*, 3874–3889. [CrossRef]
2. Amundson, R.; Berhe, A.A.; Hopmans, J.W.; Olson, C.; Sztein, A.E.; Sparks, D.L. Soil and Human Security in the 21st Century. *Science (1979)* **2015**, *348*, 1261071. [CrossRef] [PubMed]
3. Tan, Z.X.; Lal, R.; Wiebe, K.D. Global Soil Nutrient Depletion and Yield Reduction. *J. Sustain. Agric.* **2005**, *26*, 123–146. [CrossRef]
4. Gerten, D.; Heinke, J.; Hoff, H.; Biemans, H.; Fader, M.; Waha, K. Global Water Availability and Requirements for Future Food Production. *J. Hydrometeorol.* **2011**, *12*, 885–899. [CrossRef]
5. MEA. *MEA Ecosystems and Human Well-Being: Wetlands and Water Synthesis*; MEA: Beirut, Lebanon, 2005.

6. Keatinge, J.D.H.; Waliyar, F.; Jamnadas, R.H.; Moustafa, A.; Andrade, M.; Drechsel, P.; Hughes, J.A.; Kadirvel, P.; Luther, K. Relearning Old Lessons for the Future of Food—By Bread Alone No Longer: Diversifying Diets with Fruit and Vegetables. *Crop. Sci.* **2010**, *50*, S51. [CrossRef]
7. Godfray, H.C.J.; Beddington, J.R.; Crute, I.R.; Haddad, L.; Lawrence, D.; Muir, J.F.; Pretty, J.; Thomas, S.R.; Toulmin, C. Food Security: The Challenge of Feeding 9 Billion People. *Science (1979)* **2010**, *327*, 812–818. [CrossRef]
8. Waldron, A.; Garrity, D.; Malhi, Y.; Girardin, C.; Miller, D.C.; Seddon, N. Agroforestry Can Enhance Food Security While Meeting Other Sustainable Development Goals. *Trop. Conserv. Sci.* **2017**, *10*, 1940082917720667. [CrossRef]
9. Pretty, J.; Bharucha, Z.P. Sustainable Intensification in Agricultural Systems. *Ann. Bot.* **2014**, *114*, 1571–1596. [CrossRef]
10. Samadhi, N.; Mallipu, A. Reviving Local Foods, Achieving Sustainable Food System. *The Jakarta Post*, 16 October 2021.
11. Suprayogo, D.; Van Noordwijk, M.; Hairiah, K.; Meilasari, N.; Rabbani, A.L.; Ishaq, R.M.; Widianto, W. Infiltration-Friendly Agroforestry Land Uses on Volcanic Slopes in the Rejoso Watershed, East Java, Indonesia. *Land* **2020**, *9*, 240. [CrossRef]
12. Ellison, D.; Morris, C.E.; Locatelli, B.; Sheil, D.; Cohen, J.; Murdiyarso, D.; Gutierrez, V.; Van Noordwijk, M.; Creed, I.F.; Pokorny, J. Trees, Forests and Water: Cool Insights for a Hot World. *Glob. Environ. Chang.* **2017**, *43*, 51–61. [CrossRef]
13. Bernard, F.; van Noordwijk, M.; Luedeling, E.; Villamor, G.B.; Sileshi, G.W.; Namirembe, S. Social Actors and Unsustainability of Agriculture. *Curr. Opin. Environ. Sustain.* **2014**, *6*, 155–161. [CrossRef]
14. United Nations Environment Programme (UNEP). Agriculture: Investing in Natural Capital. In *Towards a Green Economy: Pathways to Sustainable Development and Poverty Eradication*; United Nations Environment Programme (UNEP): Nairobi, Kenya, 2011; p. 19.
15. Wezel, A.; Casagrande, M.; Celette, F.; Vian, J.; Ferrer, A.; Peigné, J. Agroecological Practices for Sustainable Agriculture. A Review. *Agron. Sustain. Dev.* **2014**, *2050*, 1–20. [CrossRef]
16. Ordonez, J.C.; Luedeling, E.; Kindt, R.; Tata, H.L.; Harja, D.; Jamnadass, R.; van Noordwijk, M. Constraints and Opportunities for Tree Diversity Management along the Forest Transition Curve to Achieve Multifunctional Agriculture. *Curr. Opin. Environ. Sustain.* **2014**, *6*, 54–60. [CrossRef]
17. Van Noordwijk, M.; Coe, R.; Sinclair, F.L. *Sustainable Development through Trees on Farms: Agroforestry in Its Fifth Decade*; World Agroforestry (ICRAF): Bogor, Indonesia, 2019; ISBN 9786025894039.
18. Hairiah, K.; Ashari, S. Pertanian Masa Depan: Agroforestri, Manfaat Dan Layanan Lingkungan. In *Prosiding Seminar Nasional Agroforestri 2013*; BPPTA: Malang, Indonesia, 2013; pp. 23–35.
19. van Noordwijk, M.; Speelman, E.; Hofstede, G.J.; Farida, A.; Kimbowa, G.; Geraud, G.; Assogba, C.; Best, L.; Tanika, L. Sustainable Agroforestry Landscape Management: Changing the game. *Land* **2020**, *9*, 243. [CrossRef]
20. van Noordwijk, M.; Kim, Y.S.; Leimona, B.; Hairiah, K.; Fisher, L.A. Metrics of Water Security, Adaptive Capacity, and Agroforestry in Indonesia. *Curr. Opin. Environ. Sustain.* **2016**, *21*, 1–8. [CrossRef]
21. van Noordwijk, M.; Duguma, L.A.; Dewi, S.; Leimona, B.; Catacutan, D.C.; Lusiana, B.; Öborn, I.; Hairiah, K.; Minang, P.A. SDG Synergy between Agriculture and Forestry in the Food, Energy, Water and Income Nexus: Reinventing Agroforestry? *Curr. Opin. Environ. Sustain.* **2018**, *34*, 33–42. [CrossRef]
22. van Noordwijk, M.; Bizard, V.; Wangpakapattanawong, P.; Tata, H.L.; Villamor, G.B.; Leimona, B. Tree Cover Transitions and Food Security in Southeast Asia. *Glob. Food Sec.* **2014**, *3*, 200–208. [CrossRef]
23. Jihad, A.N.; Lestari, L.D. An Overview and Future Outlook of Indonesian Agroforestry: A Bibliographic and Literature Review. *E3S Web Conf.* **2021**, *305*, 7002.
24. UNFCCC. Seventh Conference of Parties: The Marrakech Accords. Available online: https://unfccc.int/cop7/documents/accords_draft.pdf (accessed on 13 March 2023).
25. Chapin, F.S.; McGuire, A.D.; Ruess, R.W.; Hollingsworth, T.N.; Mack, M.C.; Johnstone, J.F.; Kasischke, E.S.; Euskirchen, E.S.; Jones, J.B.; Jorgenson, M.T. Resilience of Alaska's Boreal Forest to Climatic Change. *Can. J. For. Res.* **2010**, *40*, 1360–1370. [CrossRef]
26. Nair, P.R. *An Introduction to Agroforestry*; Springer Science & Business Media: Berlin/Heidelberg, Germany, 1993.
27. Achmad, B.; Siarudin, M.; Widiyanto, A.; Diniyati, D.; Sudomo, A.; Hani, A.; Fauziyah, E.; Suhaendah, E.; Widyaningsih, T.S.; Handayani, W. Traditional Subsistence Farming of Smallholder Agroforestry Systems in Indonesia: A Review. *Sustainability* **2022**, *14*, 8631. [CrossRef]
28. Viswanath, S.; Lubina, P.A.; Subbanna, S.; Sandhya, M.C. Traditional Agroforestry Systems and Practices: A Review. *Adv. Agric. Res. Technol. J.* **2018**, *2*, 18–29.
29. Duffy, C.; Toth, G.G.; Hagan, R.P.O.; McKeown, P.C.; Rahman, S.A.; Widyaningsih, Y.; Sunderland, T.C.H.; Spillane, C. Agroforestry Contributions to Smallholder Farmer Food Security in Indonesia. *Agrofor. Syst.* **2021**, *95*, 1109–1124. [CrossRef]
30. Sudomo, A.; Hani, A.; Agus, C.; Nugroho, A.W.; Utomo, M.M.B.; Indrajaya, Y. Intensification of Agroforestry Systems in Community Forests to Increase Land Productivity and Sustainable Food Sovereignty. In *Sustainable Agriculture and Food Security*; Springer: Berlin/Heidelberg, Germany, 2022; pp. 183–199.
31. Murthy, I.K.; Dutta, S.; Varghese, V.; Joshi, P.P.; Kumar, P. Impact of Agroforestry Systems on Ecological and Socio-Economic Systems: A Review. *Glob. J. Sci. Front. Res. H Environ. Earth. Sci.* **2016**, *16*, 15–27.
32. Nyland, R.D.; Nystrom, L.; Kiernan, D.H.; Bevilacqua, E. Assessing Small-Stem Density in Northern Hardwood Selection System Stands. *Can. J. For. Res.* **2019**, *49*, 237–245. [CrossRef]
33. Badan Pusat Statistik. *Hasil Sensus Penduduk 2020*; Badan Pusat Statistik: Jakarta, Indonesia, 2020. Available online: https://www.bps.go.id/pressrelease/2021/01/21/1854/hasil-sensus-penduduk-2020.html (accessed on 17 March 2023).

34. Worldometer Indonesia Population (Live). Available online: https://www.worldometers.info/world-population/indonesia-population/#:~:text=The%20current%20population%20of%20Indonesia,the%20latest%20United%20Nations%20data (accessed on 24 July 2023).
35. Abdurachman, A.; Dariah, A.; Mulyani, A. Strategi Dan Teknologi Pengelolaan Lahan Kering Mendukung Pengadaan Pangan Nasional. *J. Litbang Pertan.* **2008**, *27*, 43–49.
36. BPS. *Statistik Indonesia 2021*; BPS: Jakarta, Indonesia, 2021.
37. Mulyani, A.; Nursyamsi, D.; Syakir, M. Strategi Pemanfaatan Sumberdaya Lahan Untuk Pencapaian Swasembada Beras Berkelanjutan. *J. Sumberd. Lahan* **2017**, *11*, 11–22. [CrossRef]
38. Sarwani, M. Karakteristik Dan Potensi Lahan Sub Optimal Untuk Pengembangan Pertanian Di Indonesia. *J. Sumberd. Lahan* **2013**, *7*, 47–55.
39. Badan Ketahanan Pangan. *Roadmap Diversifikasi Pangan Lokal Sumber Karbohidrat Non Beras (2020–2024)*; Badan Ketahanan Pangan: Jakarta, Indonesia, 2020. Available online: https://badanpangan.go.id/storage/app/media/Bahan%202020/Roadmap%20Diversifikasi%20Pangan%202020-2024.pdf (accessed on 22 April 2023).
40. Bappenas Rencana Pembangunan Jangka Menengah Nasional 2020–2024. 2020. Available online: https://perpustakaan.bappenas.go.id/e-library/file_upload/koleksi/migrasi-data-publikasi/file/RP_RKP/Narasi%20RPJMN%20IV%202020-2024_Revisi%2014%20Agustus%202019.pdf (accessed on 20 March 2023).
41. Suryani, E.; Sulaeman, E.; Bachri, S.; Adhy, W. Laporan Tahunan BBSDLP 2019. 2020, pp. 29–31. Available online: https://bbsdlp-litbang-ppid.pertanian.go.id/doc/32/LAPORAN%20TAHUNAN/laptah-2019.pdf (accessed on 11 March 2023).
42. Ministry of Environment of Forestry. *The State of Indonesia's Forests 2022*; Nurbaya, S., Murniningtyas, S., Erwinsyah, E., Muttaqin, M.Z., Eds.; Ministry of Environment and Forestry, Republic of Indonesia: Jakarta, Indonesia, 2022.
43. UU pangan. *Undang-Undang Republik Indonesia Nomor 18 Tahun 2012 Tentang Pangan*; UU pangan: Jakarta, Indonesia, 2012.
44. Sibuea, S.M.; Kardhinata, E.H.; Ilyas, S.; Serdang, S. Identifikasi Dan Inventarisasi Jenis Tanaman Umbi-Umbian Yang Berpotensi Sebagai Sumber Karbohidrat Alternatif Di Kabupaten Serdang Bedagai. *J. Online Agroekoteknologi* **2014**, *2*, 1408–1418.
45. Elagib, N.A.; Al-Saidi, M. Balancing the Benefits from the Water–Energy–Land–Food Nexus through Agroforestry in the Sahel. *Sci. Total Environ.* **2020**, *742*, 140509. [CrossRef]
46. Tuturop, A.; Nugroho, J.D.; Warawarin, A. Agroforestry and Climate Smart Agriculture to Improve Food Security and Resilience Indigenous People in Teluk Patipi District Fakfak Regency West Papua Province. In *IOP Conference Series: Earth and Environmental Science*; IOP Publishing: Bristol, UK, 2022; Volume 989, p. 12005.
47. Leksono, B.; Rahman, S.A.; Larjavaara, M.; Purbaya, D.A.; Arpiwi, N.L.; Samsudin, Y.B.; Artati, Y.; Windyarini, E.; Sudrajat, D.J.; Aminah, A. Pongamia: A Possible Option for Degraded Land Restoration and Bioenergy Production in Indonesia. *Forests* **2021**, *12*, 1468. [CrossRef]
48. Altieri, M.A.; Nicholls, C.I.; Henao, A.; Lana, M.A. Agroecology and the Design of Climate Change-Resilient Farming Systems. *Agron. Sustain. Dev.* **2015**, *35*, 869–890. [CrossRef]
49. Paudyal, K.; Bhandari, A.; Adhikari, S.; Birhane, E.; Sharma, S.; Darcha, G.; Samsudin, Y.B.; Long, T.T. Framework for Assessing Ecosystem Services from Bamboo Forests: Lessons from Asia and Africa. 2019. Available online: https://www.cifor.org/publications/pdf_files/WPapers/WP255Paudyal.pdf (accessed on 10 April 2023).
50. Duque, R.L. Catherine Kohler Riessman (2008). Narrative Methods for the Human Sciences. *Forum Qual. Sozialforschung Forum Qual. Soc. Res.* **2010**, *11*, 19.
51. Sardjono, M.A.; Djogo, T.; Arifin, H.S.; Wijayanto, N. Klasifikasi Dan Pola Kombinasi Komponen Agroforestri. *Bahan Ajar Agrofor.* **2003**, *2*, 25.
52. Jemal, O.; Callo-Concha, D.; van Noordwijk, M. Local Agroforestry Practices for Food and Nutrition Security of Smallholder Farm Households in Southwestern Ethiopia. *Sustainability* **2018**, *10*, 2722. [CrossRef]
53. WFP. *Food Security and Vulnerability Atlas of Indonesia*; Food Security Council, Indonesian Ministry of Agriculture & the World Food Programme: Jakarta, Indonesia, 2015.
54. Paembonan, S.A.; Millang, S.; Dassir, M.; Ridwan, M. Species Variation in Home Garden Agroforestry System in South Sulawesi, Indonesia and Its Contribution to Farmers' Income. In *IOP Conference Series: Earth and Environmental Science*; IOP Publishing: Bristol, UK, 2018; Volume 157, p. 12004.
55. Hahury, H.D.; Prabawa, T.S.; Wiloso, P.G.; Soumokil, T.; Ndoen, M.L. Institutional Impacts on Choice of Traditional Agroforestry-Based Rural Community Livelihood Strategies in Maluku. *J. Manaj. Hutan Trop.* **2020**, *26*, 189.
56. Yundari, N.K.W.; Karmau, J.J.; Arisena, G.M.K. Kajian Kelayakan Finansial Kawasan Agroforestry. *Benchmark* **2022**, *2*, 151–163. [CrossRef]
57. Pitopang, R.; Atmoko, A.T.; Mertosono, S.R.; Ramawangsa, P.A. Plant Diversity in Agroforestry System and Its Traditional Use by Three Different Ethnics in Central Sulawesi Indonesia. In *IOP Conference Series: Earth and Environmental Science*; IOP Publishing: Bristol, UK, 2021; Volume 886, p. 12058.
58. Nandini, R. Analisis Keuntungan Usaha Tani Agroforestry Kemiri, Coklat, Kopi Dan Pisang Di Hutan Kemasyarakatan Sesaot, Lombok Barat (Benefits Analysis of Agroforestry Farming Patterned Candlenuts, Cacao, Coffee and Banana in Sesaot Community Forest, West Lombok). *J. Penelit. Kehutan. FALOAK* **2018**, *2*, 1–12. [CrossRef]
59. de Foresta, H.; Michon, G.; Kusworo, A. Complex Agroforest. In *The Everstory Book: Cultivating Connection With Trees*, 2nd ed.; Elevitch, C.R., Ed.; Permanent Agriculture Resources: Holualoa, HI, USA, 2004.

60. Gunawan, R.; Ramadhan, U.G.; Iskandar, J.; Partasasmita, R. Local Knowledge of Utilization and Management of Sugar Palm (Arenga Pinnata) among Cipanggulaan People of Karyamukti, Cianjur (West Java, Indonesia). *Biodiversitas* **2017**, *19*, 93–105. [CrossRef]
61. Bangsawan, I.; Dwiprabowo, H. Hutan Sebagai Penghasil Pangan Untuk Ketahanan Pangan Masyarakat: Studi Kasus Di Kabupaten Sukabumi. *J. Penelit. Sos. Dan Ekon. Kehutan.* **2012**, *9*, 185–197. [CrossRef]
62. Ardhana, I.P.G. Teknik Agroforestri Di Areal Hutan Kemasyarakatan Desa Pejarakan, Kecamatan Gerokgak, Kabupaten Buleleng, Provinsi Bali. *J. Ilmu Pertan. Indones.* **2011**, *16*, 81–90.
63. Sahureka, M.; Talaohu, M. Pengelolaan Agroforestry Tradisional "Dusung" Berbasis Kearifan Lokal "Masohi" Oleh Masyarakat Desa Hulaliu-Kabupaten Maluku Tengah. *J. Hutan Pulau Pulau Kecil* **2018**, *2*, 138–148. [CrossRef]
64. Umiyati, R. Diversifikasi Hasil Kegiatan Agroforestry Bagi Ketahanan Pangan Di Kecamatan Sigaluh, Kabupaten Banjarnegara. *Ilmu Taksosains* **2015**, *1*, 52–56.
65. Muliawati, E.S.; Sri Budiastuti, M.T.; Suprayogo, D.; Sutrisno, J. Agrobiodiversity in the Rural Home Gardens as the Food Reserve for Climate Change Adaptation (Case Study: Samin Sub-Watershed, Central Java, Indonesia). *Bulg. J. Agric. Sci.* **2018**, *24*, 759–767.
66. Masriah; Iskandar, B.; Iskandar, B.S.; Iskandar, J.; Partasasmita, R.; Suwartapradja, O.S. Economic, Social and Culture of Homegarden in Mekarasih Village, Jatigede, Sumedang, West Java. *Pros. Semin. Nas. Masy. Biodivers. Indones.* **2019**, *5*, 22–28.
67. Setiani, S.; Setiawan, E.; Huang, W.-C. Taneyan Lanjang Shared Home Gardens and Sustainable Rural Livelihoods of Ethnic Madurese in Madura Island, Indonesia. *Sustainability* **2022**, *14*, 5960. [CrossRef]
68. Ajijur, S.; Jette, R.; Jacobsen, B.; Robert, J.; James, H.; Sunderland, T. Finding Alternatives to Swidden Agriculture: Does Agroforestry Improve Livelihood Options and Reduce Pressure on Existing Forest? *Agrofor. Syst.* **2017**, *91*, 185–199. [CrossRef]
69. Park, J.H.; Woo, S.Y.; Kwak, M.J.; Lee, J.K.; Leti, S.; Soni, T. Assessment of the Diverse Roles of Home Gardens and Their Sustainable Management for Livelihood Improvement in West Java, Indonesia. *Forests* **2019**, *10*, 970. [CrossRef]
70. Wulandari, C.; Budiono, P.; Yuwono, S.B.; Herwanti, S. Adoption of Agro-Forestry Patterns and Crop Systems around Register 19 Forest Park, Lampung Province, Indonesia. *J. Manaj. Hutan Trop.* **2014**, *20*, 86–93. [CrossRef]
71. Belsky, J.M.; Siebert, S.F. Cultivating Cacao: Implications of Sun-Grown Cacao on Local Food Security and Environmental Sustainability. *Agric. Hum. Values* **2003**, *20*, 277–285. [CrossRef]
72. Wibawa, G.; Hendratno, S.; Van Noordwijk, M. Permanent Smallholder Rubber Agroforestry Systems in Sumatra, Indonesia. In *Slash-and-Burn Agriculture: The Search for Alternatives*; Columbia University Press: New York, NY, USA, 2005.
73. Belcher, B.; Imang, N.; Achdiawan, R. Considerations for Farmers in Kalimantan I. *Econ. Bot.* **2004**, *58*, S77–S87. [CrossRef]
74. Suyanto, S.; Khususiyah, N.; Leimona, B. Poverty and Environmental Services: Case Study in Way Besai Watershed, Lampung Province, Indonesia. *Ecol. Soc.* **2007**, *12*, 13. [CrossRef]
75. Kissinger; Pitri, R.M.N. Bioecology of Coffe Agroforestry: Vegetation Cover and Pattern of Plants Composition of Coffee Agroforestry in Pengaron District, Banjar Residence South Kalimantan. *EnviroScienteae* **2017**, *13*, 150–156. [CrossRef]
76. Madyantoro, A.; Muttaqin, Z.; Lidiawati, I. Kajian Kontribusi Sistem Agroforestri Terhadap Pendapatan Petani. *J. Nusa Sylva* **2015**, *15*, 11–16.
77. Iskandar, J.; Iskandar, B.S. Makalah Utama: Studi Etnobotani Keanekaragaman Tanaman Pangan Pada " Sistem Huma" Dalam Menunjang Keamanan Pangan Orang Baduy Ethnobotanical Study on Food Crop Diversity in Swidden Agroforestry Agriculture to Support Food. *Pros. Semin. Nas. Masy. Biodivers. Indones.* **2015**, *1*, 1265–1272. [CrossRef]
78. Iqbal, M.; Septina, A.D. Pemanfaatan Hasil Hutan Bukan Kayu Oleh Masyarakat Lokal Di Kabupaten Sanggau, Kalimantan Barat. *J. Penelit. Ekosist. Dipterokarpa* **2018**, *4*, 19–34.
79. Martin, E.; Ulfa, M.; Silalahi, A.; Winarno, B. Agroforestry Tradisional Sebagai Basis Pengembangan Hutan Rakyat: Kasus Di Kabupaten Lahat Sumatera Selatan. Prosiding Temu Lapang dan Ekspose Hasil-Hasil Penelitian UPT Badan Penelitian dan Pengembangan Kehutanan Wilayah Sumatera. 2003, pp. 133–146. Available online: https://www.researchgate.net/publication/319386585_Agroforestry_Tradisional_sebagai_Basis_Pengembangan_Hutan_Rakyat_kasus_di_Kabupaten_Lahat_Sumatera_Selatan (accessed on 7 April 2023).
80. Pujiono, E.; Sri Raharjo, S.A.; Njurumana, G.N.; Prasetyo, B.D.; Rianawati, H. Sustainability Status of Agroforestry Systems in Timor Island, Indonesia. *E3S Web Conf.* **2021**, *305*, 4003. [CrossRef]
81. Iryadi, R.; Darma, I.D.P. Multistoried Agroforestry System of Gaharu (Gyrinops Verstegii (Gilg.) Domke) in Flores Island East Nusa Tenggara. In *IOP Conference Series: Earth and Environmental Science*; IOP Publishing: Bristol, UK, 2021; Volume 648, p. 12024.
82. Wahyuningsih, E.; Lestari, A.T.; Syaputra, M.; Wulandari, F.T.; Anwar, H.; Januardi, J.; Maya, I.P.A.T.; Anggraini, D.; Aditia, G.B.D.R.; Muin, A. Pengayaan Tanaman Pakan Lebah Dengan Pola Agroforestry Home Garden Untuk Mendukung Kelestarian Sumber Pakan Lebah Madu Trigona. *J. Pendidik. Dan Pengabdi. Masy.* **2021**, *4*, 474–479. [CrossRef]
83. Antoh, A.A.; Arifin, N.H.S.; Chozin, M.A.; Arifin, H.S. Agricultural Biodiversity and Economic Productivity of the Yards in Arguni Bawah, Kaimana District, West Papua Province, Indonesia. *Biodiversitas* **2019**, *20*, 1020–1026. [CrossRef]
84. Purba, J.H.; Manik, I.W.Y.; Sasmita, N.; Komara, L.L. Telajakan and Mixed Gardens Landscape as Household Based Agroforestry Supports Environmental Aesthetics and Religious Ceremonies in Bali. In *IOP Conference Series: Earth and Environmental Science*; IOP Publishing: Bristol, UK, 2020; Volume 449, p. 12041.
85. Tefa, G.; Seran, W.; Un, P. Mamar's Contribution to Farmers' Income "A Case Study In Biau Vilage Io Kufeu District Malaka Regency". *Wana Lestari* **2022**, *6*, 22–33.

86. Njurumana, G.N.; Sadono, R.; Marsono, D. Ecosystem Services of Indigenous Kaliwu Agroforestry System in Sumba, Indonesia. *E3S Web Conf.* **2021**, *305*, 04002. [CrossRef]
87. Njurumana, G.N.; Marsono, D.; Irham, I.; Sadono, R. Konservasi Keanekaragaman Hayati Tanaman Pada Sistem Kaliwu Di Pulau Sumba (Plant Biodiversity Conservation on Kaliwu System at Sumba Island). *J. People Environ.* **2014**, *21*, 75–82.
88. Pujiono, E.; Raharjo, S.A.S.; Njurumana, G.; Prasetyo, B.D.; Rianawati, H. Kajian Aspek Ekologi, Ekonomi Dan Sosial Model-Model Agroforestri Di Nusa Tenggara Timur. *Pros. Semin. Nas. Agroforestri* **2013**, 456–461.
89. Ekarini, D.F.; Koestoer, R.H.S. Policy in Community-Based Environmental Conservation and Protection: A Comparative Study Between Brazil and Indonesia. *J. Wil. Dan Lingkung.* **2022**, *10*, 1–14. [CrossRef]
90. Prihatini, J.; Iskandar, J.; Partasasmita, R.; Nurjaman, D. The Impacts of Traditional Homegarden Conversion into the Commercial One: A Case Study in Sukapura Village of the Upstream Citarum Watershed, West Java, The Impacts of Traditional Homegarden Conversion into the Commercial One: A Case Study in Sukapura. *Biodiversitas J. Biol. Divers.* **2018**, *19*, 1926–1940. [CrossRef]
91. Widianingsih, N.N.; Schmidt, L.H.; Theilade, I. Jernang (*Daemonorops* spp.) Commercialization and Its Role for Rural Incomes and Livelihoods in Southern Sumatra, Indonesia. *For. Trees Livelihoods* **2019**, *28*, 143–159. [CrossRef]
92. Wollenberg, E.; Nawir, A. Turning Straw into Gold: Specialization among Damar Agroforest Farmers in Pesisir, Sumatra. *For. Trees Livelihoods* **2005**, *15*, 317–336. [CrossRef]
93. Feintrenie, L.; Chong, W.K.; Levang, P. Why Do Farmers Prefer Oil Palm? Lessons Learnt from Bungo District, Indonesia. *Small-Scale For.* **2010**, *9*, 379–396. [CrossRef]
94. Kusumedi, P.; Jariyah, N.A. Analisis Finansial Pengelolaan Agroforestri Dengan Pola Sengon Kapulaga Di Desa Tirip, Kecamatan Wadaslintang, Kabupaten Wonosobo. *J. Penelit. Sos. Dan Ekon. Kehutan.* **2010**, *7*, 93–100. [CrossRef]
95. Utomo, M.M.B.; Siagian, C.M.; Pietera, L.A.G. Procedure Inside Actors for Taking Activities (PIATA): An Enabling Model to Support Socio-Environmental Policy Planning Process. *Kasetsart J. Soc. Sci.* **2023**, *44*, 9–16.
96. Belsky, A.J.; Mwonga, S.M.; Duxbury, J.M. Effects of Widely Spaced Trees and Livestock Grazing on Understory Environments in Tropical Savannas. *Agrofor. Syst.* **1993**, *24*, 1–20. [CrossRef]
97. Bouchard, M.; Aquilué, N.; Périé, C.; Lambert, M. Forest Ecology and Management Tree Species Persistence under Warming Conditions: A Key Driver of Forest Response to Climate Change. *For. Ecol. Manag.* **2019**, *442*, 96–104. [CrossRef]
98. Idol, T.; Haggar, J.; Cox, L. Ecosystem Services from Smallholder Forestry and Agroforestry in the Tropics. In *Integrating Agriculture, Conservation and Ecotourism: Examples from the Field*; Campbell, W.B., Ortiz, S.L., Eds.; Springer: Dordrecht, The Netherland, 2011; Volume 1, p. 209.
99. Nguyen, Q.; Hoang, M.H.; Öborn, I.; van Noordwijk, M. Multipurpose Agroforestry as a Climate Change Resiliency Option for Farmers: An Example of Local Adaptation in Vietnam. *Clim. Chang.* **2013**, *117*, 241–257. [CrossRef]
100. Apriyanto, D.; Hardjanto; Hero, Y. Peningkatan Peran Hutan Rakyat Dalam Mendukung Ketahanan Pangan Dan Penanggulangan Kemiskinan. *J. Silvikultur Trop.* **2016**, *7*, 165–173. [CrossRef]
101. Irundu, D.; Fatmawati, D. Potensi Hutan Rakyat Sebagai Penghasil Pangan Di Desa Paku Kabupaten Polman, Sulawesi Barat. *J. Hutan Dan Masy.* **2019**, *11*, 41–48. [CrossRef]
102. Hernández-Espallardo, M.; Delgado-Ballester, E. Product Innovation in Small Manufacturers, Market Orientation and the Industry's Five Competitive Forces: Empirical Evidence from Spain. *Eur. J. Innov. Manag.* **2009**, *12*, 470–491. [CrossRef]
103. Mori, A.S.; Dee, L.E.; Gonzalez, A.; Ohashi, H.; Cowles, J.; Wright, A.J.; Loreau, M.; Hautier, Y.; Newbold, T.; Reich, P.B. Biodiversity–Productivity Relationships Are Key to Nature-Based Climate Solutions. *Nat. Clim. Chang.* **2021**, *11*, 543–550. [CrossRef]
104. Suryanto, P.; Aryono, W.B.; Sabarnurdin, M.S. Model Bera Dalam Sistem Agroforestri (Fallow Land Model in Agroforestry Systems). *J. Manaj. Hutan Trop.* **2006**, *12*, 15–26.
105. Belcher, K.W.; Boehm, M.M.; Fulton, M.E. Agroecosystem Sustainability: A System Simulation Model Approach. *Agric. Syst.* **2004**, *79*, 225–241. [CrossRef]
106. Pratiwi, A.; Suzuki, A. Reducing Agricultural Income Vulnerabilities through Agroforestry Training: Evidence from a Randomised Field Experiment in Indonesia. *Bull. Indones. Econ. Stud.* **2019**, *55*, 83–116. [CrossRef]
107. Pratiwi, D.K. Inovasi Kebijakan Pemerintah Daerah Dalam Penanganan COVID-19 Di Indonesia. *Amnesti J. Huk.* **2021**, *3*, 37–52. [CrossRef]
108. Magcale-Macandog, D.B.; Rañola, F.M.; Rañola, R.F.; Ani, P.A.B.; Vidal, N.B. Enhancing the Food Security of Upland Farming Households through Agroforestry in Claveria, Misamis Oriental, Philippines. *Agrofor. Syst.* **2010**, *79*, 327–342. [CrossRef]
109. Pieter, L.A.G.; Utomo, M.M.B.; Suhartono, S.; Sudomo, A.; Sanudin, S.; Fauziyah, E.; Widyaningsih, T.S.; Palmolina, M.; Hani, A.; Siagian, C.M. The Nexus of COVID-19 Pandemic and Rural Agroforestry Farmers' Livelihoods in Tasikmalaya Regency, East Priangan, Indonesia. *For. Soc.* **2022**, *6*, 335–354. [CrossRef]
110. Roshetko, J.M.; Rohadi, D.; Perdana, A.; Sabastian, G. Teak Agroforestry Systems for Livelihood Enhancement, Industrial Timber Production, and Environmental Rehabilitation. *For. Trees Livelihoods* **2013**, *22*, 241–256. [CrossRef]
111. Mayrowani, H. Ashari Pengembangan Agroforestry Untuk Mendukung Ketahanan Pangan Dan Pemberdayaan Masyarakat Sekitar Hutan. *Forum Penelit. Agro Ekon.* **2011**, *29*, 83–98. [CrossRef]
112. Rizki Syofiandi, R.; Hilmanto, R.; Herwanti, S. Analisis Pendapatan Dan Kesejahteraan Petani Agroforestri Di Kelurahan Sumber Agung Kecamatan Kemiling Kota Bandar Lampung. *J. Sylva Lestari* **2016**, *4*, 17. [CrossRef]

113. Belsky, J. Household Food Security, Farm Trees, and Agroforestry: A Comparative Study in Indonesia and the Philippines. *Hum. Organ.* **1993**, *52*, 130–141. [CrossRef]
114. Olson, J.M.; Berry, L. *Land Degradation in Java, Indonesia: Its Extent and Impact*; Global Mechanism with Support from the World Bank; World Bank: Washington, DC, USA, 2001.
115. Charles, R.; Munishi, P.; Nzunda, E. Agroforestry as Adaptation Strategy under Climate Change in Mwanga District, Kilimanjaro, Tanzania. *Int. J. Environ. Prot.* **2013**, *3*, 29–38.
116. Sharma, N.; Bohra, B.; Pragya, N.; Ciannella, R.; Dobie, P.; Lehmann, S. Bioenergy from Agroforestry Can Lead to Improved Food Security, Climate Change, Soil Quality, and Rural Development. *Food Energy Secur.* **2016**, *5*, 165–183. [CrossRef]
117. Toppo, P.; Raj, A. Role of Agroforestry in Climate Change Mitigation. *J. Pharmacogn. Phytochem.* **2018**, *7*, 241–243.
118. Maqoma, R.I. Mengapa Sejumlah Ahli Justru Mendukung Kelapa Sawit Menjadi Tanaman Hutan? The Conservation 2021. Available online: https://theconversation.com/mengapa-sejumlah-ahli-justru-mendukung-kelapa-sawit-menjadi-tanaman-hutan-173120 (accessed on 12 April 2023).
119. Mulia, R.; Van Sam, H.; Thi, D.; Ngoc, B.; Nguyen, D.A.; Resources, B. Earthworm Diversity, Forest Conversion and Agroforestry in Quang Nam Province, Vietnam. *Land* **2021**, *10*, 36. [CrossRef]
120. Ong, C.K.; Black, C.R.; Muthuri, C.W. Modifying Forestry and Agroforestry to Increase Water Productivity in the Semi-Arid Tropics. *CAB Rev. Perspect. Agric. Vet. Sci. Nutr. Nat. Resour.* **2006**, *1*, 1–19. [CrossRef]
121. Zhongming, Z.; Linong, L.; Wangqiang, Z.; Wei, L. *Agroforestry and Water for Resilient Landscapes*; Agroforestry Network: Stockholm, Sweden, 2020; pp. 1–8.
122. Lin, B.B. The Role of Agroforestry in Reducing Water Loss through Soil Evaporation and Crop Transpiration in Coffee Agroecosystems. *Agric. For. Meteorol.* **2010**, *150*, 510–518. [CrossRef]
123. Gusli, S.; Sumeni, S.; Sabodin, R.; Muqfi, I.H.; Nur, M.; Hairiah, K.; Useng, D.; van Noordwijk, M. Soil Organic Matter, Mitigation of and Adaptation to Climate Change in Cocoa-Based Agroforestry Systems. *Land* **2020**, *9*, 323. [CrossRef]
124. Fitri, R.; Hartoyo, A.P.P.; Simangunsong, N.I.; Satriawan, H. Pengaruh Agroforestri Terhadap Kualitas Daerah Aliran Sungai Ciliwung Hulu, Jawa Barat (Impact of Agroforestry on the Quality of Ciliwung Hulu Watershed, West Java). *J. Penelit. Pengelolaan Drh. Aliran Sungai J. Watershed Manag. Res.* **2020**, *4*, 173–186. [CrossRef]
125. Naharuddin, N. Sistem Pertanian Konservasi Pola Agroforestri Dan Hubungannya Dengan Tingkat Erosi Di Wilayah Sub-DAS Wuno, Das Palu, Sulawesi Tengah. *J. Wil. Dan Lingkung.* **2018**, *6*, 183. [CrossRef]
126. Purwiyatno, A.I.S.; Agustriani, F. Effect of Silvofishery on Ponds Nutrient Levels (Pengaruh Silvofishery Terhadap Kandungan Nutrien Di Tambak). *Ilmu Kelaut* **2014**, *19*, 81. [CrossRef]
127. Kusumaningtyas Perwitasari, W.; Muhammad, F.; Wasiq Hidayat, J. Silvofishery as an Alternative System of Sustainable Aquaculture in Mororejo Village, Kendal Regency. *E3S Web Conf.* **2020**, *202*, 06043. [CrossRef]
128. Hidayat, Y.; Lala, F.; Suwitono, B.; Aji, H.B.; Brahmantiyo, B. Implementasi Teknologi Peningkatan Produktivitas Lahan Kering Di Bawah Tegakan Kelapa Di Maluku Utara [Implementation Technology to Productivity Increase at Dryland under Coconut in North Maluku]. *Bul. Palma* **2020**, *21*, 11. [CrossRef]
129. Mbow, C.; Smith, P.; Skole, D.; Duguma, L.; Bustamante, M. ScienceDirect Achieving Mitigation and Adaptation to Climate Change through Sustainable Agroforestry Practices in Africa. *Curr. Opin. Environ. Sustain.* **2014**, *6*, 8–14. [CrossRef]
130. Garrity, D.P.; Akinnifesi, F.K.; Ajayi, O.C.; Weldesemayat, S.G.; Mowo, J.G.; Kalinganire, A.; Larwanou, M.; Bayala, J. Evergreen Agriculture: A Robust Approach to Sustainable Food Security in Africa. *Food Secur.* **2010**, *2*, 197–214. [CrossRef]
131. Grass, I.; Kubitza, C.; Krishna, V.V.; Corre, M.D.; Mußhoff, O.; Pütz, P.; Drescher, J.; Rembold, K.; Ariyanti, E.S.; Barnes, A.D. Trade-Offs between Multifunctionality and Profit in Tropical Smallholder Landscapes. *Nat. Commun.* **2020**, *11*, 1186. [CrossRef] [PubMed]
132. Nöldeke, B.; Winter, E.; Laumonier, Y.; Simamora, T. Simulating Agroforestry Adoption in Rural Indonesia: The Potential of Trees on Farms for Livelihoods and Environment. *Land* **2021**, *10*, 385. [CrossRef]
133. Palacios Bucheli, V.J.; Bokelmann, W. Agroforestry Systems for Biodiversity and Ecosystem Services: The Case of the Sibundoy Valley in the Colombian Province of Putumayo. *Int. J. Biodivers. Sci. Ecosyst. Serv. Manag.* **2017**, *13*, 380–397. [CrossRef]
134. Jonsson, M.; Bengtsson, J.; Gamfeldt, L.; Moen, J.; Snäll, T. Levels of Forest Ecosystem Services Depend on Specific Mixtures of Commercial Tree Species. *Nat. Plants* **2019**, *5*, 141–147. [CrossRef]
135. Hisano, M.; Searle, E.B.; Chen, H.Y.H. Biodiversity as a Solution to Mitigate Climate Change Impacts on the Functioning of Forest Ecosystems. *Biol. Rev.* **2018**, *93*, 439–456. [CrossRef]
136. Liang, J.; Crowther, T.W.; Picard, N.; Wiser, S.; Zhou, M.; Alberti, G.; Schulze, E.-D.; McGuire, A.D.; Bozzato, F.; Pretzsch, H. Positive Biodiversity-Productivity Relationship Predominant in Global Forests. *Science (1979)* **2016**, *354*, aaf8957. [CrossRef]
137. Claverías, R.; Quispe, C. Biodiversidad cultivada: Una estrategia. In *Perú, el Problema Agrario en Debate: SEPIA IX*; 2002; p. 180. Available online: https://sepia.org.pe/wp-content/uploads/2018/07/SEPIA-IX-PUNO-2001-completo.pdf (accessed on 25 March 2023).
138. Mori, A.S. Environmental Controls on the Causes and Functional Consequences of Tree Species Diversity. *J. Ecol.* **2018**, *106*, 113–125. [CrossRef]
139. Garí, J. Biodiversity and Indigenous Agroecology in Amazonia: The Indigenous People of Pastaza. *Etnoecologica* **2001**, *5*, 21–37.

140. Wurz, A.; Tscharntke, T.; Martin, D.A.; Osen, K.; Rakotomalala, A.A.N.A.; Raveloaritiana, E.; Andrianisaina, F.; Dröge, S.; Fulgence, T.R.; Soazafy, M.R. Win-Win Opportunities Combining High Yields with High Multi-Taxa Biodiversity in Tropical Agroforestry. *Nat. Commun.* **2022**, *13*, 4127. [CrossRef]
141. Aryanti, N.A.; Hartono, N.A.; Ramadhan, F.; Pahrurrobi, P. Hubungan Antara Aktivitas Manusia Dan Keberadaan Kukang Jawa (Nycticebus Javanicus) Di Kawasan Hutan Lindung Di RPH Sumbermanjing Kulon, Jawa Timur. *Biotropika J. Trop. Biol.* **2018**, *6*, 83–88. [CrossRef]
142. Andriansyah, A.; Supartono, T.; Nurdin, N. Gangguan Satwa Liar Taman Nasional Gunung Ciremai Terhadap Lahan Pertanian Di Desa Karangsari Kecamatan Darma Kabupaten Kuningan. *Pros. Fahutan* **2020**, *1*, 74–81.
143. Halimah, N. Faktor Yang Mempengaruhi Perilaku Dan Aktivitas Pegerakan Dari Satwa. 2022. Available online: https://osf.io/mn54w (accessed on 23 February 2023).
144. Santoso, B.; Subiantoro, D. Pemetaan Konflik Monyet Ekor Panjang (Macaca Fascicularis Raffles) Di Desa Sepakung Kecamatan Banyubiru Kabupaten Semarang. *Indones. J. Conserv.* **2019**, *8*, 138–145.
145. Wibowo, A.; KRH, I.G.A.; Sudarwanto, A.S. Implementasi Kebijakan Dalam Penanggulangan Konflik Antara Manusia Dan Satwa Liar Di Propinsi Jambi (Ditinjau Dari Hukum Dan Kebijakan Publik). *Pros. SNaPP Sos. Ekon. Dan Hum.* **2017**, *7*, 265–274.
146. Yudha, D.R.P. Persepsi Konflik Manusia Gajah Di Desa Melibur Kecamatan Pinggir, Kabupaten Bengkalis, Riau. Ph.D. Thesis, Universitas Lancang Kuning, Pekanbaru, Indonesia, 2020.
147. Duflot, R.; Eyvindson, K.; Mönkkönen, M. Management Diversification Increases Habitat Availability for Multiple Biodiversity Indicator Species in Production Forests. *Landsc. Ecol.* **2022**, *37*, 443–459. [CrossRef]
148. Pohjanmies, T.; Eyvindson, K.; Triviño, M.; Bengtsson, J.; Mönkkönen, M. Forest Multifunctionality Is Not Resilient to Intensive Forestry. *Eur. J. For. Res.* **2021**, *140*, 537–549. [CrossRef]
149. Davis, H.; Todd, Z. On the Importance of a Date, or Decolonizing the Anthropocene. *ACME* **2017**, *16*, 761–780.
150. Okumu, B.; Gwanyebit, A.; Osano, P. Current Research in Environmental Sustainability A Review of Water-Forest-Energy-Food Security Nexus Data and Assessment of Studies in East Africa. *Curr. Res. Environ. Sustain.* **2021**, *3*, 100045. [CrossRef]
151. Potapov, P.; Hansen, M.C.; Pickens, A.; Hernandez-Serna, A.; Tyukavina, A.; Turubanova, S.; Zalles, V.; Li, X.; Khan, A.; Stolle, F. The Global 2000-2020 Land Cover and Land Use Change Dataset Derived from the Landsat Archive: First Results. *Front. Remote Sens.* **2022**, *3*, 18. [CrossRef]
152. Utomo, M.; Pieter, L.; Siagian, C.M. Value Chain Structure Analysis as A Starting Point for Bamboo Enterprise Development: Lessons from Gunungkidul, Indonesia. *For. Soc.* **2021**, *5*, 405–420. [CrossRef]
153. Sabarnurdin, M.; Budiadi; Suryanto, P. *Agroforestri Untuk Indonesia: Strategi Kelestarian Hutan Dan Kemakmuran*; Cakrawala Media: Yogyakarta, Indonesia, 2011.
154. Santoso, C.E.E. Perputaran Modal Kerja Dan Perputaran Piutang Pengaruhnya Terhadap Profitabilitas Pada PT. Pegadaian (PERSERO). *J. EMBA J. Ris. Ekon. Manaj. Bisnis Dan Akunt.* **2013**, *1*, 1533 1649.
155. Alviya, I.; Suryandari, E.Y. Evaluasi Kebijakan Pelaksanaan Agrofrestry Sistem Di Indoensia. *Sosek Dan Kebijak. Kehutan.* **2006**, *3*, 1–13.
156. Torralba, M.; Fagerholm, N.; Burgess, P.J.; Moreno, G.; Plieninger, T. Do European Agroforestry Systems Enhance Biodiversity and Ecosystem Services? A Meta-Analysis. *Agric. Ecosyst. Environ.* **2016**, *230*, 150–161. [CrossRef]

MDPI
St. Alban-Anlage 66
4052 Basel
Switzerland
www.mdpi.com

Agriculture Editorial Office
E-mail: agriculture@mdpi.com
www.mdpi.com/journal/agriculture

www.ingramcontent.com/pod-product-compliance
Lightning Source LLC
Chambersburg PA
CBHW040818250926
51850CB00007BA/1200
9783725809592